Peace Parks

Global Environmental Accord: Strategies for Sustainability and Institutional Innovation
Nazli Choucri, series editor

For a complete series list, see the back of the book

Peace Parks
Conservation and Conflict Resolution

edited by Saleem H. Ali

The MIT Press
Cambridge, Massachusetts
London, England

© 2007 Massachusetts Institute of Technology

All rights reserved. No part of this book may be reproduced in any form by any electronic or mechanical means (including photocopying, recording, or information storage and retrieval) without permission in writing from the publisher.

For information about special quantity discounts, please email special_sales@mitpress.mit.edu

This book was set in Sabon on 3B2 by Asco Typesetters, Hong Kong.
Printed on recycled paper and bound in the United States of America.

Library of Congress Cataloging-in-Publication Data

Peace parks : conservation and conflict resolution / edited by Saleem H. Ali.
   p.   cm. — (Global environmental accord series)
Includes bibliographical references and index.
ISBN 978-0-262-01235-5 (hardcover : alk. paper) — ISBN 978-0-262-51198-8 (pbk. : alk. paper)
1. Transfrontier conservation areas. 2. Transfrontier conservation areas—Political aspects. 3. Conflict management. I. Ali, Saleem H. (Saleem Hassan), 1973–
S944.5.P78P43 2007
333.95′16—dc22
                                                                                                 2006037129

10 9 8 7 6 5 4 3 2

We sought to speak
Each to each in accents of trust
Dispersing ancient mists in clean breezes
To clear the path of lowland barriers
Forge new realities, free our earth
Of distorting shadows
Cast by old and modern necromancers

Wole Soyinka, *Phases of Peril*

Contents

Series Foreword by Nazli Choucri   xi
Foreword by Julia Marton-Lefèvre   xiii
Preface and Acknowledgments   xvii
About the Contributors   xix

    Introduction: A Natural Connection between Ecology and Peace?   1
    Saleem H. Ali

I Environmental Peace-Building: Theories and Histories

1 Measuring Peace Park Performance: Definitions and Experiences   23
   Anne Hammill and Charles Besançon

2 Peace Games: Theorizing about Transboundary Conservation   41
   Raul Lejano

3 Peace Parks and Global Politics: The Paradoxes and Challenges of Global Governance   55
   Rosaleen Duffy

4 Scaling Peace and Peacemakers in Transboundary Parks: Understanding Glocalization   69
   Maano Ramutsindela

5 Peace Parks as Social Ecological Systems: Testing Environmental Resilience in Southern Africa   83
   Anna Spenceley and Michael Schoon

II Transboundary Conservation in Action: Bioregional Management for Science and Economic Development

6   Connecting the World's Largest Elephant Ranges: The Selous–Niassa Corridor   109
    Rolf D. Baldus, Rudolf Hahn, Christina Ellis, and Sarah Dickinson DeLeon

7   The "W" International Peace Park: Transforming Conservation and Conflict in West Africa   127
    Aissetou Dramé-Yayé, Diallo Daouda Boubacar, and Juliette Koudénoukpo Biao

8   The Emerald Triangle Protected Forests Complex: An Opportunity for Regional Collaboration on Transboundary Biodiversity Conservation in Indochina   141
    Yongyut Trisurat

9   Conflict Avoidance and Environmental Protection: The Antarctic Paradigm   163
    Michele Zebich-Knos

10 The Waterton–Glacier International Peace Park: Conservation amid Border Security   183
    Randy Tanner, Wayne Freimund, Brace Hayden, and Bill Dolan

III Peace Parks and Regional Governance Regimes: Redefining Security and Realism

11 Bridging Conservation across *La Frontera*: An Unfinished Agenda for Peace Parks along the US–Mexico Divide   205
    Belinda Sifford and Charles Chester

12 Liberia: Securing the Peace through Parks   227
    Arthur G. Blundell and Tyler Christie

13 Preserving Korea's Demilitarized Corridor for Conservation: A Green Approach to Conflict Resolution   239
    Ke Chung Kim

14 Nesting Cranes: Envisioning a Russo–Japanese Peace Park in the Kuril Islands   261
    Jason Lambacher

15 The Siachen Peace Park Proposal: Reconfiguring the Kashmir Conflict?   277
    Kent Biringer and Air Marshall K. C. (Nanda) Cariappa

16 Linking Afghanistan with Its Neighbors through Peace Parks: Challenges and Prospects   291
   Stephan Fuller

17 Iraq and Iran in Ecological Perspective: The Mesopotamian Marshes and the Hawizeh–Azim Peace Park   313
   Michelle L. Stevens

18 Conclusion: Implementing the Vision of Peace Parks   333
   Saleem H. Ali

References   343
Index   379

Series Foreword

A new recognition of profound interconnections between social and natural systems is challenging conventional constructs and the policy predispositions informed by them. Our current intellectual challenge is to develop the analytical and theoretical underpinnings of an understanding of the relationship between the social and the natural systems. Our policy challenge is to identify and implement effective decision-making approaches to managing the global environment.

The series Global Environmental Accord: Strategies for Sustainability and Institutional Innovation adopts an integrated perspective on national, international, cross-border, and cross-jurisdictional problems, priorities, and purposes. It examines the sources and the consequences of social transactions as these relate to environmental conditions and concerns. Our goal is to make a contribution to both intellectual and policy endeavors.

Nazli Choucri

Foreword

The linkage between a healthy environment and peace is increasingly becoming apparent. Many of the roots of current and future conflicts stem from competition for scarce natural resources. In the current crisis in the Sudan, in Darfur (only one example of many), environmental factors such as desertification have contributed to competition for arable land, which has in turn led to conflicts between farmers and pastoralists. The award of the 2004 Nobel Peace Prize to Wangari Maathai, who initiated a major grassroots tree-planting scheme in Kenya, and the 2006 Prize to Muhammad Yunus and the Grameen Bank for their seminal work in helping to end poverty underscore the inextricable links between environment, sustainable development, and peace and security.

Protected areas are an integral element of strategies to achieving sustainable development. These areas safeguard biological and cultural diversity and help improve livelihoods of local communities; they also provide homelands for many indigenous people and bring countless benefits to society in general. Protected areas now cover only a little more than 12 percent of the earth's surface. It is up to nations of the world to make a major commitment to safeguard the earth's biodiversity and pass it on for future generations.

The World Conservation Union (IUCN) has been actively involved in promoting transboundary protected areas—areas that involve a degree of cooperation across one or more boundaries between (or within) countries—for many years. These areas make sound sense from the context of biodiversity conservation, since plants and animals clearly do not recognize artificial boundaries. This case was lucidly put by Dr. Z. Paulo Jordan (then South African Minister of Environmental Affairs and Tourism) in his opening address to the IUCN meeting on Parks for Peace, held in Cape Town in 1997, where he noted:

The rivers of Southern Africa are shared by more than one country. Our mountain ranges do not end abruptly because some nineteenth-century politician drew a line on a map. The winds, the oceans, the rain, and atmospheric currents do not recognize political frontiers. The earth's environment is the common property of all humanity and creation, and what takes place in one country affects not only its neighbors, but many others well beyond its borders.

The concept of transboundary protected areas, proposed by IUCN and other bodies, has taken root and is expanding rapidly. From fifty-nine in 1988, there are currently more than one hundred and seventy transboundary protected areas, indicating the widespread acceptance at all levels of this concept.

The increasing recognition that transboundary protected areas can also play a role in conflict resolution and particularly in the period following the cessation of conflicts has shown that such areas—referred to as peace parks—contribute to building bridges between nations and people. For example, the Peace Accord between Peru and Ecuador in 1998 included a peace park as a key element of the agreement between the two countries. This provided a valuable tool for assisting in the resolution of specific conflicts relating to boundary demarcation and, importantly, for building more peaceful cooperation between two countries.

Experience to date with peace parks has been more anecdotal, and there has been limited systematic analysis of the role peace parks can play in the peaceful resolution of conflict. This publication makes an important contribution to filling this void. This book is significant in that it looks at peace parks in an analytical and multi-sectoral manner. Chapters are contributed from a range of perspectives: from conservation managers, from military planners, and from experts in the fields of conflict resolution. A wealth of practical examples and guidance in the development of peace parks is provided, ensuring that this publication can assist in moving beyond theory to practical implementation. The key message is the enormous potential of peace parks in conflict resolution, particularly through building confidence and cooperation between countries. However, much remains to be done to ensure that this potential is fully realized. A number of regions where potential peace parks could be developed, as a contributing factor for conflict resolution, are also identified.

Political will and commitment as well as effective governance models will, however, be essential if peace parks are to be effective tools to contribute to peace and to environmental security. Also fundamental is the need to share experience and communicate lessons in relation to peace parks. This publication plays a vitally important role in this regard. I sincerely hope that the lessons drawn will be widely communicated and will make an effective contribution to the acceleration of peace parks around the world—both for better conservation and for better cooperation among peoples.

I have been working on these issues all my life, and certainly in my present position as Rector of the UN-affiliated University for Peace and my future position as Director General of the IUCN, I welcome this book and will do all I can to ensure that it is read and its lessons are taken to heart.

Julia Marton-Lefèvre
Rector, United Nations mandated University for Peace
Director-General designate, IUCN—The World Conservation Union
November 2006

Preface and Acknowledgments

The famed naturalist Aldo Leopold was appointed as the first conservation advisor to the United Nations soon after the end of World War II. Yet little is known about Leopold's mandate for this short-lived position that ended with his death in 1948. During my days as a student at the Yale School of Forestry and Environmental Studies, I often passed by an aging photograph of this illustrious alumnus and frequently wondered how to find a confluence between Leopold's conservation ethic and conflict resolution. I assumed, perhaps naïvely, that the appointment of a conservationist as an advisor to the United Nations would only be to serve the goals of conflict resolution.

As I proceeded through academe, I became increasingly aware that such a confluence is indeed possible, despite the onslaught of literature that diminishes the salience of environmental factors in the higher politics of war and peace. In particular, the possibility of using areas of environmental significance as a common territorial asset is beginning to show promise in various parts of the world and deserves further attention. Ideas about possible topics and authors for this book subsequently evolved through a series of seminars that I led on environmental conflict resolution for a group of midcareer conservation professionals at Brown University. I am indebted to Richard Wetzler and Steven Hamburg for giving me the opportunity to lead this seminar for three consecutive years under a grant from the Luce Foundation.

Subsequently my interactions with the staff of the Environmental Change and Security Project (ECSP) at the Woodrow Wilson Center, particularly their director Geoffrey Dabelko, provided further content and insights for this project. The Wilson Center sponsored a conference in September 2005 where several of the contributing authors to this book were given an opportunity meet and exchange ideas. At the same time ECSP partnered with the United Nations Environment Programme to launch an expert advisory group on environment conflict and cooperation of which I was a member. Interaction through such forums enriched the volume and provided greater coherence to the arguments presented here.

Preparing an interdisciplinary treatise of this size and scope requires multiple minds and processing time. I was fortunate to have such access to talented individuals, thanks in large part to the highly supportive environment at the University of Vermont. My research assistant and doctoral student Ganlin Huang patiently copyedited and formatted various permutations of the manuscript and deserves foremost gratitude for "getting the job done." Colleagues around campus, particularly Stephanie Kaza and Ian Worley, supported this effort despite the rather idealistic tone of the title.

I have also benefited from discussions and communications with numerous prominent conservationists and environmental scholars including Aamir Ali, Nigel Allan, Ken Conca, Jason Coburn, Stephen Kellert, Aban Marker-Kabraji, Julia Marton-Lefvre, Teresita Schaffer, Sheila Jasanoff, George Schaller, Jack Shroder, Larry Susskind, and Shirin Tahir-Kheli. The four anonymous reviewers for this manuscript were diligent and rigorous in their evaluation of the manuscript, and I sincerely hope to have met their expectations with the final product. All the contributing authors were prompt in responding to reviews, despite being scattered all over the planet. My acquisition editor at The MIT Press, Clay Morgan, showed remarkable patience with this diffuse project for which the authors and I are especially grateful. We all worked well as a team and met our targets. Additionally my team at home, starting with my mother Parveen, my wife Maria, and our two boys Shahmir and Shahroze, cheered me along as I sat around the house with my laptop clicking away at the edits.

Finally, this book has a futuristic tone, and perhaps the most consequential acknowledgment should be to any policy makers who are willing to realize the vision of peace parks in resolving conflicts in these troubled times.

About the Contributors

Saleem H. Ali is Associate Professor of Environmental Planning at the Rubenstein School, University of Vermont, Burlington, USA.

Rolf D. Baldus is the Community-Based Conservation Advisor at German Development Agency (GTZ), Bonn, Germany.

Charles Besançon is the Head of Programme Protected Areas at the UNEP World Conservation Monitoring Centre, Cambridge, UK.

Kent Biringer is a senior researcher at Sandia National Labs, Albuquerque, New Mexico, USA.

Arthur G. Blundell is Timber Specialist at the United Nations Security Council Panel of Experts on Liberia, New York, NY.

Diallo Daouda Boubacar is a researcher at the Abdou Moumouni University, Niamey, Niger.

K. C. (Nanda) Cariappa served as the Air Marshall of the Indian Air Force from 1995 to 2000.

Charles Chester is on the adjunct faculty of the Fletcher School of Law and Diplomacy, Tufts University, Medford, Massachusetts, USA.

Tyler Christie is Technical Director at Conservation International—Liberia, Washington, DC, USA.

Sarah Dickinson DeLeon is a graduate student of Natural Resource Planning at the Rubenstein School, University of Vermont, Burlington, USA.

Bill Dolan is Chief Park Warden at Waterton Lakes National Park, Alberta, Canada.

Aissetou Dramé-Yayé is Lecturer/Researcher of Agronomy, Abdou Moumouni University, Niamey, Niger.

Rosaleen Duffy is Senior Lecturer at the School of Social Science, Manchester University, UK.

Christina Ellis is a consultant for the Jane Goodall Institute.

Wayne Freimund is Professor and Chair of the Wildland Recreation Management Program, Department of Society and Conservation, University of Montana, Missoula, USA.

Stephan Fuller is a sustainable development Chief Technical Advisor to IUCN—the World Conservation Union, based in Sydenham, Ontario, Canada.

Rudolf Hahn is Environmental Program Director at the German Development Agency (GTZ), Dares-Salam, Tanzanio.

Anne Hammill is Project Manager in the Geneva office, International Institute of Sustainable Development, Switzerland.

Brace Hayden is Regional Issues Specialist at Glacier National Park, West Glacier, Montana, USA.

Biao Koudenoukpo Juliette is Engineer of Water and Forestry at IUCN, Dakar, Senegal.

Ke Chung Kim is a Professor of Entomology at the Department of Entomology, and Director of Center for BioDiversity Research at the Penn State Institutes of the Environment, Pennsylvania State University, University Park, Pennsylvania, USA.

Jason Lambacher is a doctoral student in Political Science at the University of Washington, Seattle, USA.

Raul Lejano is Assistant Professor at the Department of Planning, Policy, and Design, University of California, Irvine, USA.

Maano Ramutsindela is Senior Lecturer in Environmental and Geographical Science at the University of Cape Town, South Africa.

Michael Schoon is a graduate student at the School of Public and Environmental Affairs, Indiana University, Bloomington, USA.

Belinda Sifford is Director of Development at the New College of California, San Francisco, USA.

Anna Spenceley is Research Fellow at the Transboundary Protected Areas Research Initiative, University of Witwatersrand, Johannesburg, South Africa.

Michelle L. Stevens is on the faculty of the Environmental Science Program at Imperial Valley College, Imperial, California, USA.

Randy Tanner is a doctoral student at the Department of Society and Conservation, University of Montana, Missoula, USA.

Yongyut Trisurat is Assistant Professor of Forestry at Department of Forest Biology, Kasetsart University, Bangkok, Thailand.

Michele Zebich-Knos is Professor at the Department of Political Science and International Affairs, Kennesaw State University, Kennesaw, Georgia, USA.

Introduction: A Natural Connection between Ecology and Peace?

Saleem H. Ali

The announcement of the Nobel Peace Prize for 2004 surprised many leaders around the world. Hawkish commentators were shocked that diplomatic peacemakers had been sidelined, and the Nobel committee had chosen Wangari Maathai, a Kenyan environmentalist, as the recipient. In a time when armed conflict is scourging so many parts of the world, the selection seemed to many "realists" an irritating distraction. Editorials the world over questioned the choice. *The Economist* emphatically asked the question "But what does planting trees have to do with peace?"[1] The Scandinavian press was even more critical with notable academic researchers including the Norwegian editor of the *Journal of Peace Research*, Nils Petter Gleditch, questioning the rationale for the prize.[2] This line of criticism had to do with the lack of empirical evidence linking environmental issues to conflict. These critics did not ask the more salient question of whether there are certain key attributes to environmental conservation that could independently contribute to peace-building in conflict zones.

This book explores the multiple ways in which environmental conservation zones can facilitate the resolution of territorial conflicts. Such zones are often places of ecological significance or natural beauty that usually have restrictions on development activities. While environmental regulations often spark development conflicts, there may be pathways by which ecological factors in such areas can be conducive to conflict resolution. The central question we address is how environmental concerns can be transformed into cooperation between various political jurisdictions.

Our focus is on the formation of conservation zones in which the sharing of physical space can build and sustain peace. Such zones that can play an instrumental role in peacemaking or sustaining amity between communities are termed "peace parks." According to the Transboundary Protected Areas Network of the World Conservation Union there are 188 transboundary protected areas worldwide, and these are such a growing phenomenon that a separate task force has been set up under the World Commission on Protected Areas to study them.[3] We are also concerned with how such parks can be effective and how they might fail in their stated objectives. Thus the chapters in this book explore many factors such as the role of governments, the military, civil society, scientists, and conservation practitioners in negotiating effective arrangements that protect the environment, but more consequently that resolve existing conflicts which are external to the formation of a park itself. It is also important to note that while there are numerous transboundary conservation zones, many are not being effectively managed and numerous ecologically sensitive areas remain unprotected. For example, a recent geographic information systems (GIS) study conducted by Gomez et al. (2006), using a map of human influence, found 104 transboundary wild areas involving 61 countries that are not formally part of any conservation park.

The coinage of the term "peace park" can itself be traced back to the establishment of the Waterton Lakes Glacier International Peace Park between Canada and the United States in 1932. While at the time this gesture was largely meant to be symbolic, since both Canada and the United States had good relations and continue to be friendly neighbors, the term has naturally evolved to acquire many other connotations. For the purposes of this book we are concerned with any conservation zone that, by virtue of multiple jurisdictions, could either help resolve a conflict or maintain existing peace. The concept of peace parks challenges many deeply rooted historical assumptions about conservation zones, which have often been considered a source of conflict themselves due to the dispossession of land, differentiated values about conservation versus preservation,[4] and consequently ecological primacy versus political expediency. The contributors to this book accept that the establishment of peace parks will undoubtedly require political attrition. However, the argument that we present is one of possibility for making environmental factors instrumental in peace-building. Although many existing structures would require reform to prevent the cooptation of peace parks by vested hegemonic interests, with effective strategies for community engagement we can implement this novel idea for conflict resolution at an international level.

Situating the Argument

Given the large volume of environmental security discourse over the past twenty years, it is important to first put the argument presented in this book in context.

Environmentalists have often sought to highlight the linkages between resource scarcity, ecological degradation, and conflict. In this regard they have been supported by some of the earlier literature on environmental security that gained prominence toward the end of the cold war.[5] Any solutions presented in this vein tend to focus on how to improve environmental conditions as a means of addressing the conflict. While this is certainly a laudable goal where environmental factors are part of the conflict, the approach easily falls prey to critics who insist that environmental factors are a minor part of conflicts. In their view, larger ethnic, financial, or demographic factors lead to conflict and environmental issues play a subsidiary role, if any.[6]

Yet there is another way of invoking the environment in conflict resolution that would address the concerns of the skeptics. Instead of trying to tease out environmental causality in conflicts and thereby accentuate the importance of conservation, one can also try and see how environmental issues can play a role in cooperation— regardless of whether they are part of the original conflict. For example, *The Economist*'s contention that the Darfur crisis is about ethnic and political factors could still be addressed by an approach stating that desertification is a common threat to both sides, and this could be a means of bringing parties together.

Figure I.1 synthesizes the disparate strands of environmental security discourse by highlighting two key assumptions or causal links that could lead from environmental scarcity and abundance to conflict and cooperation. It is important to note that environmental quality is often differentiated from environmental quantity in resource economics. However, for our purposes, impaired environmental quality translates into default environmental scarcity of usable resources.

The environmental peace-building narrative suggests that mutual knowledge of resource depletion and a positive aversion to such depletion leads to cooperation (pathway A).[7] The Environmental Change and Security Project at the Woodrow Wilson Center has been among the few research institutions that has pursued this line of inquiry. Conca and Dabelko (2002) published the first anthology on environmental peace-making under the auspices of the Wilson Center and have also garnered interest from the World Conservation Union and the United Nations Environment Programme in this approach.[8] The main premise of environmental peacemaking is that there are certain key attributes of environmental concerns that would lead acrimonious parties to consider them as a means of cooperation. Thus environmental issues could play an instrumental role even in cases where the conflict does not involve environmental issues. As water resource theorists have frequently

4 Saleem H. Ali

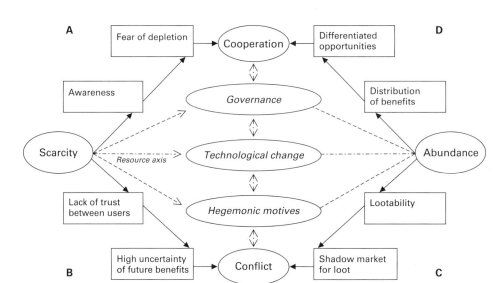

Figure I.1
Contending pathways of environmental security discourse

observed, this pathway often occurs despite perceived disputes of ownership or rights to water that may occur locally. Even adversaries who are aware of the dire impact of depletion are forced to be cooperative on water and avert any "water wars."[9]

Within political science, environmental movements have rarely been considered a direct catalyst for cooperation. Scholars such as Litfin (1998) have focused instead on understanding shifting notions of sovereignty in environmental politics. There is also emerging literature on the growth of transnational networks of civil society in diluting state sovereignty for environmental ends.[10] However, the power of such efforts in transforming the debate on environmental conservation to larger political reconciliation has eluded scholars. This may be partly because tangible institutional structures such as peace parks have so far not been empirically considered.

A reluctance to consider scholarly approaches to environmental peace-building may also be attributed to the dominance of resource scarcity and property rights discourse. By this account the classic tragedy of the commons scenario will occur when scarcity leads to conflict because of an absence of trust and relative uncertainty about accruing future benefits (pathway B). In economic terms this pathway implies a relatively high discounting rate for the future. Empirical work highlighted in a landmark volume on cooperative behavior by Oye (1984) and subsequent work by Axelrod (1985, 1997) shows that four factors are critically important in preventing parties from moving along pathway B: long time horizons for agreements, regularity of stakes, reliability of information about other's actions, and quick feedback about changes in other's actions.

At another level, resource abundance may arguably lead to conflict if there is a breakdown of governance and the resource is lootable (pathway C). This is the causality that is most often presented by environmentalists and antiglobalization activists when referring to the extractive industries. However, as the two causal links show, it is only likely to be followed if the resource in question is highly lootable and a market is allowed to exist in this lootable resource. Initiatives such as the Kimberley Process for regulating the flow of diamond revenues, linked to financing of conflicts, have attempted to circumvent such causality.[11] Additionally this is an interesting area where the "small is beautiful" approach to industrial activity that many environmentalists embrace might not be applicable either. Small-scale artisanal mining operations, for example, might be more anarchic and harder to govern than large-scale mining operations run by multinationals. Hence success stories of mining activities that have not led to conflict and helped with development and cooperation in countries such as Botswana and Chile often involve large-scale operations as opposed to the alluvial diamond mining or gold-panning in regions such as Sierra Leone or Congo.

There are of course win-win prospects of abundant resource usage leading to cooperative outcomes under ideal governance regimes (pathway D). Political ecology[12] and related literature in environmental justice movements[13] suggest that this outcome is most sustainable if power relations among players are balanced and there are enough opportunities to allow for constructive competition (that which spurs creativity but deters conflict). Good governance systems can certainly allow for such mechanisms to emerge more effectively. Increasingly such governance structures are emerging beyond usual confines of nation states and through the proliferation of deliberative global policy networks. While some of these networks operate through contentious political movements,[14] the tension they exert often cause positive transformation of inertial institutions and ultimately lead to cooperative outcomes.

Environmental Endogeneity

Social scientists trying to study causal relationships of any kind must contend with the problem of "endogeneity"—the direction of causality. Hence environmental cooperation and the resolution of larger conflicts must be considered in this light as well. Is environmental cooperation a result of conflict mitigation or is it leading to conflict reduction itself? The temporal analysis can often be so closely intertwined that the causality confounds researchers. Most politicians are quick to state that a minimal level of conflict mitigation is essential for environmental cooperation to occur. However, I would argue that the process is much more dialectical in nature. Environmental issues can be an important entry point for conversation between adversaries and can also provide a valuable exit strategy from intractable deadlocks because of their global appeal. However, they cannot be taken in strategic isolation and are usually not a sufficient condition for conflict resolution.

The key to a constructive approach in environmental peace-building is to dispense with linear causality and instead consider the conflict de-escalation process as a non-linear and complex series of feedback loops. Positive exchanges and trust-building gestures are a consequence of realizing common environmental threats. Often a focus on common environmental harms (or aversions) is psychologically more successful in leading to cooperative outcomes than focusing on common interests (which may lead to competitive behavior).

Peace Parks as Exemplars?

Now that we have the broader vision of environmental peace-building in mind, let us consider the ways in which this vision could be implemented. Conservation of environmental resources at various levels often requires the establishment of protected areas of land. Land conservation can be a contentious matter on its own because of property rights concerns and the historical misuse of such measures to depopulate areas or cause demographic shifts. The establishment of some national parks in the United States has been directly linked by historians to adverse policies toward Native Americans.[15] Similarly conservation zones for wildlife in South Africa were also linked to misappropriation of land from indigenous Africans.[16] Most recently there has been growing tension between large environmental organizations such as Conservation International, the World Wildlife Fund, and the Nature Conservancy and indigenous rights groups (particularly in South America) with accusations of corporate cooptation in the name of conservation.[17] However, all of these critiques pertain to the management and implementation of conservation plans rather than the concept of conservation itself.

There can, of course, be varying degrees of conservation and heightened local involvement in decisions between areas to address these concerns but the broader vision of bio-regionally based environmental protection remains constructive all the same. The notion of transboundary protected areas (TBPA) or transfrontier conservation areas (TFCA) was developed independent of their potential instrumental use in conflict mitigation. The World Conservation Union (IUCN) played an important role in moving this concept forward and established a task force within the World Commission on Protected Areas for this purpose. In 2001 the task force prepared a monograph that moved the idea one step forward with the suggestion that such TBPA be used for peace and cooperation—giving a renewed connotation to the term "peace park."[18] According to this publication, there were 166 existing transboundary protected area complexes worldwide comprising 666 individual conservation zones.

Previously the term "peace park" had been used to describe memorials such as the one in Nagasaki, Japan, as well as the establishment of such conservation zones for ecotourism and sustainable livelihoods. The establishment of the Peace Parks Foundation in South Africa by Dr. Anton Rupert in 1997 as a means of promoting regional cooperation (primarily in ecological-based tourism) gave greater impetus to the peace parks movement. Following the World Parks Congress in 2003, a Global Transboundary Protected Area Network was established by IUCN and based in South Africa, and a formal typology was also developed to define five different kinds of TBPA as follows:[19]

• *Two or more contiguous protected areas across a national boundary* This is what most people visualize when they hear about a transboundary protected area, but this is only one model. An example is Park "W," which is shared by Benin, Burkina Faso, and Niger and is being managed cooperatively for common conservation aims.

• *A cluster of protected areas and the intervening land* This is a more ambitious approach in that it attempts to balance strict protection with sustainable management in buffer zones and other parts of the landscape. The World Bank is currently developing such a project in the West Tien Shan Mountains of Central Asia; the project will focus first on four protected areas and later extend over parts of Uzbekistan, Kazakhstan, and the Kyrgyz Republic.

• *A cluster of separated protected areas without intervening land* In practice, it is not always politically or practically possible to include intervening land, and some successful transboundary initiatives have involved protected areas that are geographically separated but share common ecology or problems, and usually have some interchange between species. An example currently under development with support from IUCN is a transboundary initiative in the Great Lakes region of Africa involving Kibira National Park in Burundi, Virunga National Park in the Democratic Republic of Congo, and Volcanoes National Park in Rwanda, which all have common management aims but no control over intervening farmland.

• *A transborder area including proposed protected areas*   Some transboundary initiatives have started with protected areas in one country or region, with the hope of extending protection across the border, but without any formal agreement. This might be a transitional stage, with the area later becoming, for example, two or more contiguous areas across a national boundary. The Pha Taem Transborder Initiative between Thailand, Laos, and Cambodia, is developing a complex including four existing and one proposed protected areas in Thailand as well as proposed protected areas in Laos and Cambodia.

• *A protected area in one country aided by sympathetic land use over the border*   Sometimes there will be no realistic expectation (or perhaps no need) for protected areas on both sides of a border, but a need for sympathetic management in one country to safeguard a protected area in its neighbor's country. An example is in the island of Borneo, where improved forest management on the Malaysian side of the border is helping to preserve populations of large animals in the adjoining Kayan Mentarang National Park in Indonesia.

Furthermore the Transboundary Protected Areas Network formally defined the term "peace park" as follows: "transboundary protected areas that are formally dedicated to the protection and maintenance of biological diversity and of natural and associated cultural resources, and to the promotion of peace and cooperation." It is important to note, however, that some scholars prefer a definition of peace parks that does not limit them to being in adjoining border zones but rather in any zone that has endured conflict. As Gerardo Budowski of the University of Peace has argued, the border definition might exclude island states and other remote areas where conflicts might still be fought.[20]

Conservation and Conflict Resolution in the Condor Corridor

Using conservation measures as a direct means of resolving an armed conflict is the most consequential use of environmental peace-building; yet this approach is still in early stages of global acceptance. The first international peace park idea involving an armed conflict between neighboring countries was in the Cordillera del Condor region between Ecuador and Peru. This case deserves special recognition as it was the first formal effort in which conservation groups were actively involved in international conflict resolution, and the resultant peace treaty included explicit mention of conservation measures as part of the overall resolution of the conflict.

The territorial conflict between Ecuador and Peru goes back several decades. In 1995, following several failed attempts at conflict resolution, an armed conflict broke out that lasted for about three weeks. A peace agreement signed in February of 1995 committed both countries to the withdrawal of forces "far" from the disputed zone. This plan was overseen by four guarantor countries: Argentina, Brazil, Chile, and the United States. In compliance with the plan, both nations organized the withdrawal of 5,000 troops from the Cenepa Valley and supervised the demobilization of 140,000 troops on both sides.

With this much accomplished, conservation groups became very active in trying to lobby for a peace park. It should be noted that Conservation International was actively involved in biodiversity fieldwork even before the resolution of the conflict; it had worked closely with the military when fieldwork on documenting the biodiversity of the region was conducted in 1993. Therefore they were gradually able to influence more "hawkish" army officers about the collective importance of conservation and its instrumental use for conflict resolution.[21]

In November 1997, the two nations agreed in the Declaration of Brasilia to address four areas: (1) a commerce and navigation treaty, (2) a border integration agreement that would stimulate much needed development in both countries, (3) a mutual security agreement designed to prevent future conflicts, and (4) a completion of demarcation of land borders. By February 1998, they were able to agree on the first three, but that left the most important one, the demarcation of land borders. Tensions arose again in August 1998 as 300 Ecuadorian soldiers spread out along an 11 kilometer line, 3 kilometers inside Peru and 20 kilometers from the demilitarized zone. To prevent further escalation, and with pressure from conservation groups, the presidents of Ecuador and Peru both met with President Clinton on October 9, 1998, and asked that the guarantor nations make a proposal to mark the border for them. With US satellite mapping they were able to arrive at an agreeable border demarcation. As part of the peace treaty that was subsequently signed, both countries agreed that the area should be designated for conservation purposes while recognizing international borders.

Initially both countries declared national parks on their respective sides of the border. However, in 2000, Conservation International and the International Tropical Timbers Organization partnered with local conservation groups in Ecuador and Peru and with the Chimu indigenous communities (particularly the Schuar of Ecuador) to establish a bioregional management regime.[22] This has resulted in the creation of the Condor–Kutuku conservation corridor in 2004, from which many other peace park proponents can draw some instructive insights. This first test case of comprehensive conflict resolution through conservation, involving a host of organizations, has greatly helped to move the peace park agenda forward in other cases.

Book Structure and Argument Development

The chapters in this book were selected to cover a broad range of issues, while also providing in-depth case analysis. This is among the first books to integratively examine the issue of peace parks and thus must cover a great deal of territory to provide an authoritative voice. The structure reflects the multiple audiences and diverse backgrounds of the contributors. Part I presents a historical overview as well as some methodological and theoretical perspectives on the study of peace parks. Our aim here is to look at economic, political, and social theories on which the concept of peace parks may be predicated and the potential pitfalls that might arise with application of these theories. While most of the chapters in this segment are not case-specific, three cases are presented to illustrate how various theories and tools of peace-building can be applied. Game theory has been used as a tool by economists and political scientists to explore ways in which behavioral responses to uncertainty can cause or limit conflict. Chapter 2 attempts to provide analytical rigor to understanding various scenarios under which peace parks may be established with reference to case examples discussed in the volume.

The Southern African case (chapters 4 and 5) is presented via the discussion of ecological resilience theory, as a landmark effort in postconflict reconstruction and development. This case is especially instructive in developing a broader theory of peace parks because of the simultaneous attention to conservation, economic development, and conflict resolution that the Southern African experience had to contend with. Furthermore these two chapters reveal that varying perceptions of geographic and demographic scale underlie the politics of peace park formation, and may also transcend borders per se. Hence the process of peace park establishment in this context must be concurrent with broader efforts at bridging cognitive and ideological divides.

The Central American case presents an exemplification of how conservation efforts have complemented economic integration in a postconflict environment. This is often believed to be the most fertile ground for peace park formation. The emerging role of global governance structures such as international trade agreements is also placed in context by the discussion of this case. However, the challenges that peace park formation may encounter and the risk of cooptation are candidly discussed. Furthermore the formation of any conservation zone may be beset by micro-conflicts over process and resource distribution. This chapter also provides a revealing problematique that such efforts might present. Thus an effort at environmental peace-making (pathway A in figure I.1) must also contend with forces that might push the dynamics of interactions between players in the direction of pathway B, leading to conflict.

Part II presents a series of case studies of existing peace parks from around the world with two focal areas for analysis: bioregional management and economic development. Often both factors are important policy drivers for conservation efforts and hence deserve detailed case analysis. We start off with a more detailed analysis of the African experience to transition smoothly from part I to part II. An effort was made to select authors that have firsthand field experience of working in conservation zones. It is important to note here that we are largely presenting peace parks between states that have existing positive relations with each other as well as those that are in conflict with each other. While the latter category is of greater interest in terms of conflict resolution analysis, the former category can also provide some insights in terms of how cooperation can be sustained between amicable neighbors through environmental conservation.

The Selous–Niassa Wildlife Corridor exemplifies the role of foreign donors by providing mediation through conservation programs that develop better ties among African countries. This case also reveals the complex negotiating process between scientists and communities in establishing joint conservation zones from bottom-up through donor facilitation (GTZ in this case). The case also shows how such initiatives do not necessarily need formal treaties and protocols among nations to move forward. Indeed the initiatives can be formulated at grassroots and then get government acquiescence, as was the case between Tanzania and Mozambique.

The West African case of the "W" International Peace Park shows how colonial interest in conservation ironically led to a bioregional zone of interest, yet the delineation of boundaries for nation-states at the time of independence did not follow this approach. Nevertheless, the vestiges of bioregionalism from the days of French control have trumped nationalism to allow for the establishment and formalization of the park. This case also shows how asymmetric geographic territory and control between countries such as Benin, Niger, and Burkina Faso does not necessarily have to hinder the establishment of a park. Mutual gains through collaboration to form such a park can be derived even from such asymmetric players.

Nevertheless, we do not neglect the central role of a dominant power in peace park formation as exemplified by the case of the Emerald Triangle conservation zone. Here Thailand is by far the economic and demographic power in the region and has been a key force in galvanizing change with the help of international organizations such as the International Tropical Timber Organization. This chapter also shows us the mechanics of forming contemporary conservation zones and how they can play a positive role in improving relations at a regional level, particularly where riparian corridors are involved (the Mekong in this case).

The Antarctic case (chapter 9) is discussed in this segment as a historical exemplar of cooperation on science. The lessons gleaned in this case are specially revealing of the legal and logistical challenges of cooperation in harsh environments. The Antarctic case also reveals the different kinds of collaboration between both friendly and hostile partners. The role of science as a means of peace-building is also addressed using the Antarctic treaty as an important milestone in global environmental governance.

Most peace parks, so far, have been only between countries that do not have any active conflicts. This could be used as evidence to support the argument that environmental issues are low politics as compared to the high politics of war and peace, or it could be a sign of the lack of strategic vision on the part of politicians and the disconnect between ecology and politics. This volume builds on the latter argument and provides guidance on how we may move from visions of "soft peace" (between friends) to "hard peace" (between enemies). However, all the cases presented illustrate the micro-conflicts of managing a transboundary zone that can arise in various contexts.

Apart from Africa, we also cover peace parks in Southeast Asia and the last chapter of part II provides an example of ongoing bioregional management in North America—one that is often presented as an exemplary standard to many other nations. However, as the chapter shows, conservation measures and even economic development are now being trumped by yet another national imperative—border security and the "war on terror."

Part III continues with the theme of regional security and discusses some cases of peace park proposals that have not been realized on this account. This is the more "visionary" segment of the volume, laying out specific proposals for future peace parks in detail. The cases discussed here are all proposals rather than actual parks, revealing the inchoate nature of this study area and the importance of prospective work in this regard.

We transition from the US–Canadian case from part II to the US–Mexican case in part III in which regional security issues are far more crucial. The chapter provides an interesting comparison between economic cooperation versus environmental cooperation. While Mexico and the United States have managed to agree on agreements such as NAFTA and encourage industrial corridors such as the Maquiladoras,[23] despite concomitant security concerns, proposals for a peace park have come under far greater scrutiny and resistance. This case appears ripe for a public–private partnership approach to peace park formation as exemplified by initiatives such as the NAFTA Commission for Environmental Cooperation.

Postconflict reconstruction can also provide an important opportunity for peace park formation as exemplified by the chapter on Liberia. Managing the chaotic times of such transitions can be daunting such as the "invasion" of 4,000 miners into Sapo National Park, many of whom were ex-combatants. However, bringing many of these factions within the conservation community while also providing livelihoods to them is indeed possible. Instead of fueling conflict, appropriate management of the situation by the park staff mitigated potential civil strife. The formation of a Civilian Conservation Corps in this context is particularly instructive and could be emulated by other peace park aspirants.

The case of the Kuril Islands shows us how the potential resolution of a territorial dispute independent of a peace park arrangement can be environmentally deleterious. The author argues that regardless of whether the islands fall under Russian or Japanese control, the environmental management regime that best manages the ecosystem efficiently would need an arrangement similar to a peace park. Furthermore the case shows how conventional notions of rational behavior on the part of states that have much to gain from bargaining (resource-rich and cash-poor Russia versus resource-poor and cash-rich Japan) might not always materialize. Hence the proposal that Japan buy the islands from Russia (similar to Alaska) has been rejected. In such cases where irrational and intangible notions such as national pride and prestige are at work, peace parks might also provide a respectable exit strategy.

The cases of active conflict that are discussed in this section, such as between India and Pakistan, the Koreas, and Central Asia and the Middle-East, are all conducive to using peace parks as an exit strategy from conflict entrapment. Three acute international military conflicts that are now being considered as possible peace park contenders are discussed in the last chapters of the volume. While this vision may seem Utopian to some, the ongoing efforts to move these conservation zones forward from multiple channels is refreshing and reassuring to even the most pessimistic observers. For example, the formation of a peace park in the Mesopotamian marshlands between Iraq and Iran has been welcomed by all sides as a means of bringing Shi'as and Sunnis together and healing the wounds of the three wars fought in this area within the last two decades.

The chapters pertaining to these conflicts have been authored by conservation practitioners, military personnel, and strategic analysts, who are intimately familiar with the regional geographies as well as the political climate of each case. The aim here is to go beyond theory and seek means of implementation. What is striking about these conflicts is that the rationale for their perpetuation is termed more broadly as security. However, a redefinition of security to include environmental integrity may be in order. These chapters do not suggest that environmental harm will necessarily lead to conflict but rather that ecological deterioration and an inability to study it, such as the loss of water from Karakoram glacial meltwater, can reduce human development and hence make countries more vulnerable to civil strife. These cases also show the various ways by which peace park movements have arisen at an international level.

In the Korean case the demilitarized zone (DMZ) has ironically become a default sanctuary for wildlife, since there is no development activity occurring in the area. Several conservation biologists were thus attracted to using this high biodiversity characteristic of the region to develop a conflict resolution strategy between the two countries. An organization called the DMZ Forum, established in the United States in 1998, has been lobbying for this proposal to be included as part of the six-party talks. Media magnate Ted Turner has been instrumental in popularizing this effort, most recently in his visit to both North and South Korea in August 2005. In his conversations with leaders in the North, he told the media that the North Koreans were receptive to the idea but felt "preoccupied with the six-party talks"—an indication that the initiative is still perceived external to the core negotiations within the talks.[24] While this may give credence to realist assertions that environmental issues will likely be a consequence rather than a constituent of peace agreements, they certainly provide a means of trust-building and help to transform the collective psyche of conflicting parties.[25]

Conservation and Conflict Resolution: How the Twain Shall Meet?

As we endeavor to use collective environmental protection as a means of conflict resolution, let us not forget that conservation efforts are often causes of tremendous conflict. Environmental organizations and their relationship to communities have recently come under attack from both sides of the political spectrum. Since many peace park projects are often being promoted by such organizations,[26] the legitimacy of these groups is essential for meeting the goals of conflict resolution. Such conservation groups are thus often faced with the dilemma of whether or not to give primacy to their ecologically determined conservation objectives.

On the one hand, pragmatic eco-revisionists have "attributed so many of environmentalists' failures to the incuriosity about the human (read: social) sciences, like social psychology and their scientific fetishization of the 'natural' sciences."[27] At the same time some anthropologists have taken this criticism a step further by challenging conservationists about their detachment from indigenous people in their pursuit of conservation. In a much publicized article for Worldwatch magazine, anthropologist Mac Chapin recently critiqued the work of the Nature Conservancy, the World Wildlife Fund, and Conservation International by asserting that "as corporate and government money flow into the three big international organizations that dominate the world's conservation agenda, their programs have been marked by growing conflict of interest—and by a disturbing neglect of the indigenous people whose land they are in business to protect."[28]

Anthropologists and conservation scientists have encountered this debate before in various guises. An article in *Conservation Biology* by Schwartzman et al. (2000) that gave primacy to indigenous conservation practices had sparked a similar heated debate with responses from conservationists such as Redford and Sanderson (2000). Interestingly enough the disagreement here was between staff scientists at major environmental groups—some of whom were more unequivocally sympathetic to indigenous concerns over conservation priorities. Such a divergence highlights the "varieties of environmentalism" that Guha and Martinez Allier (1997) have alluded to in their work on social movements.

Yet environmentalists are collectively also accused all too often by those on the Right of the political spectrum for being too positional and uncompromising in their approach to problem-solving and not interacting adequately with free-market interests. Even Conservation International, which is often accused by more traditional environmentalists of accepting large contracts and grants from oil companies and development donors,[29] is just as much criticized by industrialists for not willing to compromise enough on extractive projects in ecologically sensitive places such as Madagascar.[30]

Environmental and human rights groups are thus often lumped together by critics of nongovernmental organizations (NGOs), such as Sebastian Mallaby (2004) or Clifford Bob (2005) who decry their unwillingness to compromise on urgent development projects. While this "damned if you do and damned if you don't" situation may lead many groups to remain sanguine, the conflicts that may be generated by conservation efforts clearly need to be addressed. The formation of peace parks must be considered part of this process of internal reconciliation as well as the extant motive of instrumental conflict resolution if it is to be sustainable.

As an environmental planner who must constantly contend with prospective and prescriptive matters, it was especially rewarding to bring together chapters that provide strategic guidance for conflict areas and show how such layers of conflict are reconciled. The proposed peace parks are presented with a measure of modest realism, while challenging realist notions of environmental issues are being relegated to "low politics." The chapters provide enough detail and instructive insights to make them plausible and appropriate for a policy audience.

When dealing with matters as emotive as environmental protection and conflict mitigation, one can't help but feel a sense of urgency and advocacy for a phenomenon that holds promise in harmonizing these two worthy goals. This book has been written at a time of transition when peace parks are being recognized as a phenomenon by some while being dismissed as a side story by others. However, there is little disagreement that if managed and implemented effectively, conservation with community consent and conflict resolution are goals worth pursuing. My aim in this volume is not to be a green or blue activist but rather to present a story of measured hope with analytical persuasion.

Notes

1. "The woman who planted trees." *The Economist*, October 14, 2004.
2. Gleditch and Urdal (2004).
3. Global Transboundary Protected Areas network Web site ⟨http://www.tbpa.net⟩.
4. Spence (2000).
5. The conventional environmental security argument presented by Homer-Dixon (1999). Instead we are positing a more nuanced human security argument offered by scholars such as Najam (2003).
6. Deudney and Matthew (1999).
7. Ali (2003) and Conca and Dabelko (2003).
9. Wolf (2002).
10. Keck and Sikkink (1998).
11. The Kimberley process was initiated by Global Witness (based in London) and Partnership Africa Canada as a means of certifying diamonds from nonconflict zones. The process has been successful in convincing governments to sign on to a stringent system of certification and compliance assurance for raw stones along the supply chain to the point at which they are polished for jewelry. For more information refer to ⟨www.globalwitness.org⟩.
12. Peluso and Watts (2003).
13. Suliman (1997).
14. Tarrow (2005).
15. Spence (2000).
16. Beinart and Coates (1995).
17. Chapin (2004). See also the response by various environmental groups and readers to this article in the February 2005 issue of *Worldwatch*.
18. Sandwith et al. (2001).
19. Typology from the Global Transboundary Protected Areas network ⟨http://www.tbpa.net⟩.
20. Budowsky (2003).
21. Conservation International published a detailed biological assessment as part of its Rapid Assessment program in 1997, and acknowledged the help of the Ecuadorian army for logistical support.
22. These indigenous communities have erroneously been called "Jivaros" by early European settlers. The process of engagement with the Schuar was managed by the Natura Foundation of Ecuador. See Ponce (2004).
23. Kamel and Hoffman (2002).
24. Ted Turner's interview with *Korea Times*, August 17, 2005.
25. Van Vugt (2000).

26. For example, the most vocal support of transboundary conservation has come from groups such as the World Wildlife Fund and Conservation International, which recently published a pictorial volume on transboundary conservation; see Mittermeier et al. (2005).

27. Schellenberger and Nordhaus (2005).

28. Chapin (2004). Sociologist Steven Brechin seconded Chapin's criticism in a subsequent comment to *Worldwatch* by stating "close relationships with wealth extractive industries like oil and gas may prevent conservation organizations from critiquing or challenging their corporate donors" (Brechin 2005). The magazine gave an opportunity for the leaders of these organizations, as well as the Ford Foundation, to respond and they collectively showed a measured exasperation with this criticism.

29. Conservation International has also established a Center for Environmental Leadership in Business, which is funded by numerous corporations.

30. Personal communication with Conservation International staff, Antananarivo, Madagascar, November 2005.

I
Environmental Peace-Building: Theories and Histories

The modern "ecological pacifist" is also awed, specifically by the biosphere and refrains to dismiss the question of its origin by a facile answer ("That's the way it was created" or "That's the way it evolved through natural selection."). He is more concerned with pursuing this question in all its complexity and most concerned with conserving the magnificence of the product.

—Anatol Rapoport, *Ecological Peace*, 2002

I know of no political movement, no philosophy, no ideology, which does not agree with the peace parks concept as we see it going into fruition today. It is a concept that can be embraced by all.

—Nelson Mandela, Public comments at Kruger National Park, December, 2001

1

Measuring Peace Park Performance: Definitions and Experiences

Anne Hammill and Charles Besançon

Developing environmental conservation zones as a means of conflict resolution is an attractive concept for conservationists and politicians alike. While peace parks may indeed contribute to a culture of peace and cooperation, their performance is as yet untested. In order to embark on a global policy project pertaining to peace park establishment, we must first be able to benchmark "peace performance" of such efforts. Empirically measuring such "soft" impacts is difficult due to the abundance of other factors that confound this type of analysis. Further complicating such assessments are experiences from the development sector that demonstrate a clear contribution to conflict of conservation groups. In particular, these environmental enthusiasts have come under considerable criticism from indigenous rights activists for sparking conflicts in remote subsistence communities.[1]

Because of the massive scale and complexity of environmental and social issues confronting the global community, new methods of understanding and responding to these challenges are being formulated, including an ever-growing geographic scale of conservation initiatives. International conservation organizations have already organized themselves around ecoregions (Nature Conservancy, World Wildlife Fund for Nature), biodiversity hotspots (Conservation International), and heartlands (African Wildlife Foundation), to name but a few.

Transboundary conservation is yet another example of this increasing geographic scale. Some of these initiatives are on a monumental scale such as the European green belt (6,800 kilometers), which follows the old route of the Iron Curtain separating Europe from the then USSR, and the Meso-American biological corridor, which is over three million hectares. Throughout this chapter we use the term "transboundary conservation" to refer to the full range of conservation initiatives that cross international borders. Many more internationally recognized definitions exist, and we wish to make special note of two of them that we will refer to throughout this chapter. First, IUCN—the World Conservation Union—defines a transboundary protected area as:

An area of land and/or sea that straddles one or more borders between states, sub-national units such as provinces and regions, autonomous areas and/or areas beyond the limit of national sovereignty or jurisdiction, whose constituent parts are especially dedicated to the protection and maintenance of biological diversity, and of natural and associated cultural resources, and managed cooperatively through legal or other effective means.[2]

IUCN goes on to define parks for peace as:

Parks for peace are transboundary protected areas that are formally dedicated to the protection and maintenance of biological diversity, and of natural and associated cultural resources, and to the promotion of peace and cooperation.

Transboundary conservation areas and peace parks have been noted to contribute to a culture of peace and cooperation between nations through the following:

Through acting as a symbol of ongoing cooperation between nations with a history of peace.

By creating an entry point for discussions between neighboring countries that may be deeply divided over economic, social, environmental, or other interests.

By increasing security and control over resources in border areas so that their rightful owners can benefit from them.[3]

By creating shared opportunities for ecotourism and sustainable development ventures on a regionwide scale, an important step in postconflict reconstruction.[4]

By developing a rich and resilient web of relationships among protected area managers from the countries involved, other government actors, including those at ministerial level or higher, local, and international NGOs and the donor community.

These relationships can be useful in providing a platform for humanitarian organizations to operate in the case of renewed conflict or natural disaster.

While the origins of transboundary conservation go back to the early part of the twentieth century, when in 1932 Waterton Glacier International Peace Park was established across the borders of Canada and United States, there has been a relatively recent resurgence and enthusiasm for the transboundary approach. The number and extent of internationally adjoining protected areas now being established are at an unprecedented rate. In 1988, there were a total of 59 complexes involving 136 countries.[5] In 2005, the number had jumped to over 188 complexes involving over 818 countries.

This recent proliferation of transboundary initiatives have been generally welcomed as a sign of goodwill and cooperation, particularly in areas with relatively recent histories of conflict. Touted by Nelson Mandela as a concept embraced by all, peace parks represent the confluence of several mutually reinforcing interests, namely those of biodiversity conservation, economic development, cultural integrity, and regional peace and security. The possibilities are impressive and attractive (especially to donors): large, contiguous ecological habitats that simultaneously protect biodiversity, create widespread opportunities for tourism venture investment, alleviate poverty, reunite previously separated ethnic groups, and promote good political relations between neighboring states. Yet their peace-building potential is rarely documented or evaluated. Cooperation and peace-building is an assumed outcome of bringing together different—and sometimes, previously opposing—stakeholders for the common purpose of managing biodiversity and protecting livelihoods. This may not always be the case, however, especially in conflict-prone settings where many transboundary conservation areas are being established.

Peace parks hold tremendous significance as arbiters for peacemaking, but in regions currently experiencing conflict or those with a history of conflict, they can inadvertently exacerbate conflict. To understand this contrary phenomenon, it is instructive to examine the links between protected areas and conflict.

Protected Areas and Conflict

As methods for protecting and maintaining biological diversity, protected areas (PAs) are central to global conservation strategies. But their overriding ecological goals do not render them socially and politically benign. Protected areas represent different things to different groups. For conservationists, they are an effective measure for protecting biodiversity; for private tourism companies, a basis for ecotourism development; for pharmaceutical companies, a source of genetic information for drug development; for oil and mining companies, an unexplored supply of revenue; for the military, a refuge and strategic target during times of violent conflict; and for surrounding local communities, PAs can signify restricted access to livelihood resources, forced relocation, or opportunities for income generation through tourism revenues. These different political understandings of PAs are a construct of broader social, cultural and economic forces at work. In many developing countries these forces include social inequality, poverty, contested resource rights, corruption, ethnic tensions, and colonial legacies. As mechanisms of resource control and power, PAs can create situations of multidimensional politicized resistance. That is to say, protected areas can be catalysts of conflict when established in economically disadvantaged regions, where surrounding communities are heavily dependent on natural resources for their livelihoods and survival. PA policies can translate into restricted access to these livelihood resources or forced relocation from traditional lands, which can undermine economic security and cultural identities. Even where provisions are made to allow for limited local resource access or to financially compensate communities, crop damage from wild animals, unequal distribution of benefits, conflicting resource rights regimes (statutory vs. customary), and exclusionary and/or nontransparent decision-making processes can continue to fuel tensions. Where PAs bring up memories of elite control and colonial power dynamics, protected areas can symbolize legacies of imperial domination.[7] The perceived imposition of unjust policies could mobilize group identities and become a rallying point for resisting authority, engendering instability and conflict.

Further, apart from instigating social conflict, protected areas can play a strategic role in sustaining ongoing military conflicts. This role is usually the result of inherent geographic conditions that make PAs valuable in the first place. The remote and relatively inaccessible location of some PAs can make them ideal refuges for military groups, as they offer physical protection, food, water, fuel, and medicine.[8] The high concentration of wildlife can provide a ready supply of bushmeat for armies.[9] Guerrilla groups in Colombia, Sierra Leone, Burundi, India, and Nepal, for example, have established bases in protected areas, with destructive impacts on PA infrastructure, management operations, and personnel.[10] Because of their strategic value protected areas can become targets in military operations. Some groups may deliberately contaminate water supplies and defoliate or burn forests in order to deprive opposing forces of shelter and resources. In 1991 the Rwandan army cut 50 to 100 meter swaths of bamboo forest that link the Virunga volcanoes in order to minimize the risk of rebel ambushes.[11]

In addition to providing physical support to military groups, resources in protected areas help finance military operations. Wildlife, timber, oil, or minerals can be plundered and sold to local and foreign markets in order to pay troops and purchase weapons. For example, the Angolan rebel group UNITA (National Union for the Total Independence of Angola) reportedly financed their military campaign through sales of ivory, teak, oil, and diamonds.[12] Similarly in Mozambique, elephant poaching and the ivory trade helped finance insurgent activities, while Charles

Taylor's coup in Liberia was made possible through revenues from timber and valuable minerals.[13] Moreover the consequences of financing wars with natural resources from protected areas extend further than immediate biodiversity loss or ecosystem degradation. Such unmitigated exploitation of valuable natural resources can drain the capital needed for development and social programs, undermining postconflict reconstruction and longer term sustainable development efforts.

Postconflict settings can give rise to new conflict issues for protected areas. Refugees, internally displaced people (IDP), and demobilized troops move into protected areas, as they contain unsettled lands and livelihood resources. In some instances resettlement in PAs can be encouraged by governments when there is no other land available and the overarching priority is to establish peace, address immediate humanitarian needs, and create some semblance of order. Following the Rwandan genocide in 1994, 50 percent of the country's population was estimated to be displaced or temporarily settled. Hundreds of thousands of refugees crossed the border into the Democratic Republic of Congo and settled in and around Virunga National Park (Lanjouw 2003), while the Rwandan government opened portions of Akagera National Park to resettlement and considered proposals for degazetting 5 percent of Volcanoes National Park to accommodate IDPs. The acute need for land, shelter, and resources that draws displaced and demobilized populations to PAs (and their immediate surroundings) have the potential for fueling further tensions and conflict. When host communities, who are also dealing with the social and environmental consequences of war, are faced with competition for livelihood resources from refugees and IDPs (sometimes of different or previously opposing ethnic groups), tensions can rise and conflicts can (re)ignite. When considered against a background of widespread arms circulation, demobilization, and general disorder and confusion in postconflict settings, the gathering of different groups in refugee camps or settlements around relatively resource-rich protected areas can become a conflict risk.

Thus protected areas are linked to the conflict *problematique* through their interaction with the complex social and political forces that traditionally fuel tension. The impacts of PAs on local livelihoods, resource rights, distribution of wealth, established management and power structures, and group identity can create grievances that, when left unaddressed, can escalate into more open forms of conflict. The more strategic and passive role of PAs in supporting militarization, warfare, and postconflict reconstruction, on the other hand, is often the result of geography, resource abundance, and a breakdown in governance and authority. Protected area management can therefore be both a contributor to and a target of local/regional conflict dynamics.

This is not a new or unexplored development—the conservation community has long searched for an optimal resolution to people-versus-nature conflict, where biodiversity protection goals would not be met at the expense of social and cultural needs. Similarly, in the wake of rising levels of local and regional violent conflicts, conservationists have been developing guidelines and management strategies for maintaining basic levels of biodiversity protection in times of conflict.[14] In addition to the PA and conflict issues described above, there are a set of other considerations unique to the transboundary context that have implications for the peace and conflict setting.

The most obvious of these considerations is the incorporation of international boundaries in transboundary conservation areas. While borders are areas with some of the world's most biologically intact ecosystems, they are also where "inequities surface and conflicts erupt."[15] Including international boundaries can therefore add a "gratuitous layer of complexity"[16] to the dynamics of PA management. Borders are political constructs that function as mechanisms of inclusion and exclusion. In many parts of the developing world, current international borders were arbitrarily drawn by colonial powers that paid little attention to the ensuing division of indigenous communities and "heritage territories."[17] This has in some cases, particularly in Africa, resulted in ambiguities about citizenship and national loyalty among border communities, fostering suspicion and political marginalization by centralized authorities. Such conditions can promote anti-national or criminal activities, including the smuggling of goods and people across borders, which can in turn contribute to the creation or escalation of tension and conflict.[18]

In addition to exacerbating political inequalities between local community and state actors, transboundary conservation can emphasize disparities between countries. Katerere and colleagues refer to this potential problem in their critique of transboundary natural resource management in Southern Africa:

The problem of distribution and access to natural resources as well as access to finance, technology, and skills is not limited to intra-state inequities. At the regional level inter-state inequities arise from differing resource endowments and the dominance of larger and economically powerful states like South Africa and to a lesser extent Zimbabwe. These differences tend to fuel economic resentment among the states in the region and conflict claims over natural resources. In some instances the inter-state inequities have fuelled xenophobic reactions towards citizens of poorer neighbors who try to seek better opportunities across their borders.[19]

Indeed national sovereignty issues can play a major role in further complicating the process of transboundary conservation area establishment and management, especially if there are outside forces driving the agenda. Formal transboundary agreements can cause more inter-state disputes than they alleviate when there is reluctance on the part of security officials, immigration, and other government representatives to cede authority. Opening up borders in remote areas can translate into increased levels of poaching and smuggling, although transboundary conservation area arrangements may allow for better monitoring and overall management presence to help curtail such activity. Because establishing transboundary areas is a lengthy process, in some cases it does not make sense to formalize an agreement at a state level. Rather, the optimum level of agreement resides at the managerial level and is often informal rather than formalized. For example, this particular arrangement has worked well in the transboundary Virunga Bwindi region that is the subject of this chapter due to tensions that have precluded high-level state meetings of Rwanda with the DRC and Uganda.

Peace and Conflict Impact Assessment

Since transboundary protected areas have the potential to either fuel tensions or foster cooperation, there is a need for analytical tools and processes that help frame the issues and examine the linkages. Specifically, we need to better understand the peace and conflict contexts in which TBPAs are established, and how TBPAs interact with these contexts to either exacerbate conflict of promote peace-building. In so doing, conservationists and policy makers can "conflict-sensitize" transboundary conservation areas and peace parks in conflict-prone or conflict-affected areas, maximizing their positive impacts and minimizing unintended negative consequences. We can turn to the development and humanitarian sectors for guidance, where the concept of conflict-sensitivity, and the various frameworks and tools for achieving it, has been gaining currency over the last ten years.

In the mid-1990s, amid allegations that humanitarian and development assistance might be generating or exacerbating violent conflicts, practitioners began developing and debating methods for anticipating and assessing the impacts of their activities in conflict zones. The impetus came from an increasing recognition and concern among NGO staff that their presence was feeding into—even prolonging—conflicts. According to Mary Anderson (1999), this was achieved in two significant ways:

through the transfer of aid resources, ranging from theft of aid supplies by armies and militias, to perceived inequalities in their distribution, and through implicit ethical messages delivered via the agency's modus operandi, such as the use of armed guards to protect staff (legitimating the use of arms to determine who gets access to aid), personal (mis)use of aid-related goods and services by agency staff (implying that those who control resources may use them for personal benefit without being held accountable), or evacuation policies that call for the removal of expatriate staff while leaving local staff behind (demonstrating that some lives are more valuable than others).

In other words, aid can unexpectedly distort social relations, entrench socioeconomic inequalities, and allow elite and/or armed groups to benefit disproportionately from unrest.[20]

With this growing understanding of the potential negative effects of aid on the conflict environment, there was a move to evaluate and restructure aid programs so at the very least, agencies could fulfill their mandates without simultaneously contributing to conflicts—that is, institute "do no harm" approaches in the design and operation of their projects in conflict zones. At the most these programs would strive to help local people disengage from conflict and effectively address the underlying causes of tension.

In response to this identified need, a number of approaches were developed for integrating the conflict perspective into the planning, monitoring, and management of development projects. Among these was the Peace and Conflict Impact Assessment (PCIA), an idea that emerged in the mid 1990s through the work by Ken Bush (1998) and Luc Reychler (1999). Although debates continue over the structure and practical application of these assessments, several northern development agencies (both governmental and nongovernmental) have seized on the idea, advocating their development and use in programming.

For the purpose of this discussion, we take peace and conflict impact assessments to encompass a broad suite of analytical frameworks, methodologies, and tools that help evaluate how an intervention (by project, policy, program, etc.) can or does affect peace and conflict dynamics in conflict-prone areas—meaning whether it increases or diminishes the prospects for long-term peace. The aim in using PCIAs is to not only reduce the unintended negative consequences of a project, but identify and optimize opportunities for peace-building. In so doing, PCIAs can lead to a discernable improvement in the quality of development and humanitarian assistance.

The end result is not necessarily a change in the types of interventions but a change in *how* they are implemented.[21]

PCIAs differ from traditional project/program evaluations in that assessments are conducted through a conflict lens. PCIAs cannot evaluate the overall success or effectiveness of an intervention—measured against stated goals, objectives, and indicators—but only their peace-added (or peace-reducing) value. Despite calls to develop a universally applicable methodology, PCIAs continue to be open and flexible in their structure and use, customized to meet the needs/objectives of different users. They can draw from a range of information sources (project documents, media reports, stakeholder consultations, etc.) and employ different analytical tools (indicators, qualitative issue-based inquiries, conflict analysis frameworks) to ascertain a project's impact on the peace and conflict environment, and vice versa. Moreover PCIAs can be carried out by different development actors and at different stages of a project cycle, again depending on the purpose of the assessment.

Generally speaking, most PCIAs involve the following components or steps:

1. *Conflict mapping*   Analysis of the causes, actors, and dynamics of a conflict affecting a project/program site.

2. *Project/program mapping*   Overview of the intervention's purpose, objectives, location, timing, beneficiaries, personnel, operational partners, and physical and financial resources.

3. *Assessing impact of conflict on project*   Examination of how the conflict has affected the design, implementation and management of the intervention, as well as any measures that have been taken to address these impacts.

4. *Assessing impact of project on conflict*   Examination of how different aspects of the intervention can or has contributed to conflict and/or peace-building in the area—that is, do they address the root causes of conflict, such as poor governance and corruption, lack of socioeconomic opportunities, inequitable access to natural resources, and lack of participation?

5. *Recommendations*   Based on the above, how can the intervention be modified so that it meets its objectives while simultaneously strengthening the structures for peace-building?

PCIAs can offer a means for understanding how interventions interact with the factors that increase the chances for peace or conflict. However, for PCIAs to be useful and effective, they must involve those individuals and groups living in conflict zones.

Relevance of PCIAs to Conservation Interventions

Besides their applicability to traditional development and humanitarian interventions, PCIAs are also relevant to conservation projects and programs in conflict-prone areas. As we explained earlier in our discussion on the links between PAs and conflict, conservation interventions are not apolitical. In fact such intervention, as one among many others, "contributes heavily to shifts in power dynamics in rural areas that are already highly politicized," as Wilshusen et al. (2002, p. 24) point out. "This is a result of [the conservation community's] relative wealth and influence compared to most local actors. In short, conservation practices are not benign. They alter the local playing field, sometimes drastically" (ibid.).

Of course, it goes without saying that conservation interventions affect more than ecosystems; they have implications for economic livelihoods, community and cultural identities, and political autonomy and control. The propensity for creating or exacerbating social or political tensions is greatest in areas where people rely most directly on access to natural resources for their survival and well-being. Many of these areas also happen to be in conflict zones, requiring a more sensitized approach to working with people and institutions for the achievement of specified goals.

Why Conduct Peace and Conflict Impact Assessments on Transboundary Conservation Projects?

Accepting that transboundary conservation is a type of conservation intervention, why does their establishment or management warrant the use of PCIAs? Apart from the afore-mentioned protected area-related conflicts that are also relevant to transboundary conservation areas, there are two additional reasons that suggest a need for conducting PCIAs. First, many transboundary conservation areas are established in conflict-prone areas in order to fulfill peace-building objectives. Unlike other traditional conservation interventions aimed at creating protected areas, transboundary conservation areas are for the most part being limited to regions with recent histories of conflict. These regions include Southern, Eastern, and Central Africa, South and Southeast Asia, and Latin America.

Second, many transboundary conservation areas are defined as peace-building projects, or at least as contributing to peace-building. Intuitively it makes sense to get previously opposing interests to come to the table to cooperate on a mutually important priority—biodiversity conservation and economic development. While this initiative could form the basis for building trust and friendly relations, given the violent histories between some of the parties, its peace-building potential could be a dangerous assumption to make. According to Bush, the first step in evaluating peace-building projects is to disregard "their self-described face value." There is growing recognition of the need to systematically evaluate them, and identify where gaps exist. Understanding how an intervention contributes to peace-building is as instructive as understanding how it contributes to violent conflict.

Given the recent proliferation and enthusiasm for transboundary conservation and the growing concern about the relationships between protected areas and conflict, it is necessary for the conservation and development community to follow the lead of the humanitarian community and ensure that their contributions "do no harm."

Conducting a Peace and Conflict Impact Assessment of the Virunga–Bwindi Region
A pilot project in the central Africa region is utilizing the Peace and Conflict Impact Assessment (PCIA) approach to the conservation of both peace and conflict in the region. At best, the peace-building goal is to keep present and future conservation projects from contributing to conflict.

Since 1991 the International Gorilla Conservation Program (IGCP) has been working in the Virunga–Bwindi region of Central Africa to protect the mountain gorillas and their Afromontane habitat. The region is comprised of a chain of forested volcanoes that straddle the countries of Rwanda, Democratic Republic of Congo, and Uganda. With an ecosystem split along national boundaries and into three contiguous national parks (and the disjunct region including Bwindi Impenetrable National Park and the Sarambwe Conservation Area), the Virungas exemplify the importance of regional collaboration and a more holistic approach in biodiversity conservation. Recognizing this, IGCP has applied a regional approach to its conservation work, emphasizing transboundary collaboration among the Virunga–Bwindi countries and their respective protected area authorities. Because violent conflict continues to threaten mountain gorillas and their habitat, as well as place conservation personnel in high-risk situations, the sociopolitical context of the region has significantly shaped IGCP's approach to conservation.

The history of conflict in the Virunga–Bwindi region is complex and deeply rooted. A detailed explanation would go far beyond the scope of this chapter. Nevertheless, much of the conflict can be traced to the establishment and polarization of group identities whose respective histories remain contested.[22] The most accepted version of the region's history, however, identifies the original inhabitants of the area as Twa pygmies, who were displaced by Hutu farmers in the eleventh century. Bantu pastoralists (Tutsis) moved into the area from either Ethiopia or Sudan 500 years later. This powerful Tutsi group created the *Ubuhake* system, whereby Hutus became indentured to Tutsis and power and land ownership was focused on Tutsi *Mwami*, or kings.[23]

The most widely known conflict in this region was the 1994 genocide in which approximately one million Tutsis and moderate Hutus were massacred over a three-month time period. This massacre caused about two million people to flee Rwanda because they either were Tutsis fearing for their lives or were Hutus implicated in the genocide. Some of the exiles, called the *Interahamwe*, have continued their fight against the Rwandan government until very recently, and they have taken refuge inside the Virunga Massif and Bwindi Impenetrable National Park, from which sites they could launch their attacks.

In 2004, after almost fifteen years of gorilla conservation and amid growing international recognition for its transboundary approach in a region beset by inter- and intra-state violence, IGCP became interested in systematically assessing the peace and conflict impacts of their work. In addition to learning the root causes of conflict, they were intent on documenting the effects of the peace-building initiative of transboundary conservation in the region and the value the PCIA has had in the process.

Conducting a PCIA of the IGCP's work makes sense for a number of reasons. The IGCP's projects are in conflict zones where there are ongoing or recent histories of violent conflict, as well as in areas where the potential for future conflict exists. Also, since peace-building is an explicit objective of some of IGCP's programming, particularly their transboundary activities, regular regional meetings are held to promote dialogue and cooperation among the protected area managers from the three countries and with the wider NGO community. There is a growing need to systematically evaluate progress, and identify where gaps exist. Finally, there is documented and/or anecdotal evidence of conservation activities contributing to both peace-building and to tension in the region.

To engage the support of consultants and the International Institute for Sustainable Development (IISD), IGCP organized a preliminary scoping mission to identify the different types and levels of conflict in the region, and how they are linked to the protected areas. Consultations with staff at conservations organizations, protected area managers, and representatives from community organizations produced a breakdown of conflict types in the Virunga–Bwindi region (see table 1.1). From

Table 1.1
Conflict typology in PCA for Virunga Bwindi Region

Conflict level	PA-related conflict issues
Inter-state conflagrations	Refugee flows to PA surroundings; subsequent environmental degradation; loss of livelihoods from armed conflict, leading to heavy reliance on PA resources; exploitation of PA resources by armed groups
Intra-state armed groups versus communities	Loss of wildlife from armed conflict; military or rebel encampments within PAs; looting, raping, and killing of local people by armed groups in PAs; destruction of infrastructure for local livelihoods, leading to further poverty and exploitation of PA resources
Local protected areas versus communities	Community displacement; tensions over access to PA-related benefits; tensions over human–wildlife conflicts; exclusionary decision-making practices; illegal poaching and resource use; encroachment on to PA lands
Hostilities among community members	Conflict over access to same PA benefits and resources

this scoping exercise, an in-depth PCIA research process was developed with the following main goals:

Assist the IGCP in achieving their conservation goals by integrating conflict-sensitivity into their activities so that peace-building opportunities are maximized and threats to biodiversity conservation are addressed.

Develop knowledge about the relationship between conservation and peace/conflict dynamics that can contribute to conservation thinking and practice beyond the central African region.

To achieve these goals, the following program was proposed:

Analyze how IGCP activities address—both positively and negatively—the root causes of conflict in the Virunga–Bwindi region.

Integrate the results of the PCIA analysis into the management and monitoring plans of IGCP activities (including the development of peace and conflict indicators).

Build the capacity of IGCP staff to design, conduct, and implement the results of PCIAs.

Develop general guidelines for PCIA processes in other transboundary conservation initiatives.

Communicate the value of conflict-sensitivity in transboundary activities to the broader conservation community.

These directives are expected to be carried out over an 18-month period. Using a combination of desk-based analysis, field research, and stakeholder consultations, the IGCP is gaining insight into how their conservation activities are linked to local and regional peace and conflict dynamics. Four specific IGCP conservation interventions were selected for in-depth analysis, one of which focuses on its transboundary work. Attempts to understand this work's contribution to peace-building drew from the emerging literature on environmental peacemaking, as Conca and Dableko (2002) explain:

> The basis for this [environmental peacemaking] claim lies partly in the general conditions understood to facilitate cooperation, partly in the issue characteristics common to many environmental problems, and partly in the kinds of social relations that are engendered by ecological interdependencies (p. 10).

They go on to describe two ways in which environmental cooperation may occur (pp. 10–13):

Change the strategic climate   Environmental problems can be treated as opportunities in conflictual situations. That is, discussions over environmental issues can be a means to create a modicum of trust, cooperation, and transparency among peoples, thereby improving the "contractual environment" in the bargaining process.

Strengthen post-Westphalian governance   Use environmental concerns to deepen trans-societal linkages, strengthen regional identities, and transform state institutions to become more open, democratic, and accountable. That is, look outside of formal, state-sanctioned negotiations to broader social dynamics.

Transboundary conservation initiatives, such as those supported by the IGCP, have the potential to promote environmental cooperation in both such ways. For example, the technical cooperation needed to establish and manage PAs across borders could create an opening to other forms of cross-border cooperation. Then the opening of borders could allow for animal migrations and personnel exchanges could deepen transnational relations (personal and economic) and regional identities, thereby lessening the incentives for conflict. Conservation could thus prove to be one of the more viable opportunities for peace-building in a conflict-prone or conflict-affected setting such as in the Great Lakes region.

The IGCP's own transboundary work emphasizes collaboration at all levels, from the protected areas to governments in the three countries, and the importance of effective communication and sharing of resources. Specific regional activities include ecological monitoring and surveillance, tourism development, joint training, planning, and community initiatives and management planning. The IGCP also organizes regional meetings every three months with the protected areas authorities in the three countries, NGOs who are active in the region, and consultants. The purpose of these meetings is to coordinate park management, share experiences, and receive some sort of education or awareness about certain topics, ranging from veterinary health to conservation enterprises, and to conflict resolution. The meeting locations rotate among the three countries.

Analysis of this case revealed that the IGCP is perceived as contributing to peacebuilding in the region. The regional meetings are an important relationship-building mechanism, encouraging dialogue and cooperation over shared interests. The meetings have provided both a structure and venue for regular interaction among Ugandans, Rwandans, and Congolese, as well as between government actors and civil society.

Given the highly charged sociopolitical context in the Great Lakes region, the relative neutrality of the forum is a welcome relief. Participants have expressed appreciation for the opportunity to meet "as conservationists, not as politicians," and share information on technical issues such as monitoring.

This is not to imply that all of the issues discussed at these meetings are apolitical, as control over natural resources is among the root causes of conflict in the region. Since gorillas represent significant tourism revenue for all three countries, issues related to their management and control cannot help but become politicized and sometimes confrontational. But the relationships that have been created and fostered through regional meetings—through regular discussions over predominantly technical issues—have provided a foundation upon which the trickier, more political issues can be addressed. Frustrations can be expressed, and clarification over miscommunication obtained. In a region where political relations are characterized by mutual suspicion, this is important. No one can say that these meetings are transforming regional conflict dynamics, but their role in promoting a network of relationships is proving that increasing regular dialogue and information sharing is a small step in the right direction.

Conclusion

The context in which both conservation and development work is done is extraordinarily complex, involving numerous actors and multiple competing demands for resources. This level of complexity is exacerbated at international borders where protected areas meet and conflict exists over the use of natural resources, and so can confound the understanding of these relationships. Peace parks and other transboundary initiatives are being developed at a dizzying rate with claims that they not only promote biodiversity conservation and provide a context for increased revenue through ecotourism and other sustainable development activities, but that they also promote a culture of peace and cooperation. While not criticizing these claims, we are calling for an reassessment that reconciles the optimistic claims of peacebuilding with those that say conservation can contribute to conflict. In this chapter we drew together a broad range of literature and experience from the conservation and development sectors to explore the relationships among transboundary conservation, peace parks, and conflict. We then demonstrated how one approach, the Peace and Conflict Impact Assessment, is being tested in one of the premier examples of transboundary conservation in the world, the transboundary Virunga–Bwindi region at the border of Rwanda, Uganda, and the Democratic Republic of Congo, a region with a considerable history of conflict.

New robust methods of analysis including assessment and evaluation tools for transboundary conservation must be made to figure prominently in the work of conservation organizations and protected area authorities, funders, and so on. By thoughtfully and systematically evaluating the progress in achieving transboundary conservation and peace parks objectives, we hope to be able to chart some course corrections and improve overall conservation practice.

Acknowledgments

The authors wish to thank the Howard G. Buffett Foundation for providing funding for the assessment phase of this project. Additional thanks to the staff of the International Gorilla Conservation Programme for their continued support, especially to Eugéne Rutagarama, the Director, and to Annette Lanjouw, the International Technical Advisor.

Notes

1. See Chapin (2004).
2. Sandwith et al. (2001, p. 3).
3. Van der linde et al. (2001, p. 111).
4. Ibid (p. 112).

5. Zbicz and Greene (1997).
6. Zbicz and Greene (1997).
7. Wilshusen et al. (2002).
8. Austin and Bruch (2003) and Shambaugh et al. (2001).
9. Yamagiwa (2003).
10. Austin and Bruch (2003), McNeely (2002), and Shambaugh et al. (2001).
11. Austin and Bruch (2003), McNeely (2002), Shambaugh et al. (2001).
12. Austin and Bruch (2003).
13. *Global Witness* (2004) and Smillie (2002).
14. Shambaugh et al. (2001).
15. Katerere et al. (2001).
16. Westing (1998).
17. According to Singh (1999), "heritage territories are areas that have been established though well-established historical use, such as migratory patterns of indigenous peoples."
18. Singh (1999).
19. Katerere et al. (2001, p. 21).
20. Gaigals and Leonhardt (2001).
21. Bush (1998).
22. Mamdani (2001).
23. *East Africa Living Encyclopedia* (2005).

2

Peace Games: Theorizing about Transboundary Conservation

Raul Lejano

The present discussion concerns the extent to which we can analytically model the efficacy of peace parks. We begin with a fundamental notion that whatever the merits of peace parks, their employment would correspond to hitherto unseen shifts in position of the contending parties. For example, in Cyprus a diplomatically created buffer zone separates Greeks from Turkish Cypriots. The buffer zone has, for decades, been ostensibly simply an unutilized area. However, in recent times there have been social and cultural exchanges that occur within the buffer zone, which is a hopeful sign of change vis-à-vis the protracted conflict.[1] Another example to consider is the Limpopo Transfrontier Conservation Area in South Africa where proponents have begun envisioning as not just an ecological reserve but an experiment in peace.[2] A third example is the peace treaty of 1994 between Jordan and Israel; it included, as part of the compact, the Binational Red Sea Marine Peace Park, which they jointly manage.[3] On a vastly different stage is the Siachen glacier. India and Pakistan are discussing plans for a transboundary peace park on the Siachen glacier, notwithstanding more than half a century of hostility between the two parties.[4] While only time and experience can prove the value of these peace parks, it is clear that on some plane and even in this early juncture, they do seem to matter. From these developments we have a unique opportunity to study the dynamics occurring on and around these exciting laboratories for peace.

This chapter takes up the alternative constructions of territory. Throughout history, territoriality has been the subtext for violent conflict.[5] So being, it is ironic that territory is now being turned on its head as an instrument for peace.[6] We will explore the manner by which this reversal happens.

Specifically, what is the mechanism by which the park acts on a situation of conflict? How does the park alter the priorities of the contending parties? A secondary question concerns alternative constructions of space even within the category of the peace park.

We employ the simple notion of rationality as a way to depict the mode of reasoning of each contending party. Rationality, in the radically simplified model of the decision sciences, will be taken to simply mean the penchant for each party to pursue the greatest gain to itself. In our model we will treat each nation as a single, rational agent. We will see how even this admittedly overly reductionistic model can give us new insights into the action of peace parks. The specific model derives from noncooperative game theory. It will be seen how this model can depict the manner by which the park might act as a buffer zone in a manner consistent with observations in the field.[7]

The Game-Theoretic Model

Some researchers talk about peace parks as if these were pure and simple buffers spread between parties in a conflict.[8] So we will start with the simple notion of the park as a physical barrier to hostilities. Let us say that it corresponds to the classic model of the commons where conflict is sparked by the physical encounter of one party with another.[9] In the model of the commons the metaphor of cows belonging to different owners who compete with each other for pasture space can be applied to an infinite number of present-day social situations (cars on a highway, fishers in a lake, even outright military clashes). Conflict can arise because of pure congestive circumstances, and even the dynamic of religious, symbolic, or ethnic conflict can be heightened by mere physical proximity.

Figure 2.1 shows the situation as a two-person, noncooperative, non–zero-sum game. The two players in this game correspond to the two neighboring territories, each maintaining its territory of area X on either side of a common border. Each player uses its land for its highest benefit, as is a player's prerogative, but this creates conflict along the border. We express the zone of conflict as a smaller strip of land with area $2\Delta x$, and we simply assume that Δx or half of this total area belongs to each player.

A peace park amounts to essentially having each player set aside exactly the Δx area of conflict for a proposed buffer zone. How can this neutral zone be implemented? One way might be to simply preclude or minimize crossover effects in the neutral zone, meaning to curt one party's incursion into the other's zone and thus

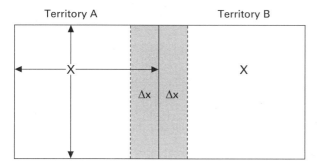

Figure 2.1
Conceptual model of a border park (Reprinted by permission of Sage Publications Ltd. from R. P. Lejano, Theorizing peace parks: Two models of collective action, *Journal of Peace Research* 43(5): 563–81. Copyright © International Peace Research Institute, Oslo, PRIO 2006.)

prevent a violent response. Setting aside the land, however, means that both players give up the beneficial use of their set-aside land (area Δx). Using the two-person noncooperative model, we show how, under certain conditions, the park adjusts the utilities of each player to such a degree as to foster a cooperative outcome. Generally speaking, Δx can be as narrow as a small strip of land with some kind of boundary (the limit being a mere fenceline), or as large as the entire territory itself, X. For the purpose of this analysis, let us assume that Δx lies between these two extremes.

As is usual in these game-theoretic treatments, players (i.e., states) are understood to be atomistic, autonomous, rational agents, whereby "rational" means individually utility-maximizing. Policies and actions are understood by these agents in terms of decisions that consist of choices between competing alternatives. For simplicity's sake, we can treat the two states as identical, each owning identical areas of land.

We will assume that each unit of area can be put to beneficial use by each state, such as for agriculture, and that beneficial use gives each party a return equal to a total present value of utility per unit area Y. The total benefit experienced by each state from the full use of its land is estimated as $Y(X)$ without the peace park and $Y(X - \Delta x)$ with the peace park. The implicit assumption is that the peace park solution will entail leaving this buffer zone unused. We let the two costs in this model be Δy, which expresses the total present value of loss to each party from conflict with the other, and Δc, which expressed the total present value of cost to each party

	Player B	
	Use X	Don't use X
Player A Use X	[γX−Δy, γX−Δy]	[2γX, 0]
Don't use X	[0, 2γX]	[0, 0]

Figure 2.2
Payoff matrix without buffer. The first entry in each quadrant is the payoff to player A, and the second is the payoff to player B. (Reprinted by permission of Sage Publications Ltd. from R. P. Lejano, Theorizing peace parks: Two models of collective action, *Journal of Peace Research* 43(5): 563–81. Copyright © International Peace Research Institute, Oslo, PRIO 2006.)

from creating and maintaining the peace park (including vacating the land, setting up boundaries, etc.). The game-theoretic model allows us to engage in the following sequence of reasoning:

Proposition A (No-agreement alternative) When the land is productive, then the equilibrium condition consists of both parties using their land and incurring the cost of continued friction with the other. Let us call the no-agreement alternative the default.

The payoff matrix is depicted in figure 2.2. In this case each party decides either to actively use its area X or to abandon it. While it may seem extreme for each party to abandon a large tract of land, this is part of what we want to illustrate using the model. From the model in figure 2.2, we see that if productive use of the region X is considerable, and if it is greater than the cost of continued conflict Δy with the other party, then each party will continue to use its land to the detriment of ongoing friction with the other state. We can obtain a single Nash equilibrium for this simple game as (Use, Use)—meaning each player continues to use its land. Let us refer to this as the default, or no-agreement scenario. Note that there exists an aberrant case, which is not so important for this discussion, when the land is of low productivity, namely when $\gamma X < \Delta y$. Then there would arise two pure strategy Nash equilibria, (Use, Don't Use) and (Don't Use, Use).

From this starting model we can proceed to analyze two alternative propositions: First, each party sets aside a small strip Δx on either side of the boundary to act as

	Player B	
	Maintain buffer	Don't maintain buffer
Player A Maintain buffer	[γ(X−Δx)−Δc, γ(X−Δx)−Δc]	[γ(X−Δx)−Δc−Δy, γ(X+Δx)−Δy]
Don't maintain buffer	[γ(X+Δx)−Δy, γ(X−Δx)−Δc−Δy]	[γX−Δy, γX−Δy]

Figure 2.3
Payoff matrix with buffer. (Reprinted by permission of Sage Publications Ltd. from R. P. Lejano, Theorizing peace parks: Two models of collective action, *Journal of Peace Research* 43(5): 563–81. Copyright © International Peace Research Institute, Oslo, PRIO 2006.)

an empty buffer; second, each party uses the said strip as a zone of cooperative activity.

Proposition B (Active, cooperative use of buffer zone) When Δc, the cost of conflict, is high, then the outcome of the game has both parties setting aside Δx and maintaining an empty buffer zone.

We illustrate the cooperative case in figure 2.3. In the payoff matrix of the figure, each party has the option of either actively respecting a buffer zone agreement or reneging on the buffer zone agreement and continuing to use all of its land, including Δx. Let us compare the payoff matrix in figure 2.2. The lower right-hand quadrant contains the payoff for the no-agreement alternative (Use, Use)—the first number in the parenthesis corresponds to the payoff to player A, and the second number is the payoff to player B. Notice that if the cost of conflict Δy is high enough that it exceeds the gains of reneging, equal to $2\gamma\Delta x + \Delta c$, then the game produces two single-strategy Nash equilibria: (Maintain Buffer, Maintain Buffer) and (Renege, Renege). Essentially, either both parties maintain the buffer or both renege on the agreement. On the other hand, if the cost of conflict is low, meaning $\Delta y < 2\gamma\Delta x + \Delta c$, there is a single, pure strategy Nash equilibrium, (Renege, Renege). The first equilibrium dominates the second, however, since both parties are better off with the first solution than the second. So being, we can reason that both parties would choose to comply with the buffer arrangement.

At this point let us take a step back and consider how proposition B's buffer zones work as more than a simple physical means of separation to prevent parties from encountering one another in the contested area. There are quite a few examples that we can cite where buffer zones do seem to work, as simple physical separation, to reduce conflict. In a statistical analysis of the outcomes of such buffer zones, Fortna[10] suggests that full demilitarized zones or DMZs (defined as buffer zones of 2 kilometers in width or more), such as the zone between El Salvador and Honduras or between North and South Korea, have reduced the probability of war by an average of about 90 percent. Even the narrower strip of land between India and Pakistan, established after the second Kashmir war, seems to have contributed to a marked drop in incidents in the area. One consideration is that it may be less the actual, physical effect of the boundary that is most salient, since it is easy for one party to cross the neutral zone and infringe onto the other. The real mechanism of peace in this case is each party's making a conscious, rational decision to comply with the institution of the neutral zone. A more interesting proposition follows, in which we ask whether the neutral zone should be used as a vacant buffer or an active zone of cooperation.

Proposition C (Neutral, noncooperative use of buffer zone) When the cost of (or losses due to) conflict is high, then the Nash equilibrium is the use of the buffer zone as an empty neutral zone and not an active site of cooperative activity.

Figure 2.4 shows a slightly modified game where the agreement being considered is the use of the median zone as an active site of cooperation. Let the expression Δb be the total present value to each state of beneficial, cooperative activities in the peace park. As we have seen before, when Δy is high enough, the parties can be

		Player B	
		Maintain joint use	Don't maintain joint use
Player A	Maintain joint use	$[\gamma(X-\Delta x)-\Delta c+\Delta b-\alpha\Delta y, \gamma(X-\Delta x)-\Delta c+\Delta b-\alpha\Delta y]$	$[\gamma(X-\Delta x)-\Delta c-\Delta y, \gamma(X+\Delta x)-\Delta y]$
	Don't maintain joint use	$[\gamma(X+\Delta x)-\Delta y, \gamma(X-\Delta x)-\Delta c-\Delta y]$	$[\gamma X-\Delta y, \gamma X-\Delta y]$

Figure 2.4
Payoff matrix with joint use. (Reprinted by permission of Sage Publications Ltd. from R. P. Lejano, Theorizing peace parks: Two models of collective action, *Journal of Peace Research* 43(5): 563–81. Copyright © International Peace Research Institute, Oslo, PRIO 2006.)

expected to agree around the peace park alternative—the upper left-hand quadrant. In this formulation we allow for the (small) possibility that the cooperative interaction will instead lead to more hostility, represented by the probability α, that the agreement will instead trigger the cost of conflict Δy. So, for the different equilibrium payoffs obtained in figures 2.3 and 2.4, the question is, Will the parties choose to create an active zone of cooperation, or will they opt for the neutral, barren buffer zone? To be sure, if Δc is higher than the net gains of reneging (i.e., $\Delta y > 2\gamma\Delta x + \Delta c - \Delta b + \alpha\Delta y$), then the parties should choose the vacant buffer over the alternative zone of cooperation. Figure 2.5 summarizes all of the outcomes

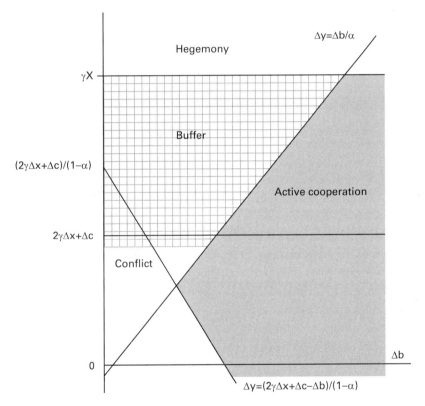

Figure 2.5
Graph of game-theoretic outcomes. (Reprinted by permission of Sage Publications Ltd. from R. P. Lejano, Theorizing peace parks: Two models of collective action, *Journal of Peace Research* 43(5): 563–81. Copyright © International Peace Research Institute, Oslo, PRIO 2006.)

discussed. (There is one other case that is of little interest to us. When Δy is so great that it exceeds the total possible utility from beneficial use of the land, equilibrium is obtained when one of the parties forgoes its use of the land altogether. This is referred to as a hegemonic situation; see figure 2.5.)

At this point we are ready to consider how relevant these propositions are to the real world. The most complex case is that of proposition C, by which vacant buffer zones would serve to physically separate contending parties when cost of conflict is high. In regions where we find a long history of protracted (and even armed) conflict, we can suppose that the costs of conflict are extraordinarily high. Already we see evidence of the reasonableness of proposition C in the case of the Korean demilitarized zone (DMZ), which continues to be a vacant, neutral zone. As an aside, we might add that after decades of non-use the DMZ is now a thriving nature reserve.[11] Some authors maintain that the DMZ has been instrumental in maintaining peace between the two parties for what has been almost half a century.[12] It is possible for parties to move boundaries so as to shrink the buffer zone to where it is no more than a narrow strip of land or simply a fence, as is found in the Carpathian Biosphere Reserve.[13] However, something of an "out of sight, out of mind" attitude takes over when a thin stretch of land virtually separates warring parties.

But not all real-world borders between conflicting parties end up as buffer zones with cessation of activities. In some cases the boundary zone can become a venue for joint activities. For example, Jordan and Israel are two parties that chose to forge a joint relationship around a common ecological preserve, the Binational Red Sea Marine Peace Park, which has become a joint eco-reserve.[14] In 1994 the eco-reserve was part and parcel of the Agreement on Special Arrangements for Aqaba and Eilat. In the example of Cyprus, what could have remained an empty buffer zone is slowly evolving into a site for bi-communal peace efforts between Turkish and Greek Cypriots.[15] Both examples go against the grain of more typical game-theoretic formulations that favor juridical solutions. By juridical, we mean the setting of rules and standards (regarding ingress and egress, ownership, borders, neutrality, etc.). Ostrom's model of common property resources (CPRs) is an example that depends on rule-making. Ostrom shows that rules on boundaries and membership are essential for successful establishment of CPRs.[16] However, in our formulation, under certain conditions, we allow the parties to choose nonjuridical solutions to establishing relations (and the numerous other activities that can emerge from social interactions).

Other possible outcomes might result if we were to change the conditions of the game or use a more complex model. A variation might involve a highly asymmetric game, where one or some players might have significantly more resources or power—for example, we may find a hegemonic power championing the status quo in which it receives a disproportionate share of the total payoff.[17] Nevertheless, as is evident from the international relations literature, the stronger player in the asymmetric game, by virtue of its power, can provide incentives (or threats) to induce the other to cooperate. Thus an asymmetric game can increase the likelihood of a stable outcome.[18]

If we further model a game as a repeated game, we can have an infinite number of outcomes. This is stated by the Folk theorem, that any set of feasible payoffs that are as good as or better than any Nash equilibrium for each player can be an outcome of (i.e., can be a Nash equilibrium, or more exactingly subgame perfect Nash equilibrium, of) the repeated game (as in Friedman 1971). As we suggested above, buffer zones might shrink over time, or they might do the opposite: start as small, "trial" buffer zones and over time expand into large, elaborate and cooperative zones. To be more explicit, we take Δx to be an endogenous variable and allow players to vary the size of the buffer zone according to some rational calculation. We can treat the gains to each party from having a buffer zone as accruing simply by its mere existence (regardless of size), while assuming that the opportunity costs of forgoing use of Δx is proportional to the size of the zone. We let the individual utility functions to be of the following form:

$u_i = \alpha_1 - \alpha_2 \Delta x$ for each player i

provided that $\Delta x > 0$ and the other player also agrees to $\Delta x > 0$.

We then construct a sequential bargaining game where players take turns proposing and counterproposing different values for Δx, with the noncooperative result as the disagreement point. Here the players will seek a Nash equilibrium where Δx becomes exceedingly small, approaches zero but does not reach zero (e.g., it might end up simply being a wall). In the repeated game the bargaining could end in the very first round or, depending on the exact specification of rules of play, could proceed in an iterative manner. To consider another possibility, suppose that the risk of friction in the neutral zone is inversely related to Δx and that such risk decreases with each round of play. These are the conditions that can lead to an extended, even indefinite, period of play.

We could also seek other possible outcomes by testing variations on the utility functions. As another example, assume a polynomial utility function such as

$u_i = \alpha_1 - \alpha_2 \Delta x + \alpha_3 \Delta x^2 - \alpha_4 \Delta x^3$ for each player i.

Then there are two local maxima or equilibrium outcomes that the players might tend to—either Δx approaches zero, as before, or a larger, finite sized buffer zone is equal to

$$\Delta x = \frac{\alpha_3 + (\alpha_3^2 - 3\alpha_2\alpha_4)^{1/2}}{3\alpha_4}$$

(and note that solving the polynomial gives a local minimum, which neither player will opt for).

We can extend this discussion to the model where the players had different utility functions. Suppose that they differ such that expanding Δx is beneficial to one player but not to the other, as is often seen in allocative bargaining games.[19] Another, related model that might conceivably be relevant to peace parks is that of signaling games where the sequence of play allows a player to gain limited information about the other's preferences before making a decision.[20] What we will find is that the exact outcomes of a game is sensitive to the exact "rules of play" involved in the bargaining model. Further, a sequence of play can dictate the outcome, thus conferring a first-mover advantage, as in the models of Rubinstein (1982, 1985).

Some other variations are possible. For example, we could specify conditions where each party's strategy cannot be pinned down except to probability distributions over a range of actions. These specifications will lead to the so-called Bayesian Nash equilibria. However, more complex game specifications call for an unrealistic exactitude to which players know and quantify utilities, specify exact strategies, calculate outcomes, and so forth, none of which hold true in real life. Thus it often makes more sense to use simple models that provide reasonable, but not exact, propositions about what happens in real negotiations.

In general, then, a game-theoretic formulation specifies ideal conditions that we cannot expect to exist in the real world. To be sure, the possibility of a complete specification of utilities requires a smoothness in the ability of each party to strike the most infinitely fine trade-offs or balancing of costs and benefits.[21] Such smooth negotiations are countered in the real world by the situations we mentioned earlier, namely parties with long histories of conflict, sometimes involving violent, hair-trigger upheavals that take us far away from the margin where fine deliberations are thought to occur. Moreover what constitutes "moves" in real situations can encompass an equally innumerable set of actions, formal and informal, taken by many parties. So, if an analyst is willing to accept the strict assumptions of a model with a grain of salt, then the game-theoretic model can actually help explain certain situations while not predicting or fully modeling them per se.

The propositions A, B, and C we have formulated in this chapter are three rules of thumb. They are not useful for making sharp predictions about real situations but nevertheless point to tendencies in real-world negotiations that we can seek to understand.

To these propositions, we might add a fourth one that speaks to the need for other than a game-theoretic formulation to deepen our analysis of peace parks. This is necessary if we are to consider human agents as something more than rational, utility-maximizing automatons. In fact, if contending parties are to begin forging relationships with the other, it may be useful to think outside the box, and instead of rational individuals imagine a joint identity that both parties might desire to share. This last proposition is left for future research to elucidate.

Proposition D The game-theoretic model should be complemented by other, more qualitative analyses, that consider elements of history, context, religion, culture, identity, and other sociological phenomena that game theory assumes away.

Conclusion

This chapter has analyzed the action of peace parks by employing a model of rational choice. In this model peace is obtained through computations of individual utility values, as carried out by individual, utility-maximizing autonomous agents. The game-theoretic model used is that of the simple two-player, noncooperative game. This allows the formulation of three general propositions about how contending parties might jointly resolve hostilities through the institution of the park. The most interesting of these solutions is the one that posits under what conditions the park might function as not just a buffer zone but an active zone of cooperation. This idea leads us to a fourth proposition that suggests employing other, qualitative models grounded in the real world to complement the game-theoretic model. Such a qualitative model should allow us to distinguish among situations where peace is the outcome because the parties rise above thinking as individual, autonomous agents and instead begin to constitute themselves in terms of the other and in union with the other. Some outlines of such a model may be found in Lejano (2006), wherein institutions are seen as structures built on relationships—but this is left for future research. The peace park works in such a relationship-based model because it allows relationships between parties to guide future actions as an embodied virtue. Actions, in this second model, arise not out of individual calculations of ultility but in conjunction with constituted relationships.

The game-theoretic model does provide some guidance with respect to the practice of peace-building. It suggests that we look more closely at the structure of incentives experienced by each party, the cost of conflict to these, and the degree to which peace-building institutions affect their profile of utilities. It provides some testable propositions that we can begin to assess empirically, such as the possibility that the efficacy of peace parks might hinge on the severity of conflict. It also provides some guidance on how to begin reasoning with contending parties about the joint gains to be had by employing this policy instrument. Last, we see how our analyses can be affected by a specific construction of space built into the peace park—for example, as a physical buffer or as a site for collaborative exchange. It is hoped that future work will bring out more and richer conceptualizations of space that can be used to guide the formulation of the peace park.

Acknowledgment

Figures 2.1 through 2.5 are taken from R. Lejano (2006), Theorizing peace parks: Two models of collective action, *Journal of Peace Research* 43(5): 563–81, and used with permission from Sage Publications.

Notes

1. Broome (2004).
2. Singh (1999).
3. Sandwith et al. (2001).
4. Ali (2002).
5. See Kocs (1995), Huth (1996), and Vasquez and Henehan (2001).
6. See Thorsell (1990), De Villiers (1999), and Sandwith et al. (2001).
7. Singh (1999) and Broome (2004).
8. For example, Thorsell (1990), Brock (1991), Singh (1999), and Broome (2004).
9. Hardin (1968).

10. Fortna (2003).
11. Westing (1997).
12. Lindley (2004).
13. Fall (1999).
14. Sandwith et al. (2001).
15. Broome (1998) and Anastasiou (2002).
16. Ostrom (1990).
17. For example, Katerere, Hill, and Moyo (2001) and Wolmer (2003).
18. For example, Keohane (1984) and Snidal (1985).
19. For example, Rubinstein (1982, 1985).
20. For example, Cho and Kreps (1987).
21. Herstein and Milnor (1953).

3

Peace Parks and Global Politics: The Paradoxes and Challenges of Global Governance

Rosaleen Duffy

Today's rising interest in transfrontier conservation can be regarded as part of a broader set of changes in global politics. In particular, the commitment to transfrontier conservation areas (TFCAs), often called peace parks, can be seen as a reflection of the shift way from nation-states as the key actors in international politics. The end of the cold war and ensuing globalization has changed the international system so that new and important transnational networks of governance have developed. Global politics is no longer defined by the bipolar system of the cold war years. The end of the cold war opened up new opportunities for international cooperation, but it also raised questions about the changing nature of conflict in the world system. The development of such a multipolar rather than a bipolar international system has meant that numerous organizations and actors can now be regarded as sites of power and authority; the growing importance of these non-state organizations is enhanced by the ways that they operate together through complex networks, and these are clearly not contiguous with nation-states.[1] For example, international financial institutions (e.g., the World Bank and IMF), international and local NGOs, donor organizations, and international human rights institutions (e.g., the United Nations) have increased their influence in the post–cold war system.[2] In terms of environmental management, changes in the international system have often been identified as offering greater potential for global cooperation to tackle transboundary problems. This shift toward a multicentric and networked international system is nowhere more apparent than in the environmental arena. Environmental problems are by their very nature, transboundary. As a result there is increasing interest in developing forms of environmental management which go beyond the confines of the nation-state.

This chapter will focus on one example of this in the form of transfrontier conservation areas, or peace parks, in Central America. Peace Parks can be briefly defined as conservation areas that cross one or more international borders, and they are intended to have common management practices, often to conserve a single transnational ecosystem.[3] While peace parks have recently become fashionable, they do have a long history, with the first bi-national park established in 1932 between the United States and Canada. However, peace parks have become especially important in global conservation policy since the mid-1990s, and are a key part of environmental policy-making within Central America. In many ways peace parks can be regarded as not just reflective of but are also constitutive of transformations in the global system brought about by the end of the cold war and the onset of globalization.

This chapter will analyze the rationale for peace parks in the Belize–Guatemala–Honduras border regions, to provide a broader view of why peace parks have been promoted as a form of global environmental governance. It will examine the politics of peace parks to explore how they reflect the wider shifts in global politics, including the rise in international forms of management, the use of technical justifications for policy interventions, the shift toward complex networks of governance that stretch through international organizations, NGOs, national governments and local communities. Finally, this chapter will investigate the ways that illicit activities impact ambitious schemes to implement peace parks.

Global Governance and Peace Parks

Notions of global governance are especially important for understanding new and changing forms of transnational environmental management. The increasing interest in global governance is related to the broader shift in global politics that occurred at the end of the cold war. The change from the bipolar global system to a more multicentric and globalized world at the beginning of the 1990s prompted a fundamental change in the ways scholars have conceptualized and described international relations. The more traditional realist framework that emphasizes inter-state relationships has proved inadequate for understanding the complicated network of actors involved in global politics in an era of globalization.[4] In particular, global governance can be regarded as a useful framework for examining the ways that multiple interest groups operate together to govern and regulate. Global governance refers to forms of control, regulation, and management that are different or encompass more than bureaucratic forms of state government.[5]

The increasing debate about new forms of global politics and regulation has focused on notions of global governance, which differ significantly from national government. A standard definition was provided by the 1995 UN Commission on Global Governance, which defined governance as including the formal institutions and regimes empowered to enforce compliance, as well as the informal arrangements that people or institutions have agreed or perceive to be in their interest.[6] While there are differences in the ways that it is interpreted and understood, there is some general agreement about what it means. Selby suggests it is a definitively liberal idea, conveying a pluralistic, post-national and post-ideological conception of the world. In essence, the global governance project is normatively about dispersing power away from geographically defined nation-states.[7]

Furthermore global governance also reflects the extension of neoliberal and market-oriented forms of economic management in the post–cold war era.[8] In sum, global governance is about extending liberal democratic values and procedures, and it focuses on ordering people and things through recourse to reason, knowledge, and expertise. In this way global governance is ultimately a project for rationalizing global social relations.[9] Global governance also raises questions about the changing nature of national sovereignty. In many ways, in terms of understanding the workings of global governance, it is more useful to think of sovereignty as increasingly held by a wide range of actors and not just nation-states. These new sovereign powers include a broad range of actors and institutions that depend on local communities authorized to manage natural resources to the regulations and conditionalities placed on countries and organizations by the World Bank. This does not negate sovereignty as an issue but changes our understanding of it. Furthermore peace parks are a classic example of the new problems facing the conceptualization of sovereignty in international politics: they require national governance to cede some power to transnational ecosystem managers that have been empowered to enforce compliance with the new regulations surrounding peace parks. Nevertheless, it is clear that peace parks exist in a system where national governments retain significant levels of power, but their sovereign power is modified and challenged by their engagement with the powerful global actors engaged in promoting and implementing these parks.

This chapter will now turn to an examination of global governance in relation to peace parks in Central America. It will provide an analysis of how the rationale for peace parks reflects broader debates centered on global governance; furthermore it will examine the politics of peace parks implementation and the challenges it faces from multiple interest groups.

Peace Parks in Central America

Over the last ten years transboundary conservation has become a prominent feature of global conservation policy. The environment is a key area of global governance in practice because ecosystems cross human-created national boundaries, and so environmental change is rarely bounded by nation-states. Furthermore, in line with global governance theory, it is clear that managing peace parks requires a range of innovative mechanisms, through which the previously dominant role of the state is supplemented by and possibly displaced by governance through networks of international organizations, NGOs, and local community groups. These networks are designed to cut across international boundaries and ensure cooperative and complementary management of single ecosystems. In addition the peace park movement reflect changes within the conservation community. By the late 1990s there was a sense within the global conservation community that the community-based natural resource management (CBNRM) rationale was out of date; a new set of arguments and rationales were raised to justify continued conservation activity. The emergence of the transboundary natural resource management (TBNRM) discourse was a direct result of the feeling that CBNRM was no longer the fresh and exciting argument to use to draw in global funders.[10] The rationale behind transboundary management was complex and had to be justified as a better form of environmental conservation. This chapter will examine the ways that peace parks are justified and then analyze the politics surrounding those rationationalizations and the attempts to implement peace parks in Central America.

In line with global scientific discourses about environmental management and protection of biodiversity, peace parks are presented as having a scientific imperative. This is in part related to global debates about universal scientific prescriptions for environmental management that are relevant in a wider context than that of a locality.[11] Global environmental governance relies on positivist and uncontested scientific ideas that can be used to draw up universally applicable forms of environmental management. This notion of neutral scientific management translates into a universalist conservation policy practice.[12] The appeal to neutral science is also linked with notions of bioregionalism, where boundaries are re-drawn in environmentally appropriate ways to encompass whole ecosystems.[13] These rationalizations are then used to justify and legitimate highly political management policies.

The tri-national park planned for the Gulf of Honduras is such an experiment to encompass a single integrated ecosystem that stretches from rainforest to mangroves and coral reefs in Honduras, Belize, and Guatemala. Part of the rationale for the park is that the endangered West Indian manatee is found in the Gulf of Honduras. Currently Belize has one of the largest manatee populations, and its manatee conservation legislation is the strictest in the region. However, while manatees are protected in Belize, they are subject to much less stringent conservation legislation and practice in the neighboring states of Honduras and Guatemala. Moreover manatee conservation laws in Honduras and Guatemala are more weakly enforced than in Belize. As a result it is difficult for Belize to maintain its manatee population when the policies of neighboring countries undermine conservation efforts.[14]

In an attempt to create a common manatee conservation program, two local NGOs have joined together to assist cooperation between Guatemala and Belize. In Belize the Toledo Institute for Environment and Development (TIDE) and Fundeco in Guatemala have joined a tri-national alliance of NGOs aimed at establishing cooperative manatee conservation policies for the three countries.[15] Harmonized conservation practice between neighboring states is a very important part of the justification for peace parks.

Nevertheless, applying a scientific and politically neutral rationale to peace parks is problematic; in particular, peace parks are not due to neutral conservation strategies but highly politicized creations. As Neumann argues, global conservation strategies tend to gloss over the magnitude of political change involved; they instead invest international conservation groups and states with increased authority over resources and often local communities. For example, many new global conservation schemes require land registration and the creation of buffer zones. Both are highly political interventions that are likely to stir up serious challenges.[16]

The bioregional and biological reasons for large transfrontier protected peace parks can upset the commitment to internationally recognized borders. The boundaries that demarcate political space are also not always conspicuous on the ground. Contrary to artificially separating or human-created national political boundaries, peace parks are intended to restore connections between parts of an ecosystem through migration corridors for wildlife and common management policies for single ecosystems. Since political frontiers are not ecological boundaries, often key ecosystems are divided between two or more countries and are subject to a variety of contradictory management and land use practices.[17]

An important part of global governance is the extension of neoliberalism as the most effective and appropriate political and economic methods of managing the environment. In line with this, the supporters of peace parks in Central America have consistently argued that they have an economic rationale through the development of ecotourism. Broadly, ecotourism is promoted by governments of the North and South, international lending institutions, and private businesses as an ideal development strategy that combines economic growth with environmental conservation.[18] Peace parks are intended to be not only economically self-sustaining but to provide revenue to the state and its conservation agencies, and to local communities that live within or adjacent to the transfrontier schemes. Peace parks also intersect with more established debates about the need for conservation to pay its way. Therefore peace parks are expected to be largely self-financing through the development of market based economic activities.[19]

For peace parks the core economic activity is tourism, and more specifically ecotourism. The transfrontier conservation area planned for Sarstoon–Temash on the Belize–Guatemala border has taken account of its ecotourism potential from its inception. Natalie Rosado, of the Forest Department, has suggested that the local communities that inhabit the Sarstoon–Temash were keen to develop ecotourism, but that currently the area has no road. Since the Peace Parks would cover 41,000 acres and require tourists to cross an international border, ease of access is considered a critical factor in the ability to develop ecotourism to secure the financial viability of the protected area.[20] These extensive economic justifications indicate that peace parks conform to notions of extending global governance through the proliferation of market-based systems, in this case to pay for conservation.

Nevertheless, the promotion of ecotourism as an ideal means of providing a financially sustainable form of conservation has been misplaced.[21] One of the difficulties associated with such regional plans is that some partners stand to gain more than others. Paradoxically for transnational forms of management, it is the continued importance of state power and different levels of power and economic development between states involved in peace parks that can be most problematic. For example, Belizean government officials refused a plan from the Mexican government that the Belize barrier reef be renamed and marketed as the Maya reef or El Gran Arrecife Maya. The Belize barrier reef is part of a much larger reef system that stretches from Honduras in the south to Mexico in the north. While the reef is marketed as part of the Mundo Maya experience, the Belize government was concerned that Mexico had already degraded many of its reefs (especially around Cancun), and so the Mexican tourism industry would benefit disproportionately from claiming that the Maya reef was in Mexico. In effect the Mexican tourism industry would make financial gains from giving the impression that less environmentally damaged reefs of Belize were within Mexican borders.[22] The often complex political and economic relationships between neighboring states means that the reliance on ecotourism is highly problematic despite the assumption that it provides an important economic rationale for conservation programs.

The economic justification for peace parks is also closely linked with the use of rural communities as partners in the schemes to provide "local legitimacy" for them. In line with global governance theories, local communities have been identified as key stakeholders in peace park initiatives. This reflects two interrelated themes in global politics. The first is that local communities are increasingly drawn into transnational networks of governance. In particular, peace parks have been intimately bound up with notions of community conservation, where responsibility for environmental management is devolved to local communities.[23] Supporters of peace parks see communities as vitally important actors in ensuring that TFCAs are socially as well as environmentally sustainable.[24] For example, peace parks promise cross border cooperation that can assist in reinvigorating traditional cultural practices such as dances and spiritual rituals. Conejo Creek in the Sarstoon Temash area has re-established its deer dance with funding from the Kekchi Council of Belize. This funding allowed them to rent costumes from Mayan communities in Guatemala that still practiced the dance.[25] The promise of cultural re-connection and revitalization through the extension of cross border cooperation is a vital part of the justification for peace parks.

However, as Neumann argues, demands from local communities for the power to control, use, and access environmental resources are not the same as plans for local participation in externally driven conservation schemes and commitments to local benefit sharing.[26] The difficulty is that local participation is far from politically neutral and has often assisted dominant economic, political, and social groups within communities to further their interests at the expense of others. The presentation of communities as single units with common interests that support peace parks is a clear oversimplification. Local communities affected by or involved in peace park schemes are organizationally complex and contain many different interest groups and are stratified by age, gender, income, and so on. In addition the costs and benefits of peace parks are not evenly distributed across different interest groups within communities. In the case of the peace parks for Belize–Guatemala, the communities that surround the Sarstoon–Temash ecosystem, which crosses the international border, have been involved in lobbying for the park and for a meaningful role in its conservation and management. Gregory Ch'oc of the Kekchi Council of Belize stated that he had been elected by the twelve communities in the Sarstoon–Temash area to represent their interests and conduct negotiations with relevant outside agencies, such as the World Bank and the Inter-American Development Bank.[27] Peace parks clearly draw together transnational networks of governance, since communities are not expected to manage or participate in peace parks in isolation. Rather, they are intimately interlinked with global institutions. In Central America, discussions over the proposed parks are often conducted between local communities, local and global NGOs, and international financial institutions such as the World Bank. For example, the Nature Conservancy and the United Nations Development Program have given financial backing to the Meso-American Biological Corridor Project and the transfrontier parks initiatives in southern Belize.[28] Peace parks have attracted enthusiastic financial backing from other organizations regarded as supports and implementers of neoliberal forms of global governance, such as donors and international NGOs. A number of peace parks have received funding from the World Bank's Global Environmental Facility (GEF), and the Meso-American Reef System Project, received US$10 million from it. Belize is set to benefit significantly from this new source of funds because it has been chosen as the regional headquarters of the transfrontier reef management scheme.[29] The announcement of World Bank funding brought further pledges from donors (including the Canadian Trust and the Central American Commission on Environment and Development) to assist the stakeholders in getting the transfrontier reef project up and running.[30]

Still this appearance of decentralization and linking up with local communities has been heavily criticized. Wolmer suggests Peace Parks are the latest in a line of top-down, market-oriented environmental interventions by international bureaucracies such and the World Bank, bi-lateral aid donors, and international environmental organizations.[31] The anxiety surrounding the importance placed on global–local partnerships is linked to this concern about external dominance. Wolmer suggests that the fashionable language of "stakeholders," "partnerships," and "capacity-building" has allowed for an unhelpful and depoliticized discussion of the role and dimensions of community involvement in peace parks. Instead, the new focus on transboundary natural resource management (TBNRM) and the shift away from community conservation are directly linked and neatly serve the interests of numerous stakeholders, not necessarily communities, in peace parks.[32] It is clear that notions of decentralization and community management fit in with global governance theory but are highly problematic in practice.

In line with theories of global governance that point to the growing power of nonstate actors, peace parks have attracted the attention of international and local NGOs seeking to be involved in the latest trends in conservation. The role of NGOs, in particular, has been criticized by persons active in the conservation sector. For example, Gregory Ch'oc of the Kekchi Council of Belize has suggested that national and international environmental NGOs that are involved in shaping the Sarstoon–Temash management plans hardly have room for the local community perspective. He points out that community viewpoints must be properly addressed in a practical way to ensure that the peace park is workable.[33] The interplay between local and global NGOs has proved to be extremely complex, and has challenged attempts to govern environments from a regional or global level. On the one hand, the bargaining power of communities can be significantly enhanced through their relationships with the international NGOs. On the other hand, the needs and political power of communities can be severely undermined through their participation in transboundary conservation schemes that incorporate a number of globally powerful actors.

Finally, supporters of peace parks have used arguments about national security, environmental security, and conflict resolution to justify the schemes. Conflict resolution has become a key component of global governance, and it is claimed that this can be achieved through better management, resources, and market-oriented economic development and international involvement. The World Bank and the Peace Parks Foundation argue that transfrontier conservation encourages regional integration and fosters peaceful cooperation between countries that have been or may be engaged in conflict with one another. Peace parks are being promoted as a means for reducing or eliminating the impact of violence over natural resources, and for cooperatively encouraging sustainable economic development.[34] This idea, of course, intersects with debates about environmental security and resource scarcity as a cause of conflict in the developing world. Such a rationale is in line with Homer-Dixon's definitions of environmental security and the need to understand the dynamics of scarcity and conflict.[35] Similarly Matthew, Halle, and Switzer suggest that the environment can be regarded as a security issue because resource scarcity is a significant cause of conflict; the implication of this is that better resource management practices can contribute to peace and stability.[36]

In accordance with this conflict resolution rationale, regional links at the highest government levels over peace parks are intended to increase cooperation and reduce the possibility of regional conflict. Peace parks are also promoted as a means for reducing or eliminating the impact of violence in or over natural resources, and for cooperatively encouraging sustainable economic development and peace.[37] In the case of the Belize–Guatemala peace park, the security issue is compounded by the longstanding border dispute between the two countries. Since the Guatemalan government does not recognize Belize as a sovereign state, but considers it to be part of Guatemala, there is a question over the legality of transfrontier initiatives and concerns about national security.[38] Therefore, while peace parks are intended to be a means of conflict resolution and peace-building between neighbors, the creation of transfrontier reserves can raise new conflicts in areas where there is a border dispute.

The assumption that peace parks reduce competition over scarce resources needs more refined analysis in relation to regional politics. Indeed an examination of peace parks indicates that in some areas the assumption that resource scarcity leads to conflict, as Homer-Dixon[39] might argue, is highly problematic; cooperation in the environmental sector does not necessarily lead to a reduction in violent conflict. The politics of peace parks indicate that an abundance of resources (rather than scarcity) can even create new forms of conflict over who has access to and control over the resources.[40] The big problem is that areas that have been identified as sites for transfrontier conservation initiatives already contain networks that are equally interested in gaining access and control over the resources held within peace parks. These interests often challenge and resist the plans of other stakeholders to implement peace parks. The areas earmarked as peace parks are not "empty lands" but have already been transnationalized in ways that are closely associated with other processes of globalization. The central problem in establishing peace parks is that they are already "transnationalized" by illicit networks.

Peace parks are also often planned in areas that provide key resources for those interested in illegally harvesting flora and fauna for either local use or sale into the international trade. For example, the Sarstoon–Temash area is more heavily used by various groups from Guatemala, and is largely underutilized by Belizeans. In Sarstoon–Temash people from Sastun village cross the international border from Guatemala into Belize to collect orchids in the protected area. The orchids are used in local medicine as a blood tonic, and there are concerns that they are also trafficked into the international trade in rare plants (Belize is one of the few places where the extremely rare black orchid can be found, which is prized by international collectors).[41] The problem for law enforcement is that the orchid collectors simply disappear over the international border to avoid capture and prosecution. This is compounded by an understandable unwillingness among the local *alcades* (local councilors and representatives) in Sastun to admit to the problem. The *alcaldes* fear reprisals from the illegal harvesters who are often heavily armed and have lucrative business interests to protect.

Similarly local people on the Belizean side of the Sarstoon–Temash area are afraid of the orchid collectors, whom they see as dangerous criminals.[42] These struggles over control and access to natural resources in border areas are in direct conflict with the ambitions of agencies interested in implementing peace parks. In turn peace parks constitute a direct threat to illicit networks, and so far they have resisted them often through the threat or use of violence. Then there is the international trade in narcotics, which raises more critical issues for peace park planners because the trade is sustained by much wider clandestine global business networks. The transborder trade in narcotics has proved to be a central focus of any discussion regarding a policy of opening borders between states. The US Department of State has identified Belize as a significant drug transit country. Since Belize lies between the producing countries of South America and the consumer countries of Europe and North America, its position marks it out as an ideal route for smugglers.[43] It has also been claimed that elements in the formal state apparatus have been complicitous, leading to the state's incorporation into global trafficking networks. For example, the US Bureau for International Narcotics and Law Enforcement Affairs stated that the ability of the government of Belize to combat trafficking was severely undermined by deeply entrenched corruption that reached into senior levels of government. In addition there is evidence that ministers in the government as well as police officers are complicitous in the drug trade in Belize.[44] This means that these interest groups engaged in the illicit narcotics trade are highly resistant to new forms of control over borderlands. In essence they regard the levels of local, state, and global control represented by peace parks as a significant threat to their own globalized business interests. Therefore it is important to recognize how illicit networks present a core challenge to implementation of peace parks. The supporters of peace parks often overlook or ignore such illicit networks, and this can lead to significant problems when conservation agencies attempt to implement them.

Conclusion

In the environmental arena, peace parks are clearly linked to wider changes in the global system, and especially proliferating forms of global governance. Global governance is a particularly important factor because ecosystems cross human-created national borders and environmental change is transboundary. In addition, in line with theories of global governance, peace parks have a scientific rationale, rely on market-based principles, and connect complex networks of people and institutions, from international and state level agencies to local NGOs and rural communities. This way they are not just reflective of the shifts in the global system since the end of the cold war, they are also constitutive of such changes. Peace parks have contributed to growing process of global governance in theory and practice.

An examination of the politics of peace parks in Central America, however, demonstrates that global governance in the environmental arena is very uneven and problematic. It is clear that peace parks are highly political interventions that are far from the neutral conservation strategies that their supporters might imagine. The commitment to scientific justifications, appeals to ecotourism as a means of making them financially self-sustaining, and the use of communities as partners or stakeholders are not neutral. Furthermore the failure to recognize the ongoing and often longstanding illicit activities and the networks that support them make it even more difficult to implement peace parks successfully. In short, peace parks are ideologically related to neoliberal forms of management that are promoted through appeals to global governance, which is in itself a political choice.

Notes

1. Rosenau (1995); also see Wilkinson (ed., 2005).
2. Cox (2005) and Cammack (2003).
3. Wolmer (2003: 2).
4. Hewson and Sinclair (eds., 1999: 5–11), Rosenau (1990: 10–12), McGrew (1992: 13), Rosenau (1999: 290–96), Linklater (1998), Buzan and Little (1999: 89–104), and Duffield (2001).
5. Rosenau and Czempiel (eds., 1992), Reinicke (1989), O'Brien et al. (2000), and Wilkinson and Hughes (eds., 2002).
6. Commission on Global Governance (1995: 2–4).
7. Selby (2003: 6); also see Wilkison (ed., 2005), Murphy (2000), Keohane (2005), and Weiss (2000).

8. For further discussion, see Fukuyama (1992); also see Cox (2005) and Cammack (2003).

9. Selby (2003: 6); for further discussion of the debates on global governance theory, also see Wilkinson and Hughes (eds., 2002) and Gill (1998).

10. IUCN-ROSA (2002: 2).

11. Also see Peace Parks Foundation, *Key TFCA Issues Tackled at Maputo Workshop*, ⟨http://www.greatlimpopopark.com/press_release.php⟩ (accessed January 14, 2004), Wolmer (2003: 4), and Hutton, Adams, and Murombedzi (2005).

12. See Litfin (1998).

13. See Berthold-Bond (2000).

14. Interview with Nicole Auil, manatee researcher CZMP, Belize City, November 23, 1998; and interview with Roberto Echevarria, tour guide, Toledo Institute for Development and Environment, *Punta Gorda*, May 23, 2000; and *Amandala*, October 10, 1999, Manatee Week '99: Get to Know the Manatees.

15. Interview with Roberto Echevarria, tour guide, Toledo Institute for Development and Environment, *Punta Gorda*, May 23, 2000.

16. Neumann (2000: 220–22), and see Neumann (1998).

17. Griffiths et al. (1999: 11).

18. Duffy (2002).

19. Boo (1990: 1–3); also see Duffy (2002).

20. Interview with Natalie Rosado, Conservation Division, Forestry Department, *Belmopan*, May 17, 2000.

21. Duffy (2000: 549–65).

22. *Amandala*, May 18, 1997, Airlines in bed together: Tourism Minister Henry Young.

23. For further discussion, see Hulme and Murphree (eds., 2001).

24. Neumann (2000: 220–42), Barrett and Arcese (1995), and Cooke and Kothari (eds., 2001).

25. Interview with Gregory Ch'oc, Kekchi Council of Belize, *Punta Gorda*, May 23, 2000; also see Duffy (2002: 112).

26. Neumann (2000: 235–37), also see Sivaramakrishnan (1999), Peluso (2001), and Cooke and Kothari (eds., 2001).

27. Interview with Gregory Ch'oc, Kekchi Council of Belize, *Punta Gorda*, May 23, 2000 and interview with Natalie Rosado, Conservation Division, Forestry Department, *Belmopan*, May 17, 2000.

28. Interview with Natalie Rosado, Conservation Division, Forestry Department, *Belmopan*, May 17, 2000; Amandala, September 26, 1999, Meso American countries meet to discuss reef sustainability; and *Amandala*, July 11, 1999, Toledo Eco-Politics: Maheia vs Espat.

29. *Amandala*, June 13, 1999, Belize designated international barrier reef headquarters; and *Amandala*, July 18, 1999, Ministry of Natural Resources and PACT: seeking to improve relationship with the public.

30. Interview with Natalie Rosado, Conservation Division, Forestry Department, *Belmopan*, May 17, 2000; *Amandala*, September 26, 1999, Mesoamerican countries meet to discuss reef sustainability; and *Amandala*, October 17, 1999, Regional Meso American Barrier Reef System Project planning workshop complete; *Amandala* (Belize), September 26, 1999, Meso American countries meet to discuss reef sustainability; *Amandala*, September 5, 1999, Fishers lead in planning for sustainable reef resources management; and *Amandala*, October 17, 1999, Regional Meso American Barrier Reef System Project planning workshop complete. The Meso American Biological Corridors Project aims to re-establish wildlife migration corridors, and create unbroken bioregions that cover rainforests, mangroves and coral reefs stretching from Mexico, through Belize and Guatemala, to Honduras.

31. Wolmer (2003: 7) and Hutton, Adams, and Murombedzi (2005).

32. Wolmer (2003: 19).

33. Interview with Gregory Ch'oc, Kekchi Council of Belize, *Punta Gorda*, May 23, 2000.

34. Interview with Werner Myburgh, Peace Parks Foundation, Somerset West (South Africa), March 16, 2000; also see World Travel and Tourism Council (1999: 48); Griffiths, Cumming, Singh, and Metcalfe et al. (1999: 25).

35. See Homer-Dixon (1999); also see ⟨http://www.homerdixon.com⟩ (accessed April 22, 2005).

36. Matthew, Halle, and Switzer (2002: 5); also see Homer-Dixon (1999) and Bannon and Collier (2003).

37. World Travel and Tourism Council (1999: 48), Weed (1994: 175–90), and Litfin (1998).

38. Shoman (1995: 206).

39. Homer-Dixon (1999).

40. See Peluso and Watts (2001), Fairhead (2001), Richards (1996, 2001), and Nordstrom (2004).

41. Interview with Roberto Echevarria, tour guide, Toledo Institute for Development and Environment, Punta Gorda, May 23, 2000; and interview with Gregory Ch'oc, Kekchi Council of Belize, *Punta Gorda*, May 23, 2000.

42. Anonymous interviewee, Belize City; and anonymous interviewee, *Punta Gorda*.

43. Bureau for International Narcotics and Law Enforcement Affairs (1998: 1); *Guardian* (UK) June 17, 1997, Whose Colony is this Anyway?; *Amandala*, January 19, 1997, US Pressure in drug fight rankles some; *Amandala*, February 9, 1997, Colombia and Mexico sign agreement to Cooperate in anti-drug fight; and *San Pedro Sun*, December 12, 1997, *Drug traffic increasing in the Caribbean*. For further discussion of the global nature of the drugs trade, see Stares (1996) and Nietschmann (1997: 193–224).

44. Bureau for International Narcotics and Law Enforcement (1998: 2).

4

Scaling Peace and Peacemakers in Transboundary Parks: Understanding Glocalization

Maano Ramutsindela

The recent explosion of interest in peace parks across the globe reveals not only the internationalization of peace-building efforts, but also suggests the attempt to build new sites and to create new scales that can be deployed toward re-ordering conditions that underpin much of the violence and conflict we encounter today. Clearly, the development of the world's first international peace park, Waterton-Glacier, between Canada and the United States, accounts for the dominant trend toward creating peace parks on borderlands (See Tanner et al. chapter 10 in this volume). Indeed the Rotary Club International, which was the force behind the establishment of the Waterton–Glacier International Peace Park in 1932, saw the park as "a way to cement harmonious relations between [nations], while providing a model for peace for nations around the world."[1] By combining high mountains and deep canyons, forest belts and prairie grasslands, and deep glacial-trough lakes and rivers into an expanded protected area, the site of the park also symbolized the importance of nature conservation. In that way the first peace park cemented the relationships between nature and peace. It also provided the dominant model for peace parks, which is based on protected areas, especially national parks, across state borders. Ecologists are increasingly supporting the model, largely because they see state borders as disruptive of ecological systems. This is not to suggest that all peace parks are established across international borders. For example, the International Institute for Peace through Tourism (IIPT) in Vermont has embarked on the mission to establish peace parks throughout the United States by the year 2010,[2] and peace parks have been established as commemorative parks within the boundaries of states such as the peace park next to Christchurch Cathedral (Ireland) and in Nagasaki (Japan). Others are established to honor certain individuals as it has been the case with the establishment of the Dag Hammarskjold International Peace Park in Zambia in 2000 to mark the 40th anniversary of the death of the United Nations

Secretary General, Dag Hammarskjold, who died in a plane crash in Ndola (Zambia). In Canada, each of the 400 cities and towns dedicated a park to peace to commemorate Canada's 125th birthday as a nation in 1992.[3] There is also an idea that preserving pristine areas as peaceful areas that contrast sharply with conflict areas might serve to promote the importance of eco-based peace parks. Notwithstanding the existence of these nontransborder peace park projects, the establishment of peace parks is generally focused on borderlands as sites for peace, as the various chapters of this volume imply. Accordingly this discussion is concerned with cross-border peace parks. The borderlands have a peculiar appeal to proponents of peace parks because the demarcation of state borders have often led to disputes, most of which erupted into bloody wars. It is therefore logical to assume that borders as sites of conflict can be brought under new cross-border regimes, with ecoregions acting as a catalyst for peace. In this sense the establishment of peace parks on borders represents a change of scale from the national to the supranational. This new scalar arrangement transcends traditional areas of state sovereignty while relying on the participation of states and the cooperation of local populations for the success of peace parks. And that scalar arrangement is underscored by the changing functions of the state, and by the belief in the reconstitution of states into a supranational or regional entity.

Proceeding from the view that peace parks are promoted as part of the process of regionalism, this chapter presents three arguments that are mutually reinforcing. First, it argues that in peace parks, we witness a process of re-scaling in which the role of the nation-state is scaled up and down with relation to the regional's scale. Second, it suggests that while peace parks emphasize the regional scale as a locus of peace and as a sound ecological basis for biodiversity protection and management, the regional complex in which they operate is often undertheorized. Agents of regionalism are not necessarily limited to the regional scale; they are part of a constellation of practices and a mix of actors.[4] Third, it argues that the borderlands as sites for peace do not necessarily diminish their roles as a locus of political struggles and contestation. My aim in making these arguments is neither to deny the essential importance of peace parks nor to dampen the spirit of those working hard to encircle the world with peace parks. Rather, I seek out the assumptions behind peace parks in order to highlight other possibilities that can emerge in the process of establishing peace parks at the regional scale. Regions cannot, and should not, be treated as unproblematic containers of peace processes. The re-signification of a supranational scale has the potential to unleash new sites of struggles or to allow old patterns and processes to thrive under "new conditions." These conclusions can more appropriately be appreciated by looking closely at the debates arising around the scale question and the burgeoning literature on borders and regions. This in turn can help establish the links between peace parks and scalar narratives and the implications of transforming scales on borderlands and the citizens. My main defense of the choice of scale analyses, border studies, and regions is that the establishment of peace parks creates a supranational scale, which is believed to be important for peace, and peace parks seek to engineer new meanings of borders, particularly, in existing and potential war-torn or conflict zones.

Scalar Transformation

Critical questions of scale lie therefore in unraveling the complex processes that work to produce scalar arrangements and how, once produced, scale works to define, delimit, and enable possibilities for human action.[5]

To claim that the establishment of peace parks is an act of transforming scales, as I do in this chapter, warrants a closer examination of the analyses of the scale. Scale analyses have, at least within the discipline of geography, been found useful for making sense of social change. That is to say, scale analysis enables us to construct theoretical vocabularies through which to analyze sociopolitical and economic changes, and the processes underpinning those changes. Collinge is of the view that the scale analysis is important for capturing important trends of social change, including the restructuring of nation-states and the revaluation of regions as an outcome of globalizing tendencies.[6] In support of this view, Newstead concludes that "while there is growing diversity of literature on scale, the empirical forces of much of the work has been on the particular sorts of conflicts, struggles, and scale rearranging arising from the challenge of globalization to the territorially defined nation-state."[7] For peace parks, the construction of scale is a temporary and contingent sociospatial production. Decisions on scale flow from the different but related logics of ecology, international relations, and economic development, all deriving from supranational/regional considerations in an increasingly globalizing world.

Much of the discussions about scales proceed from the propositions that geographical boundaries are socially constructed. Moreover it is widely held that scale arrangements do not operate in isolation nor are they fixed. Instead, they are interlinked and enmeshed into sociospatial processes. As will be shown below, this observation is important for understanding the implication of privileging particular scales as important for the establishment and operation of peace parks. For the moment, we should note that the political economy perspective suggests that the rearrangement of scales is a product of globalization. Thus globalization is viewed as a process that rearranges scales by challenging and altering the territorially defined nation-state. The extreme view of this process is that globalization will lead to the disappearance of territoriality as an organizing principle for social and economic life,[8] the implication being the creation of a society without borders. A counterview is that globalization provides new avenues in which states, groups, and individuals create new borders as a way of resistance.

Drawing from a study of urban culture in Amsterdam, Nijman concluded that globalization contributes to "a proliferation of national, ethnic and revivalist movements and involves complex tendencies towards both cultural convergence and inter-group differentiation."[9] Of significance to our discussion is the view that the grouping of states into regions is often ascribed to globalization, and that regions are a feature of the post–cold war world.[10] Grant and Söderbaum are of the view that many forms of current regionalism are manifestations of economic globalization and prevailing forms of hegemony.[11] In fact it has been argued that the Soviet Union and the Eastern European communist republics established the Council for the Mutual Economic Assistance (CMEA) as a socialist alternative to the European Economic Community; the Asian Pacific Economic Cooperation (APEC) emerged as a response by Australia, New Zealand, the United States, and Canada to a Pacific cooperation that threatened to exclude them; and Japan promoted the South-East Asia region as a way out of its predicament in the 1930s.[12]

However, regions cannot be ascribed to economic forces alone, because they are, as Paasi has argued, "historically contingent social processes emerging as a constellation of institutionalized practices, power relations, and discourses."[13] And the relationships between regionalization (as region formation) and globalization can be ambiguous. For example, although regions are generally associated with economic globalization, the proliferation of regional communities in the mid-1990s was viewed as the greatest threat to global "free trade."[14] Without wanting to overelaborate on regions, another general comment on regions is in order. Hettner differentiates "regionness" according to five stages of growth: the delimitation of the geographical unit, the development of social systems, transitional cooperation, the organization framework of civil society, and the development of the personality of the region.[15] These stages are relevant for analyses of regions as conduits of peace.

If Hettner's typology of stages of region formation is acceptable, it follows that the stage of regionness is crucial for peace processes.

It would be unrealistic to expect that the delimitation of the geographical unit would be conducive for peace while the social and organizational systems that are important for holding the constituent parts of the region together are least developed or, worse, nonexistent. The views on regions raise serious questions for peace parks, not least because peace parks delimit certain geographical units (through the establishment of bioregions) and promote cooperation while giving inadequate attention to social systems and organizations operating in those units.[16] The stages of regionness, where other forms of regions already exist, are useful for understanding the environment in which peace parks operate and to which they contribute. As I will argue below, bioregions on which peace parks are anchored are an expression of particular views of regions and bioregions are not insulated from other regional complexes. This is to confirm the obvious: peace parks are neither formed in a political vacuum nor operated in smooth sociopolitical and economic milieus.

As I suggested above, debates about the question of scale focused on regions and limitation of scale. First, two influences on the limitations of scale are the Western and North American dominance in explaining scalar processes. As a result other narratives of spatial transformation from other parts of the world are being marginalized, and particularly those of the global South. In Britain the debate about scales followed the popularity of the locality concept in the 1980s. There, locality studies were rekindled by the emergence of projects such as the Changing Urban and Regional Systems and the Economic Restructuring, Social Change and Locality. These projects sought to grapple with profoundly different socioeconomic changes in different parts of the country.[17] More recently the new regionalism awareness in the United States has raised the importance of urban regions in their redistributive impact within metropolitan areas.[18] Notwithstanding the importance of experiences from the global North, Nagar et al. claim that the postcolonial experiences of marginalized groups provide alternative areas of insights into the operation of global power relations.[19] It has even been argued that human security and development in Africa provides a good theoretical perspective to discourses of regionalism.[20]

Second, scalar analyses overemphasize economic and social processes. This is understandable since scales are socially produced through struggles over resources and power, and the struggles themselves are mediated by scale.[21] In emphasizing the primacy of capital in scalar transformation, Swyngedouw maintained that "scale articulations are profoundly reconfigured in the perpetual struggle of capital to control place through its command over space."[22] These limitations can vary, depending on the extent to which peace parks can provide sites for understanding the reconfiguration of scale. Nevertheless, in the process of the scalar transformation of peace parks, local contestations are sure to be submerged under new supranational formations or regions.

Peace Parks: A Glocalizing Experience

Glocalization is the moment of re-scaling that involves the double re-articulation of political scales: the scaling up of some state regulatory functions and the scaling down of others.[23] As I indicated above, peace parks can contribute to the production and reconfiguration of regions. The obvious questions are how to construct peace parks for different regions and how to determine if a region has an appropriate regulatory framework for an eco-based peace park. The questions are critical because with globalization the role of the nation-state is being reworked, opening up space for transnational regions. Regions are starting to "take on new importance in the regulation and organization of economic, political, and social processes."[24] However, whereas border narratives have played an important part in the establishment of bioregions as the basis for peace parks, the conventional function of state borders is to mark the limits of sovereignty. In conflict zones the border is seen a line of truce in a battlefield.[25] So conflicting adjacent states tend to perceive the border between them from a military perspective. Traditionally this led to the fortification of borders, as in the cases of the Maginot Line (France), Golan Heights (Israel), the Berlin Wall (Germany) and the electrified fence (South Africa). Ironically, such fortification of borders as a response to perceived threats have served to deepen conflict, with the borders erroneously symbolizing the conflict. Indeed border conflicts are by and large multilayered contestations that have their roots elsewhere; that is, the conflict over borders resides within and is about the borders. As Anderson and O'Dowd have argued, it is the "lack of congruence between state borders and other types of boundaries [that] remains a perennial source of border disputes and conflicts."[26] Borders set by colonial powers have been the main cause of such disputes.

The borders superimposed by foreign powers have led to the control of resources being contested and political influence in distant places. A lot of the disputed territories are in countries emerging from cold war politics that tore apart states and continents along capitalist and communist ideologies. This was because the cold war, as declared by Europe and North America, was actually fought in these subjugated countries. Equally momentous was the end of the cold war, which with the collapse of the Soviet Union in 1989 created secessionist movements that wracked havoc in Eastern Europe. To be sure, the causes of conflict in Eastern Europe were much larger than border disputes. Nevertheless, Nigel and Antonia Young have proposed the establishment of a peace park on the border of Kosovo, Montenegro, and Albania as part of the solution to political instability in the area.[27]

Likewise it should be noted that many postcolonial governments in Asia and Africa have failed to change the conflictual nature of the inherited borders. In Asia, the India/Pakistani conflict over the Kashmir is inextricably linked to the act of independence and postindependence power struggles at the national and international levels.[28] In postindependence Africa, borders account for much of the tension between and among states, and the bloody war that broke out in mid-1998 between Eritrea and Ethiopia only served to remind us of the continuing effects of the colonial legacy on the continent. The fact that the defunct Organization of African Unity (OAU) and the newly formed African Union have endorsed the sanctity of inherited colonial borders complicates attempts toward changing state borders for unity and peace in the continent.

Against the background of these examples, in which ways can peace parks contribute to conflict resolution? For the moment, the answer seems to lie in changing the meanings and significance of borders rather than their physical position. This is in line with broader conceptions of border change, which refer to, among other things, changing the symbolic meanings and or the material functions of existing borders in situ.[29]

The underlying motive behind the peace parks initiative is therefore to re-define state borders as sources of peace rather than conflict. The aim is to change borders from their roles as physical barriers to, as Nugent and Asiwaju put it, "theaters of opportunities."[30] This concept demands answers to a number of questions: What processes can be put in place to redefine borders for purposes of peace? How does such peace-building impinge upon other equally important meanings of national and subnational borders? In the event of a redefinition of borders succeeding, will the new meaning of a state's borders be in line with the understandings and activities of ordinary citizens whose lives are directly affected by the state borders? Presently, of course, these questions cannot be answered definitely. I raise them simply to suggest a way of reflecting on the possibilities that can arise from attempts to change borders as part of a peace-building process. The immediate historical past indicates that state borders cannot, by their very nature, lead to the development of a common good. Historically state borders have emerged from political contests and the meanings assigned to them by politicians and the people living adjacent to them vary. The evidence we have from a number of examples is that a reformulation of the traditional meanings of a state's borders is an act of contesting the definition of that state. Peacemaking solutions must regard the meanings of existing borders and understand the implications of designating borderland areas as peace parks. That is to say, we should be aware that although peace parks can infuse new meanings in borderland landscapes, they can also create situations that are beyond the rhetoric of peace.

For peace parks to exist across state borders, states must willing to participate in a process that holds promise of bringing about a harmonious coexistence of states. Peace parks mainly can offer change to the symbolic meaning of border while respecting the territorial integrity of a state. Indeed the following excerpt from the Treaty governing the Ais-Ais/Richtersveld Transfrontier Park between Namibia and South Africa validates this point: "The Government of the Republic of South Africa and the Government of the Republic of Namibia; [recognize] the principle of sovereign equality and territorial integrity of their states; . . . [and desire] to promote ecosystem integrity, biodiversity conservation and sustainable development across international boundaries."[31]

The ascendancy of concerns with, and the outpouring of assistance toward the protection and management of transborder natural resources, together with the international nature of environmental problems, could encourage states to participate in peace parks. Nevertheless, the very nature of peace parks implies that states must be willing to rethink the nature of their sovereignty over borderlands. Commenting from a different but relevant context, Krasner has argued for the creation of shared sovereignty, which would then involve the engagement of external actors in some of the domestic authority structures of the target state for an indefinite period of time.[32] His view is that the rules of conventional sovereignty—recognition of juridically independent territorial entities and nonintervention in internal affairs of other states—no longer work and "their inadequacies have had a deleterious consequence for the strong as well as the weak."[33]

Such governance of peace parks demands that states cede some of their powers to a supranational body. To put it differently, agreements between or among states over the establishment of peace parks must be based on the notion of limited land use options, with biodiversity protection as the primary goal. This point is clearly stated in the Memoranda of Understanding (MoUs) and treaties under which peace parks are governed. Once the signatures are exchanged, the states can no longer dictate land use in areas designated for peace parks. For example, article 6(e) of the treaty for the Great Limpopo Transfrontier Park on the Mozambique–South Africa–Zimbabwe border states: "the Parties (i.e., the governments concerned) shall synchronize related development actions in areas bordering each other."[34] In most developing countries, the challenge for creating peace parks is on negotiating land use and land rights. As Spenceley and Schoon show in chapter 5 of this volume, the establishment of a peace park can seriously undermine the land rights of local residents and thereby create tension on a supposed site for peace. That is to say, peace parks created at the regional scale such as in the Great Limpopo Transfrontier Park do not necessarily address challenges that occur at the subnational level. The land reform in this case proved controversial in Zimbabwe, and approximately thirty land claims in South Africa's Kruger National Parks have been peripheral to the process of creating the park. More crucially, there turned out to be the human settlements on the Mozambican side (see chapter 5) that did not feature on the earlier maps used to gain the support of the governments and the private sector for the Great Limpopo Transfrontier Park. This calls into question of the extent to which MoUs and treaties translate into scaling up functions of the state that are necessary for both the establishment and operation of peace parks. In my view, the scaling up of functions and authority is one of the basic requirements for the establishment of a long-term regulatory framework for peace parks. It is in that regulatory framework that traditional spaces of conflict can be radically transformed. The decided framework should in turn be informed by current realities on the ground, as is already the case in moves toward strengthening legislative instruments governing natural resources in peace park sites. Institutionally it is held that the governing framework should be localized, and operate, within the region. Indeed the nature of peace parks is the shift away from nationally bounded protected areas toward regionally protected areas. In Southern Africa the proponents of peace parks, particularly the Peace Parks Foundation, have made peace parks an initiative of the Southern African Development Community (SADC).[35] As elsewhere it is intended that the region become a repository of legal instruments that provide an avenue for funding and governing peace parks. This initiative has remarkably taken place despite a thick record of dismal performance of Africa's regional blocs.

Within the history of region formation (through the amalgamation of states), states constituting a region have been reluctant to cede their authority to a supranational body. Historically the interest of the nation-state has often come before that of the region. The recent rejection of the European Union's constitutional amendments by the French public is a case in point. Apart from contesting state interests and powers, regions can also become a source of conflict, as the example of the Great Lakes Region of Africa clearly shows. State disintegration in the former Zaïre, and the subsequent unfolding of genocide in Rwanda in 1994, plunged the Great Lakes Region in a war that has been dubbed Africa's First World War. Mamdani[36] considers Rwanda as an epicenter of the crisis in that region. What started as a local conflict developed into a complex political milieu involving Rwanda, Burundi, Uganda, Chad, Namibia, Angola, and Zimbabwe.[37] The Zimbabwe–Angola–Namibia–Democratic Republic of Congo alliance challenged SADC as a regional catalyst for peace, more especially because SADC members were divided into two main opposing camps in the conflict. Observers have also commented that South Africa's attempt to bring stability to the region through diplomacy further complicated the resolution of conflict in the Great Lakes Region. There is a view that the war in that region was about redefining the position and role of postapartheid South Africa.[38] Two lessons can be drawn from this example. First, regions have their own peculiar dynamics that can be a source of, rather than a solution to, conflict. Second, political instabilities manifested at a regional level also have domestic roots. Most conflicts in Africa mirror internal rather than regional conditions. These include the collapse of the state,[39] colonial legacies, and the nature of governance. This is not to suggest that Africa's political problems are by nature domestic, since the boundaries between domestic and external factors are often blurred in the context of Africa. Rather, it is to suggest that understanding conflict at the subnational level is important for any peace-building exercise, including the establishment of peace parks. There is a need in establishing peace parks to regard various scales, as conflict is not limited to a particular scale.

Experiences from political and economic regions raise pertinent questions about the advent of new ecological regions that are formed through the participation of states. At issue here is the extent to which peace parks can be vehicles for engineering a new regional order that can become a conduit for peace among participating states. In my view, the bioregional approach to peace-building parks must allow for the expansion of areas (designated for peace parks) beyond the limits of state sovereignty. However, this can only be possible where the contours of bioregions do not follow state borders so that state borders arbitrarily cut across bioregions. By using bioregions as instruments for region formation, the peace park initiative advances a scientific argument that underplays political interests. Nevertheless, denial of access to natural resources is the highly contended consequence, particularly in developing countries. Unlike the joint management of biodiversity, which is the vortex around which peace parks revolve, the livelihoods of local populations are left to the whims of nation-states[40] and are not treated as priorities of a region. This is precisely where the double re-articulation of the scale takes place. To be sure, whereas the need for joint protection and management of biodiversity forces the state to scale up its functions and authority to a regional entity, the local demands and struggles for resources result in the scaling down of the functions of the state. Individual states must remain responsible to their citizens in terms of their constitutions, the organizational structure of population groups, the nature of civil society, and the like. It is when these national and subnational scales fall apart that discontentment arises, challenging the peaceful existence of cross-border parks, as in the case of Mozambique (see chapter 5 in this volume). That is to say, the success of peace parks, particularly in developing countries, lies with national and subnational authorities, and the needs and demands of the citizens should be addressed at these levels. There should be a harmony of purpose in the re-scaling of state functions in order to deal effectively with regional complexes and national demands and aspirations.

Conclusion

An analytical look at peace parks must not fail to appreciate the resurgence of debates over scales, borders, and regions. As I have shown in this discussion, these debates challenge the assumption that peace parks can be established across state borders. Implicit in this assumption is that border changes (symbolically) can be achieved through the establishment of peace parks. On the contrary, history has proved that a broader perspective of borders should be recognized and acknowledged. Borders have multiple meanings that cannot easily be subsumed into one common goal. Historically borders have been the loci of conflict. The reality is that the sources of interstate conflict are by and large much deeper than the border that separates states.

A supranational scale may be the most appropriate for peace parks. Scale analyses suggest that scales do not exist in isolation; they are nested and hierarchically interrelated parameters. An understanding of the kinds of scales that peace parks produce and/or transform is necessary.

Regions are unproblematic containers of peace processes. Regions emerge from historically contingent socio-spatial processes, and border regions are generally incubators of conflict. They even serve as nodes in the shadow economy—economic activities that are conducted outside state-regulated frameworks.

In my view, it is important to understand that propositions for peace parks are grounded on assumptions about strategies for achieving peace. Currently the bioregionalism underpinning peace parks does not insulate the propositions for peace parks from socio-spatial processes affecting and regulating people's lives.

Notes

1. See ⟨www.peace.ca/rotarypeaceparks.html⟩ (accessed May 24, 2005).
2. International Institute for Peace through Tourism ⟨http://www.iipt.org⟩.
3. Ibid., 3.
4. Beauregard (1995: 239) and Brenner (2001: 592).
5. Smith, as cited in *Newstead* (2005: 47).
6. Collinge (2005: 189).
7. *Newstead* (2005: 47).
8. Waters (1995: 3).
9. Nijman (1999: 150); see also Featherstone (1990: 10) and King (1995: 221).
10. Sidaway (2002: 5).
11. Grant and Söderbaum (2003: 7).
12. Sidaway (2002: 24).
13. Paasi (2004: 540).
14. *The Economist*, cited in Sidaway (2002: 30).
15. Cited in Poku (2001: 71–72).
16. Ramutsindela (2004a: 132).
17. Massey (1991: 269–70).
18. Sites (2004: 767).
19. Cited in *Newstead* (2005: 46).
20. Poku (2001: 2) and Grant and Søderbaum (2003: 9).
21. Collinge (2005: 194).
22. Swyngedouw (1992: 61).
23. Swyngedouw (1996: 1499–1500).
24. *Newstead* (2005: 45).
25. Glassner and Fahrer (2004: 81).

26. Anderson and O'Dowd (1999: 595).
27. See ⟨www.acfnewsource.org/environment/peace_park.html⟩ (accessed May 24, 2005).
28. See Cariappa and Biringer, chapter 15 in this volume.
29. Anderson and O'Dowd (1999: 595).
30. Nugent and Asiwaju (1996: 11).
31. Treaty of the Ais-Ais/Richtersveld Transfrontier Park (2003: 4–5).
32. Krasner (2004: 85).
33. Krasner (2004: 85).
34. Treaty of the Great Limpopo Transfrontier Park (2002: 10).
35. Ramutsindela (2004b: 70).
36. Mamdani (2001: 18).
37. The last three states are members of SADC.
38. Baregu (1999: 37).
39. Some call it the rebirth of the state.
40. See Baldus et al., chapter 6 in this volume.

5

Peace Parks as Social Ecological Systems: Testing Environmental Resilience in Southern Africa

Anna Spenceley and Michael Schoon

Peace parks must inevitably balance natural constraints for conservation and social choices for conflict reduction. In particular, this analysis of peace parks has come from Southern Africa, where the fall of apartheid in South Africa and the end of the civil war in Mozambique heralded a tremendous push toward transboundary conservation.

This chapter provides detailed descriptive analysis of the Great Limpopo Transfrontier Park and suggests that applying the framework of a "social-ecological system" can help one understand peace park functioning and lead to more constructive management regimes. The role of intergovernmental, governmental, and nongovernmental bodies to synergistically drive the emergence of the transfrontier park is considered. These institutions devise appropriate policies, and form decision-making structures with management mechanisms. Their management decisions can facilitate more balanced interactions at local, national, and international scales, and between natural and societal choices. In particular, this chapter considers the utilization of transfrontier natural resources for nature-based tourism, not only because of its potential to generate revenue to finance biodiversity conservation but also because of its role in local economic development and its capacity to improve the livelihoods of local people. Therefore the linkages between commercial tourism developments and people living within and around the Great Limpopo Transfrontier Park are discussed. However, the divergent experiences of communities interacting with protected area authorities in different regions of the park are also illustrated.

The linkages include government-led public-private partnerships that incorporate contractual corporate social responsibility obligations; joint ventures between the private sector and historically displaced populations on their restituted land; land invasions; and the negotiated resettlement of residents outside the transfrontier area. We use a social-ecological framework that addresses issues of resilience and robustness can enhance transfrontier management, particularly where there are different policies and processes of land reform in participating countries.

The Southern African Case

For the past several decades, discussions of a transfrontier or transboundary park have been raised in South Africa, but these movements gained little ground until the 1990s. At this time several political occurrences coincided to increase issue salience. First, South Africa was re-emerging from under the apartheid regime, and the government was looking for a popular movement to improve its international image. Second, Mozambique's divisive civil war came to an end. Third, Zimbabwe's government, like South Africa's, needed a means of recuperating its international image due to Mugabe's dictatorial regime and its conflict over land use and ownership. Finally, the South African National Parks Board was struggling to deal with balancing its controversial policy of hands-off natural resource management, including its refusal to cull animal populations, with potential environmental disaster due to elephant populations in excess of the land's carrying capacity. A transfrontier park would help to alleviate some of the pressures faced by these four actors.

The governments of South Africa and Mozambique, spurred by the lobbying efforts of public entrepreneur, Dr. Anton Rupert, the president of World Wildlife Fund (WWF)–South Africa, met in 1996 to discuss the creation of transfrontier conservation areas (TFCAs) between the two countries. In response to these efforts, the Peace Parks Foundation (PPF) was established by the WWF–South Africa and the Rupert Nature Foundation in February 1997 to coordinate management of these TFCAs. The PPF became one of many nongovernmental organizations to play key roles in the development of a peace park between these countries. In November 2000, the ministers for the environment of Mozambique and South Africa, accompanied by the Zimbabwean representative, signed a memorandum of understanding for one of six peace parks along the South African border—the Gaza–Kruger–Gonarezhou (GKG) Transfrontier Park (Ford 2002). The name of the GKG Transfrontier Park came from the three national components of the park—the Gaza province of Mozambique, Kruger National Park of South Africa, and Gonarezhou National Park in Zimbabwe. In December 2002, the prime ministers of these three nations signed a memorandum of understanding to establish the Great Limpopo Transfrontier Park (GLTP) by officially combining the individual parks of the GKG Park into one jointly managed area. The newly created park currently covers 35,000 km² and will encompass over 100,000 square kilometers when complete (figure 5.1). To put this size in perspective, it is larger than Portugal or roughly a third of the size of Germany.[1]

Underlying this TFCA initiative is the Southern African Development Community (SADC) Wildlife Policy. The aim is to establish TFCAs as a means for interstate cooperation in the management and sustainable use of ecosystems that transcend national boundaries. In addition the SADC protocol on Wildlife Conservation and Law Enforcement promotes regional cooperation in the development of a common framework for the conservation of natural resources, enforcement of laws governing these resources and their sustainable use, and it serves as the framework for transfrontier development and management.[2] The institutional structure of stakeholders working on the GLTP has a hierarchical arrangement that incorporates a Trilateral Ministerial Committee (TMC), a Joint Management Board (JMB), a coordinating party, and management committees.

The TMC consists of the ministers designated and mandated by the participating countries, and it is responsible for overall policy guidance in the development of the transfrontier park. The committee is chaired on a two-year rotational basis and meets at least annually.[3] The JMB includes senior representatives of the pertinent authorities of each country. The four representatives of each member country are seconded from the public sectors of planning and finance, security/home affairs, provincial/regional institutions, and the national agency responsible for protected area management.[4] The board is responsible for interpreting the political directives of the TMC into a set of operational guidelines and policies. It is also in charge of approving development and management plans, harmonizing the expectations of the various parties with respect to the transfrontier park, and monitoring the implementation process of the establishment, development, and management of the transfrontier park. As with the TMC, the JMB is chaired on a rotational basis and meets bi-annually.

The JMB is supported by four management committees, which advise and assist in the implementation and day-to-day management of the GLTP. They are composed of representatives appointed by the relevant authorities of the participating countries and/or representatives delegated by the relevant ministries. The management committees address legislation, finance and human resources, conservation, tourism, safety and security. They are responsible for implementation of the JMB's action plan and need to ensure full participation by all appropriate stakeholders when preparing policy recommendations and resource management plans.[5]

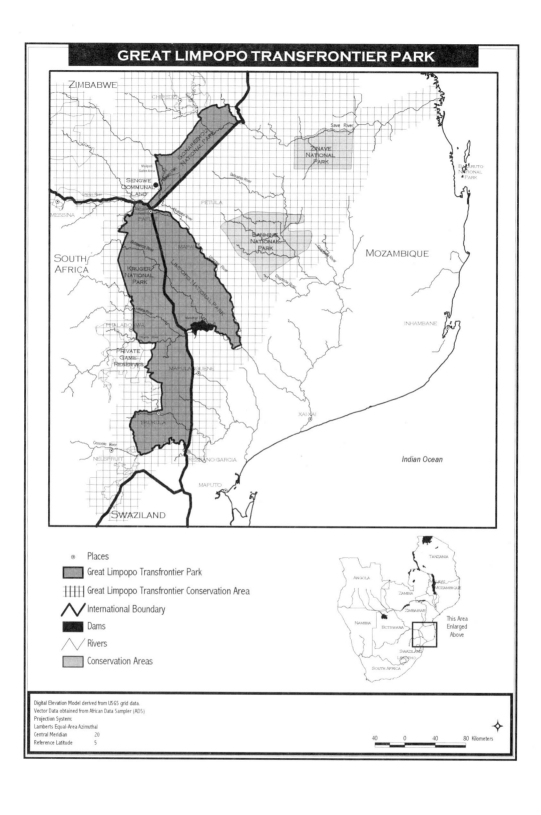

Table 5.1
Tourism estimates in the GLTP countries

	South Africa	Zimbabwe	Mozambique
Nature tourism arrivals ('000s)[a]	4 634.5	1 494.4	42.0
Total tourism arrivals ('000s)[b]	5 872.3	1 966.6	400.0
Indicative % nature arrivals[c]	78.9%	76.0%	10.5%
Nature tourism income (US$ mil)[d]	2 298.8	143.5	8.4
Tourism income (US$ mil)[e]	4 561.7	184.1	—
Indicative % income from nature tourism	50.3%	77.9%	—
Travel and tourism industry GDP (US$ mil)	3 563.9	158.8	—

a. Scholes and Biggs (2004). Data for 2000 for African, non-African, and domestic tourists.
b. South Africa data: SA Tourism (2004); Zimbabwe data: ZTA (2001). Data for 2000 used. Mozambique data for 2001 drawn from Christie (2004).
c. The proportion is indicative only, as the figures compared are for different years and were drawn from different techniques: border entry counts, estimates, and satellite accounts.
d. Scholes and Biggs (2004). Data for 2000 for African, non-African, and domestic tourists.
e. WTTC (2003). Data for 2003, but not available for Mozambique.

During 2000 nature-based tourism was estimated to generate an aggregate of $3.6 bn from Africans and non-Africans, and to have contributed 9 percent of the total gross domestic product (GDP) for the SADC region in 1999. The nature-based tourism sector was responsible for an estimated 2.8 million non-African arrivals and 6.1 million African arrivals[6] (see table 5.1).

The nature-based tourism markets in South Africa, Zimbabwe, and Mozambique are characteristically strong in comparison to other market niches. Therefore the promotion of nature-based tourism in the GLTP to derive revenue for conservation and poverty alleviation is understandable. To probe the significance of these developments, a formal theoretical base is useful for exploring the complex interactions between and among actors in the development of a peace park.

Conceptual Framework Linking the GLTP as a Social-Ecological System

One means of contextualizing the creation and development of a peace park lies in examining the park as a linked social-ecological system. The initial impetus of WWF President Rupert, and the explicit goal of environmental organizations to create a

Figure 5.1
Map of the Great Limpopo Transfrontier Park

robust and dynamic ecological landscape intricately connected to the fabric of the surrounding society, tie directly to the theoretical foundation of resilience and the concept of a social-ecological system. The concept of resilience as originally defined by Holling (1973) is "a measure of the persistence of systems and of their ability to absorb change and disturbance and still maintain the same relationships between population or state variables." In later studies Holling modified his concept by differentiating between two types of resilience: engineered and ecological. Engineered resilience is about the return time of a system to equilibrium after a disturbance. Because there are multiple stable domains in many ecological systems, the resilience of a system can be measured by the amount of change the system can undergo without changing to a different basin of stability. Ecological resilience, on the other hand, is about how big a disturbance a system can absorb and still retain the same characteristics of function and structure.

Anderies et al. (2003) use the term robustness, rather than resilience, when applying such equilibrium disturbance concepts to a social-ecological system (SES) as a way to distinguish between a designed system and an evolved system. In their concept of robustness, a SES loses its robustness when a collapse of both the social and the ecological system occurs. More formally, a SES refers "to the subset of social systems in which some of the interdependent relationships among humans are mediated through interactions with biophysical and non-human biological units."[7] In the case of the GLTP, the social systems would include governmental and nongovernmental service providers, local resource users, and nature-based tourist and operators, all interdependent on the ecosystem(s) encompassed by the transfrontier park. In such a system with complex relationships, a conceptual framework enables us to mentally grasp the multiple linkages inherent in a SES. The framework, shown in figure 5.2, provides such a tool.

The framework of figure 5.2 was originally developed for assessing SES robustness, defined here as the ability of a system to maintain itself despite external stresses, shocks, and perturbations. In its present application, the framework will provide a structure for assessing the relationships of various actors in the system and the impacts of decisions in one part of the system on its other parts. In what follows, the entities and linkages will be explored in an attempt to capture the richness of park development in the GLTP.

The primary resources (see figure 5.2, position A) include wildlife and land. The major tourism draw of the park stems from wildlife and the allure of Africa's big five—lion, leopard, elephant, rhinoceros, and buffalo. The allure of charismatic

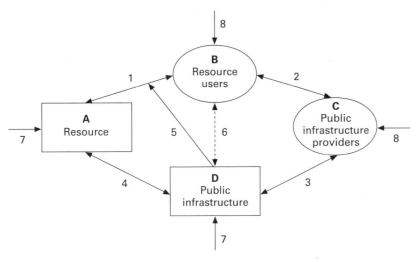

Figure 5.2
A conceptual framework of a social-ecological system

mega-fauna provides a compelling draw for conservation groups and tourist enterprises to lobby for governmental support of protected areas. In turn governmental agencies create policy governing the rule infrastructure of protected areas, animal-culling practices, and veterinary practices in neighboring agricultural communities. While wildlife management may be difficult, land tenure arrangements often prove the more contentious resource. Each of the countries involved in the GLTP has struggled with land and resource tenure arrangements, as noted earlier in the section on local communities—South Africa with land claims and contractual parks, Zimbabwe with squatter communities in Gonarezhou, and Mozambique with ongoing resettlement plans.

The resource users (position B) can be divided into several sets. Three of the main groups include tourists interested in wildlife viewing and trophy hunting; community members using land for growing crops, grazing animals, hunting, and general livelihood; and businesses supporting any of these activities. Much of the complexity of protected areas arises from mutually exclusive resource usage. For instance, increases in hunting may reduce the quality of wildlife viewing, which in turn causes declines in tourism. Changing land tenure arrangements to create national parks provides increased wildlife viewing while decreasing land available for agriculture, timber harvesting, and ranching. Overlapping claims on resources and diverse resource usage brings up the ongoing debate of conservation and development and whether parks serve to improve local livelihoods.[8] The purpose of the framework is not to resolve these debates but rather to provide a forum where such issues can be raised in an organized and thorough manner.

Inherent in these discussions are the main goals of the various public infrastructure providers (position C) involved in the transboundary park process—the governmental agencies, the conservation NGOs, and development NGOs. These goals include encouraging economic development, fostering peace between nations, and conserving biodiversity. While many infrastructure providers agree on these three overarching goals, different prioritization may lead to divergent individual efforts which in turn lead to failure in collaborative efforts. One example of this would include the development of community-based natural resource management plans for a community that other infrastructure providers intend to relocate.

Understanding the roles of infrastructure providers is further complicated by the multilevel governance of transboundary parks.[9] As discussed previously, the GLTP is governed by the JMB, which coordinates between each participant country at the national level. Their work is complicated, however, with the need to coordinate multinational donor organizations and international conservation groups, on the one hand, and local communities and community development organizations, on the other. Balancing at scales ranging from the global to the local while negotiating between three sovereign states makes joint management a tricky task.

The public infrastructure (position D) of a transboundary protected area includes both the physical and social infrastructure of the park. While the physical infrastructure builds on that existing in the Kruger National Park, much development is still needed in the Limpopo National Park and Gonarezhou National Park as discussed in the following sections. Further development is contingent on raising funds and resolving such issues as sharing tourist revenues for cross-border sightseeing, sharing infrastructure development costs, and coordinating efforts with local communities. Even more challenging than the work on physical infrastructure is the coordination of social infrastructure—the rules, regulations, and policies—among the three nations. Each country has specific policy tasks that it must resolve. South Africa's park board continues to tackle the overpopulation of elephants, attempting to improve the situation without culling, as well as dealing with indigenous land claims. Mozambique faces the arduous job of formulating a resettlement policy within the Limpopo National Park. Zimbabwe's park service has the difficulty of dealing with an ambivalent national government.

Linkage one, the connection between resources and resource users, profiles the disputes highlighted earlier of the various incompatible forms of resource usage. These disputes may increase in the future without proper institutional frameworks obtaining acceptance from all resource users, including local communities. The association between resource users and public infrastructure providers, linkage two, has historically been weak in the GLTP development. To date many of the decisions in the GLTP have been influenced by powerful donors, NGOs, and government sponsors. Resource users with less influence, including indigenous communities, have had less involvement. Linkage three, between the infrastructure providers and the public infrastructure is quite strong at the constitutional and collective level, with the Trilateral Ministerial Committee established, the Joint Management Board in place, and international treaties signed. At the operational level, many day-to-day management decisions must still be resolved. It is too early to tell how effective joint management of the GLTP would be and what level of transboundary cooperation will emerge from the interplay, linkage four, between the public infrastructure and the governance of the resources. In addition to the management challenges already discussed, it is unclear what ramifications the rules of the jointly managed park will have on issues ranging from border control and illegal migration to poaching and inappropriate harvesting of natural resources. Linkage five denotes the influence of the public infrastructure on resource dynamics and usage. This is where issues of equity may arise with rules constraining certain users to benefit other groups. In this case traditional local resource usage may be hindered for the improvement of the ecotourism experience. This, coupled with linkage six, tying resource users to public infrastructure, may prove the biggest challenge of the JMB. Some resource users are not directly involved in infrastructure development or enforcement, yet these rules govern and transform their legal resource usage and economic livelihoods. Many efforts are being made to strengthen this linkage, but it is unclear how circumstances will play out in the coming years. External disturbances to the system are incorporated into the system through arrows seven and eight. Arrow seven denotes the effects of biophysical disturbances on resources and public infrastructure ranging from floods and droughts to disease outbreak. Arrow eight addresses social changes influencing the resource users and public infrastructure providers including population changes, fads in ecotourism destinations, and political upheaval in one of the member nations.

The indigenous community, however, is undergoing a forced change from an agriculture-based system to a service-driven economy. A major concern of this shift in economies is that people who formerly provided their own source of sustenance may lose the ability to feed themselves and become reliant on external sources of food. If the service economy provides enough of a monetary reward, this problem can be overcome. However, during a rapid transition period the requisite skills may not exist, and a sudden catastrophic collapse of the local economy may occur. If direct linkage between the ecological system and the social system is damaged, the robustness of the entire SES may be lost. This could occur if there is an increase in poaching either due to less local monitoring or due to an increased need for food, if agricultural practices are changed from traditional manners in boundary land surrounding the park, or for a host of other human-ecological changes. The current governance structure in the park creation is one of centralization and hierarchy. The diverse and complex interaction within and between social, ecological, and economic systems at multiple levels, however, limit the knowledge and information that any one person or group can comprehend regarding the entire system. In turn, the consequences of decisions by one set of actors on the system as a whole may not be as intended. The geographic scale and remoteness of the park and the sociological diversity of race, tribe, and nationality also contribute to making centralized rule-forming difficult.

By further expansion of the basic SES framework presented here, this chapter intends to provide a theoretical lens applicable for improving understanding of peace park functioning and the roles of a diverse range of actors. The following sections will expand upon the specific role of governmental agencies, NGO actors, and local communities in each of the three countries in the GLTP. Using the concepts of resilience/robustness and of social-ecological frameworks, these agents can improve their management capacities.

South Africa

Similar to the early National Parks in North America, the origins of Kruger National Park (KNP), which comprises South Africa's portion of the GLTP, are beset by a tragic tale of human displacement. In 1969 the Makuleke people were forcibly removed from an area of 24,000 hectares that they inhabited in the north of KNP by the state. They were finally compensated for their relocation in 1998 with the restitution of their land and the creation of a contractual park[10] in the Pafuri Triangle (see figure 5.1). A 25-year agreement was forged between the Makuleke and SANParks to return the ownership and title of the land to the people, although the title specifies that the land may only be used for wildlife conservation.[11] The contract that governs the incorporation of the Makuleke land in KNP enables them to make sustainable use of specified natural resources.[12] They have the option to construct six small camps with a cumulative occupancy of 224 beds.[13] However, when compared to the full scale of KNP, this is a relatively meager amount. KNP has well developed tourism infrastructure, with 25 rest camps of various sizes providing 4,056 beds, in addition to 405 caravan and camping sites.[14] Although the overall standard of accommodation is not considered high by international standards, accommodation ranges from luxury tented-camps to self-catering guesthouses and cottages. KNP has approximately 1 million visitors per year and generates an annual income of around US$40 million.[15] Trophy hunting has been able to deliver quick revenues to local people during the period when tourist lodges and camps were being built. Between 2000 and 2003 the Makuleke used trophy hunting to generate both income and "venison" for households. Four hunts earned the Makuleke the equivalent of roughly $230,000 during this period.[16] The Communal Property Association (CPA) wants to phase out hunting as the lodges and camps begin to bring tourists and revenue to their area. Makuleke's CPA has a contract with a private sector partner, Matswani Safaris, who developed a luxury 24-bed lodge called The Outpost on the confluence of the Luvhuvhu and Mutale Rivers. The lodge development began in 2002 and cost between R10 and 15 million (ca. US$1 million to $1.5 million) to construct.[17] The lodge is an up-market establishment, designed to have low environmental impacts. An initial fee was paid to the Makulekes on signing of a contract, and Outpost Lodge has provided a bond to guarantee its performance for the duration of the agreement. In addition the Outpost pays the CPA 8 percent gross earnings generated by the business, a traversing fee for game drives, and 2 percent of gross earnings into a social development fund earmarked for the education of Makuleke youth. They also pay a monthly traversing fee for every vehicle based in the Makuleke region.[18] Projections have indicated that when running at 60 percent occupancy the lodge will pay an annual rent of US$75,000 to the Makuleke community, and around US$150,000 to the 30 local employees. This would generate an estimated US$400 per family through the initiative, which is significant in relation to the average annual wage of around US$750.[19] In July 2003 the CPA announced an R45 million concessionary agreement with Wilderness Safaris to build three luxury low-impact tented lodges.[20] Wilderness Safaris propose to invest R20 to R30 million and are currently undertaking an EIA process (costing ca. R4 million).[21] Wilderness Safaris state that the Makuleke will receive a percentage of turnover, rather than a flat rental fee, community members will receive training and employment (up to 200 jobs), and some scholarships will be provided.[22]

Both tourism partnerships have similar payment systems, and are built–operate–transfer arrangements. This means that their partners build and operate for a specific number of years and then they transfer the ownership of the lodge to the CPA. The Makuleke will then decide if they want to run it themselves or invite the partners to operate on different terms (as they will then own the facilities).[23]

Until 2000 all tourism infrastructure inside KNP was developed, owned, and operated by South African National Parks (SANParks). At this time SANParks embarked on a commercialization process that allowed the parastatal to grant concessionaires rights for the use of defined areas of land and infrastructure within National Parks, coupled with the opportunity to build and operate tourism facilities over twenty-year concession periods.[24] The aim of the process has been to increase the net revenue that commercial activities contribute to SANPark's core function of nature conservation. The program led to the transfer of management of tourism operations to commercial operators, who were considered to be more qualified and equipped to such facilities than SANParks.[25] Major objectives included the promotion of economic empowerment of the formerly disadvantaged, the promotion and provision of business opportunities to emerging entrepreneurs (in particular, local communities adjacent to national parks), and the application of SANPark's environmental regulations and global parameters to all concessions.[26]

Seven accommodation concession contracts in KNP were agreed in December 2000 that guaranteed SANParks a minimum income of R202 million[27] over a twenty-year period. Three of the concessionaires were black-controlled consortia, and all of the others had significant percentages of shareholding by historically disadvantaged individuals (HDIs[28]). The average percentage of HDI shareholding within the seven concessionaires (either immediately or contractually bound to be in place within three years) was 53 percent.[29] The concessions added approximately 250 beds to KNP's existing accommodation facilities.[30]

However, the concessionaires have not performed as well as they had anticipated over the first three years. In 2004 SANParks undertook an investigation of the concessionaires' financial progress and found that concessionaires had effectively bid too much for their sites. For the first three years occupancies were 22 percent lower than the bid forecast, and the fixed concession fees amounted to R29 million despite concessionaires incurring losses of R34.9 million. The total actual capital investment for the KNP concessions was R218.8 million: 65 percent more than the anticipated

R132.3 million. The companies had invested between 22 and 118 percent more than they had predicted in their bids.[31] The up-market game lodge industry in South Africa is highly volatile and risky, characterized by sensitivity to market fluctuations, a highly competitive market, high capital requirements, high fixed costs, and a requirement for predictable and high-quality game viewing. Apparently the lack of information regarding the market for concessions inside national parks at the time of the tender process coupled with concessionaires bidding in a competitive environment led to them overestimating their potential performance. Tourism demand in South Africa has become increasingly price driven, and growth is being experienced in the three-star market due to the high-end five- and four-star markets being less affordable. SANParks is currently undergoing a consultation process with concessionaires to address and ameliorate some of their constraints. The problems faced by the concessionaires in KNP need to be understood in order that future concession programs in Mozambique and Zimbabwe build on the South African experience.

Other tourism investment in KNP to support the TFCA has included money from the national Department of Environmental Affairs and Tourism (DEAT). DEAT's Poverty Relief Fund funded two major projects within KNP between 2002 and 2004: upgrading staff accommodation (R10 million) and partial upgrading of the western boundary's foot-and-mouth fencing (R2.5 million). Poverty Relief Fund money has certain socioeconomic conditions attached to it, such as the following:[32] between 2 and 4 percent of the project budget must be dedicated to training and capacity-building, and all training must be accredited to enable the trainee to either manage the funded project or seek employment elsewhere with the skills or knowledge gained; projects must promote the creation of SMMEs and favor previously disadvantaged individuals (PDIs); 90 percent of temporary jobs created must go to local people, and 60 percent of the temporary and permanent jobs should be reserved for women, 20 percent for youth, and 2 percent for disabled people.

DEAT also used R40 billion granted by the National Treasury to invest in tourism infrastructure within the GLTP at the end of 2002. This investment was used to improve the quality of roads in KNP, to construct a new border post, to develop research infrastructure, interpretation facilities, and picnic sites, to build bridges, and to drop a section of fence between KNP and LNP.[33]

Zimbabwe

In Zimbabwe the GLTP consists of Gonarezhou National Park (GNP), and is proposed to include the Sengwe communal area. Gonarezhou has one rest camp,

Mwenezi, providing rondavels with 16 beds. The transport infrastructure, although available in and around GNP, is in a state of disrepair due to poor funding and maintenance. Roads are untarred in the park, poorly maintained and mainly accessible by 4-by-4 vehicles.[34] Between 1996 and 1998, there were approximately 6,000 visitors to Gonarezhou annually, of which around 20 percent were foreigners. In 2000, this number declined steeply (along with other areas in the country) to just over 2,000 per annum, of which more than half were day visitors.[35]

The 1998 to 2004 GNP Management Plan sought to create 13 "undeveloped" campsites and 15 developed sites, 20 day-visitor picnic sites in addition to new hides and viewing points.[36] Reports from the ZTA indicate that during 2004 the Ministry of Finance allocated Z$17.9 billion (R26 million) to upgrading international airports, including refurbishing the Buffalo Range Airport that is used to reach Gonarezhou. The state also allocated Z$2.2 billion (R3.2 million) to upgrade existing tourism facilities in GNP. It was indicated that they had begun with three main areas of road development, electrification and communications to provide for accessibility for interested investors to develop campsites. However, the ZTA indicates that a lack of funds has put de-mining programs along the borders on hold.[37] There has been no evidence of tourism investment in Gonarezhou since 2000.

In the runup to the 2000 general election, various coalitions of actors, gathered under the banner of "war veterans," stepped up a previously low-level campaign of occupying commercial farms and some state-owned land. ZANU(PF), the ruling party, fought the election under the slogan "Land is the economy and the economy is land."[38] In February 2000 a national referendum was held regarding a new draft constitution for Zimbabwe, and this was followed in June by a nationwide parliamentary election. These two processes dramatically changed the country's political environment. The changes were traumatic and reported as such, resulting in a poor international image for Zimbabwe, which had adverse implications for the whole economy, most particularly the tourism industry.[39] Farm occupations and a "fast-track" land reform process picked up momentum after the election, underpinned by a policy emphasis on the importance of small-scale peasant agriculture at the expense of white-dominated commercial agriculture in general, and the wildlife industry in particular.[40]

When Gonarezhou was declared a National Park back in 1975, the inhabitants of the park were forced to resettle outside the park's boundaries.[41] In 2000, reoccupations began as the Chitsa people used the invasion of commercial farms in Zimbabwe to pursue their historical claim for the land, driven by the need for access to grazing and hunting.[42] By November 2000, areas of the park were being occupied, cleared, and burned by residents of neighboring communities.[43] By May 2001, Agritex, the Agricultural and Rural Extension Department, had planned 10 villages in Gonarezhou and had also allocated separate arable plots and a communal grazing area. Provisions were made for 750 settlers on 520 plots covering 11,000 hectares. Although there was an immediate increase in settlers and bush clearing, a combination of drought and elephant crop raiding let to most settlers returning to communal areas by June 2002.[44]

Since November 2002 the governor of Masvingo Province has reportedly encouraged families of those previously evicted and opportunitisic "war veterans" to take over 5,000 hectares of the park, north and south of the Runde River. Cattle are grazed in the park, and cattle fencing has been pulled down. Although the incursions have only taken place in some parts of Gonarezhou, poaching and decline in tourist numbers have affected the GLTP.[45] However, Wolmer et al. (2003) indicate that it is not only cattle and agriculture that are forming the mainstay of enterprise development in Gonarezhou. There appear to be unique cases of local entrepreneurs allocating themselves land in a former veterinary corridor, with the objective of developing commercial wildlife tourism. Apparently 50 hectares of self-contained plots have been allocated to 56 people, who are members of a relatively wealthy and politically well-connected elite including, councilors, war veterans' leaders, army personnel, policemen, and National Parks staff. Although none of the people have physically moved back into the veterinary corridor, they propose to operate it as a mini-conservancy, where revenues from a safari concession would be disbursed to the new landowners.[46] More recently, in January 2004, in a reversal of previous actions, the government of Zimbabwe began to relocate illegal settlers out of Gonarezhou and its buffer zones, because of the proposed TFCA.[47]

Mozambique

The GLTP in Mozambique is proposed to include the Limpopo National Park (LNP).[48] LNP was a hunting zone called Coutada 16 until November 2001 when it became a national park.[49] When the government of Mozambique (GoM) proclaimed the Coutada 16 as Limpopo National Park in November 2001, it did so despite the fact that people were living within it. The World Bank reported that "... despite their efforts to resolve the fate of the communities living in Coutada 16, and the assurances given by GoM that the issue would be adequately resolved prior to any action, in 2001, the LNP was gazetted as a national park and some animals were allowed to enter the park area. This created a lot of media attention and controversy around the project."[50]

In all, it is estimated that there are 6,500 people living along the Shingwedzi River and close to Massingir Dam, with an additional 20,000 people living along the Limpopo River.[51] The people living in the Shingwedzi River basin inhabit an area of approximately 370,000 hectares. These people are clustered in nine villages: the biggest of them amounting to about 2,000 inhabitants and the smallest to less than 150. Each homestead cultivates an average of about 3 hectares of land, and there are more than 5,000 head of cattle within the park. Communities living along the Limpopo River as well as along the Olifants River in the southeast of the Park number approximately 20,000 people, settled in the support zone throughout about forty villages.[52]

A resettlement process is underway in LNP to determine whether the inhabitants of LNP will remain there, or whether they will be relocated to areas outside the park and compensated. LNP has aligned its resettlement policy with the World Bank's safeguard policy on involuntary resettlement (Operational Policy 4.12). The first principle of this policy is that "Resettlement must be avoided or minimised."[53] A Resettlement Working Group (RWG) has been appointed that is composed of district administrators, provincial government officials, local leaders, and community representatives. The RWG will draw up the Resettlement Action Plans.[54] The resettlement policy considers the "no resettlement" options with regard to creating fenced enclaves within the park, or remaining unfenced. Huggins et al. suggest that in the enclave scenario, land available for farming and grazing would be limited, and villagers would not be able to leave the enclave on foot but be dependent on arranged transport.[55] If homesteads remained unfenced, there would be competition among people, livestock, and wild animals over resources. Living with wildlife can increase the danger of spreading diseases between domestic and wild animals. Although homesteads would be free to move, they would live in a more dangerous environment. However, the LNP Tourism development plan also states that tourism may have positive impacts on local communities through job creation and other opportunities.[56]

Aside from tented camps constructed by Gaza Safaris, the previous hunting concessionaire, there is no tourism infrastructure inside LNP.[57] Tourism in LNP is predominately at the planning stage; a tourism masterplan, zoning plans, and an infrastructure development plans have been drawn up. The Tourism framework identifies five forms of tourism that can be accommodated in LNP: recreational, adventure, consumptive, cultural-historical, and ecotourism.[58] The German Development Bank (KfW) has helped finance the construction of a network of safe roads, basic infrastructure for the establishment of private tourist camps and lodges, fences to protect the villages from roaming wildlife, and the construction of health stations and village schools in the fringe areas of the park.[59]

The Peace Parks Foundation (PPF) summarizes progress in Mozambique as follows:[60]

1. An anti-poaching drive has been launched leading to the confiscation of forty firearms. The anti-poaching unit in LNP works closely with its counterpart in KNP.

2. A de-mining contractor has completed 70 percent of the de-mining.

3. A total of 1,987 animals have been translocated into a wildlife sanctuary as a first phase of restocking LNP.

4. Seventy-three field rangers have been trained and deployed in the park, and workshops between communities and field rangers have been held to enhance collaboration and cooperation.

In addition to facilities for administration, game rangers and environmental education, the LNP Tourism Development Plan proposes four entrance gates: at Massingir, the Giriyondo border post, Mapai, and the Pafuri border post. A road from Massingir to Giriyondo is proposed to establish access from the Kruger National Park into LNP, in addition to a series of loop roads in the moderate-use zones of the PNL for tourists. Tracks for 4-by-4's are proposed for safaris and park management around Limpopo, Shingwedzi, Makandezulo, Mapai, and in the Sandveld area. Viewpoints are proposed along these routes at suitable points. This management infrastructure will be developed in phases over a period of time, depending on the availability of funding. It is predicted that in addition to the 2,184 potential overnight visitors, 160 day-visitors will also be able to enter LNP. At capacity, and with a two-night average stay, it is anticipated that the park may have a capacity of 486,180 visitors per annum.[61]

A community-based tourism enterprise near the Massingir dam has been facilitated by the Swiss NGO Helvetas. Covane Community Lodge is located about 7 kilometers from the Massingir township, and it opened in May 2004. The lodge has two 5-bed chalets, three 3-bed tents (19 beds), and additional space for people to bring their own tents. Besides accommodation the lodge offers traditional dances, traditional food, hiking trails, village visits, and boat viewing, and the opportunity to purchase local crafts. There are self-catering facilities, a restaurant, ablutions, and a seating area overlooking the Massingir dam. Between July and September 2004 the lodge had 95 domestic and foreign visitors.[62]

The lodge was financed jointly by USAID ($50,000) and Helvetas ($20,000) in 2002 but is now owned by the Canhane community. After presenting the concept of a community lodge to the provincial ministries of tourism and agriculture, the district level administration, and the departments of agriculture and tourism, the concept was raised with the Canhane and Kubo communities. There was a low level awareness of tourism at the time, but Helvetas talked to the communities about CBT, potential locations for a lodge, the land law, and land delimitation. The community identified a suitable location and, at the suggestion of Helvetas, organized a steering committee of ten local volunteers. Community members chose members of the steering committee during a community meeting. Helvetas facilitated the delimitation of an area of 7,024 hectares so that the community "owned" the land the lodge was to be built on. The community chose twenty local people to work on the construction between February 2003 and November 2003, without assistance from the private sector. Helvetas organized a constitution for the association, which addresses the financial responsibilities of the Canhane community and Helvetas.[63]

Nine members of the community are now employed at Covane lodge. Helvetas suggested the type of qualifications that employees would need, and the community searched for these qualifications within the community. Helvetas sent these people on tourism and hospitality training courses, and the manager is about to attend a hotel management course. The NGO also organized visits for community members to visit community-based tourism enterprises in Swaziland and at Xai-Xai.[64] The lodge accrued 8 million Metacais (ca. R260,000) from the accommodation and services sold to tourists between May and October 2004, and plans to have a community meeting to decide how to spend the money within the 3,033 strong population of Canhane. The agreement currently states that 50 percent of money should be spent on community infrastructure and 50 percent on investment for the camp. This later proportion will eventually be spent on salaries when Helvetas completes their involvement.[65]

Conclusions

The GLTP is in the early stages of development. South Africa has a well developed tourism industry and infrastructure within the park. Mozambique, however, has only minimal infrastructure and is still in the planning stages for tourism. At the same time Zimbabwe's investment has been heavily constrained by a lack of funding, ensuing from political unrest and land instability. Although this study has revealed the level of tourism development existing and planned in the participating countries, the implications for sustainable tourism development cannot yet be evaluated. As the TFCA develops, as tourists return to Zimbabwe, as tender processes for concessionaires evolve in Mozambique, as socioeconomic problems are addressed in addition to conservation management, and as linkages with other destinations in regional circuits are realized, the implications of the area for sustainable development will become clearer.

The active engagement of local communities, working in concert with transnational efforts of both governments and NGOs, will help to ease the economic transition for the subregional markets between the old and new economies. Perhaps the biggest risk to social-ecological resilience is through a societal collapse due to a poor economic transition. Such an event could reverberate through the ecological system and cause irreparable harm to the social-ecological system. In anticipation of the challenges of the economic transition, two steps can be taken to mitigate these risks.

What is obvious is that through the process of planning and development, stakeholders are learning more about the complex issues involved in catalyzing a sustainable nature-based tourism industry. They are learning more about the market demand for nature-based tourism (e.g., SANParks concessioning process), more about how to address local community issues—especially with regard to people living within the TFCA (e.g., Sengwe Communal Area, Limpopo National Park)—more about developing viable community-based tourism ventures (e.g., Makuleke, Covane Community Lodge), more about the environmental impacts of nature-based tourism (e.g., SANParks concessioning process), and more about the fragility of the industry in relation to political instability and uncertainty of land tenure (e.g., Zimbabwe). An adaptive approach to the TFCA development is compatible with sustainable tourism development, and Hunter[66] argued that it must address widely divergent situations, different goals, and different mechanisms of utilization. The extent to which tourism in the GLTP will develop in such a way that it does not degrade the environment to the extent that it prevents the successful development of other activities, and the extent to which it can alleviate poverty in the region, remains to be seen.

Acknowledgments

This chapter is largely drawn from A. Spenceley, *Tourism Investment in the Great Limpopo Transfrontier Conservation Area: Scoping Report* (2005). This was a report to the Transboundary Protected Areas Research Initiative presented on March 30, 2005. For more information on the conceptual framework see *Do Parks Harm More Than They Help? The Role of Peace Parks in Improving Robustness in Southern Africa*. Presented at the USGS-Sponsored Conference on Institutional Analysis for Environmental Decision-Making, January 29, 2005. This is available in the USGS Scientific Proceedings of the Conference.

This research was supported by the Workshop in Political Theory and Policy Analysis, Indiana University, the BMW Chair for Sustainability, University of the Witwatersrand, and the Transboundary Protected Areas Research Initiative (TPARI), an IUCN-SA program that is funded through the Centre for the Integrated Study of the Human Dimensions of Global Change, by way of a cooperative agreement with the National Science Foundation (SBR-9521914).

Notes

1. Ford (2002).
2. Mtisi and Chaumba (2001).
3. Codex (2001–2004) and Schuerholz (2003).
4. Schuerholz (2003).
5. Codex (2001–2004).
6. Scholes and Biggs (2004).
7. Anderies et al. (2004: 4).
8. Brockington (2002).
9. Janssen et al. (2004).
10. Elliffe (1999).
11. Steenkamp (1998) and Steenkamp and Grossman (2001).
12. Koch (2002), personal communication.
13. Grossman and van Reit (1999).
14. UNEP/WCMC (2004).
15. JMB (2002).
16. Ford Foundation (2002), Collins (2003).
17. Ford Foundation (2002).

18. THETA undated.
19. Koch (2001).
20. Turner (2004).
21. C. Bell (2004), personal communication.
22. Turner (2004).
23. Makuleke (2003) and Collins (2003).
24. SANParks (2001).
25. van Jaarsveld (2004).
26. SANParks (2000).
27. In real net present value terms.
28. HDIs are described as socioeconomically disadvantaged by the apartheid system. HDIs include black, colored, and Indian people as well as women and the disabled.
29. SANParks (2001).
30. KMPG (2002).
31. van Jaarsveld (2004).
32. DEAT undated.
33. Theron (2004) and Swanepole (2004).
34. KPMG (2002) and JMB (2002).
35. JMB (2002).
36. KPMG (2002).
37. Chikanga (2004).
38. Wolmer (2003).
39. de la Harpe (2001).
40. Wolmer (2003).
41. Ferreira (2004).
42. Chaumba et al. (2003).
43. Sharman (2001).
44. Chaumba et al. (2003).
45. Ferreira (2004).
46. Wolmer et al. (2003).
47. *Staff Reporter* (2004).
48. JMB (2002).
49. Huggins et al. (2003).
50. World Bank (2004: 11).
51. Huggins et al. (2003).
52. GFA Terrasystems (2002).

53. Huggins et al. (2003).
54. Limpopo Tourism Consortium (2004).
55. Huggins et al. (2003).
56. Limpopo Tourism Consortium (2004).
57. G. Vincente (2004), personal communication.
58. Limpopo Development Consortium (2004).
59. KFW (2004).
60. Codex dds undated.
61. Limpopo Development Consortium (2004).
62. G. Palane (2004), personal communication.
63. Helvetas Moçambique (2004).
64. G. Palane (2004), personal communication.
65. G. Palane (2004), personal communication.
66. Hunter (1997).

II
Transboundary Conservation in Action: Bioregional Management and Economic Development

Will governance challenges be met by mere extensions of the traditional legal cliche of permanent sovereignty over natural resources? Alternatives are on the horizon, and among them is the concept of public trusteeship: the idea that the Earth's natural heritage might not be the property of nation-states after all, but a kind of public trust for the benefit of all peoples, with governments as trustees accountable for their diligent exercise of mere fiduciary rights and obligations.

—Peter Sands, Carnegie Endowment for International Peace, 2001

Rather than idolize wilderness as a nonhuman landscape, where a person can be nothing more than a visitor, parks might provide important new lessons about the degree to which cultural values and actions have shaped the natural world.

—Mark David Spence, *Disposessing the Wilderness*, 1999

6

Connecting the World's Largest Elephant Ranges: The Selous–Niassa Corridor

Rolf D. Baldus, Rudolf Hahn, Christina Ellis, and Sarah Dickinson DeLeon

Starting from the Grassroots

Transboundary projects in southern Africa have usually started from the top with the signing of protocols by the heads of state. This chapter challenges the assumption that peace parks projects require high-level political movement to be initiated. In the case of the Selous–Niassa Corridor, cooperation developed from the grassroots. We also consider the role of development agencies and donors in facilitating such processes without spurring microconflicts and resentment of external interference.

Conservation policies usually require local communities to alter their way of life. Although these changes appear to affect only superficial aspects of an individual or community's lifestyle, the impact can often be more profound. Frequently these changes alter resource use behaviors that are connected to traditional practices, cultural beliefs, and the structuring of social relationships within a community. Moreover the hidden values underlying new policies, such as the value of nature, its exploitation and role in human cultural practices, can be alien and undermine existing beliefs. As Douglas and Wildavsky (1983) noted, these ontological values and beliefs represent a key component to individual and group identity and can be the firmest and most defended beliefs that cultural groups hold.[1]

The importance of basic human identity needs, as suggested in the food security theme, extend beyond biological requirements to include the satisfaction of social, physiological, and spiritual needs. Human needs theory can be a useful framework to better understand the dynamics driving the intensity of identity conflicts; it is outlined here as a tool for practitioners to become more sensitive to the role of identity in conflict. Situations or conflicts that evoke strong emotions from participants usually represent conflicts that threaten the basic identity needs of those involved.

The term "deep-rooted conflict" describes situations where basic human identity needs of one or more of the actors involved is threatened. Conservation, by proposing to organize, use, and exploit the environment in different ways, threatens the traditional integrity of a group and, consequently, its basic human needs. Because our identity, and the values and beliefs underlying it, define our experience and understanding of the world, it cannot simply be negotiated away. Addressing deep-rooted conflict is understandably intimidating. It requires that all actors acknowledge the identity needs at stake and reconcile the opposing values and beliefs. This cannot be accomplished through negotiation, but it is a longer term process based on a respectful focus on process and relationship building, joint experience, and dialogue. Understanding deep-rooted or identity-based conflict and its impact on conservation requires a broader awareness of identity needs and their satisfiers.

Human needs theory, as explained by Burton (1990), suggests that there are universal human identity needs but that the particular satisfiers of these needs are defined by culture, beliefs, and experience. Although each individual's specific identity needs might be unique, language and culture provide a common framework to which most of the people within its range share a great deal of their world. Yet a particular conflict might combine different identity needs that can be linked to more tangible issues that are more amenable to compromise. Navigating this complexity can often feel overwhelming. Examining the five categories of basic human needs—meaning, connectedness, security, action, and recognition—is a first step in making practitioners more sophisticated in their awareness of identity and conflict. Although writers have defined different human needs typologies, we briefly summarize Redekop's (2002) categories below because it serves as a good overview and is directly linked to deep-rooted conflict theory.

Meaning
The human identity need for meaning contains not only our worldview but also our sense of justice. Despite observing similar facts, individuals with different meaning structures will understand and interpret the information differently. Engaging these underlying structures of meaning requires an extended dialogue and must be done in a nonthreatening manner. The emotion most often connected to an individual's sense that their "meaning" is threatened is anger.

Connectedness
Humans create a "web" of connections during our lifetime: personal relationships in a social network and geography in a physical environment. Connectedness repre-

sents the need to belong. Yet it is the profoundly different ways in which cultures or individuals satisfy that need for connectedness that often generates deep-rooted conflict within conservation settings. Culture can serve as a common language of belonging, including common stories, values, and experiences. Changes in cultural practices therefore affect an individual's or community's sense of belonging or connectedness and lead to feelings of separation, alienation, and ostracism.

Security

The need for security incorporates the physical, emotional, economic, and spiritual dimensions of human experience. An individual's sense of security can extend to incorporate family, community, ethnic group, or way of life. For this reason conflicts that appear to only involve individuals might take on larger group implications because underlying many disputes are the internalized beliefs and values linked to group identity. By binding the project goals to the security needs of the entire community, the project mirrored the individual sense of security that extended beyond the family unit. Fear is most often the emotion associated with a lack of security.

Action

The identity need for action centers on the impetus to take meaningful, significant action in relation to the actor and his or her environment. This is often interpreted as a strong need for control or independent action, but different cultures manifest behaviors differently. Systemic discrimination or strong government control (i.e., colonialism) destroys the sense that meaningful action that influences their environment can be taken because the ability for creative, independent actions (even as a group) has been removed. The emotional state that reflects the lost of agency is depression.

Recognition

The identity need for recognition in individuals and groups can be simplified to the concept of respect. Treating individuals with dignity or respect not only meets identity needs, it also affirms an individual's place within a community or social context. Again, what is deemed as respect or recognition can vary significantly among cultures and individuals. But for individuals or groups who have been historically undervalued, recognition takes on greater meaning. Even the initial interactions with communities or individuals might be distorted by preexisting, yet unrelated, deep-rooted conflicts.

Addressing identity-based conflicts are generally outside the standard conservation practitioners toolbox. While conservation projects would most likely solicit the help of a community reconciliation expert to address identity-based conflicts, there are some basic points that will increase the practitioners awareness of the depth and power of identity needs:

- It is important to distinguish between conflicts that may involve a change of the power relationship between implicated parties and those that will not.
- It is important to identify if there exists a mutually acceptable institution to deal with changes in power between the parties.
- Standard tools such as joint problem-solving, and its use of negotiation and mediation are often not appropriate in situations with large power inequalities between the parties. Changes might be needed to make sure that what indigenous peoples believe is at stake for them is better heard, a willingness to meet in settings where indigenous peoples are comfortable, or the adoption of indigenous peoples' discussion styles.
- Conflicts that involve core values such as identity are often not negotiable. Raising awareness of commonalities and shared concerns may increase the parties' perceived relationship and responsibility for resolving the issue at hand.
- Individuals and groups cannot negotiate how they feel about their identity. They can, however, negotiate how they think and speak about their identity; how they formulate it in public discourse. Social identity is a construction and can be reconstructed through new and shared experiences.
- It is most important to adopt reconciliation principles that "dignity and respect foster truth" and to fervently "listen."

With these guidelines, it should be possible to frame a worthwhile conservation effort from the grassroots perspective. In this chapter we will see this approach succeed in a region of eastern Africa recovering from protracted civil conflict and ethnic complexity.

The beginnings of communications within the Selous–Niassa corridor were rather down to earth. In 1998 wildlife and conservation managers in southern Tanzania and from the Niassa Reserve of northern Mozambique started to discuss their common interest in reduction of poaching as a cross-border issue. The process was facilitated by a German development cooperation program in Tanzania that had assisted the Selous Game Reserve and the communities in the buffer zones since the 1980s. Communication on security matters and anti-poaching slowly developed into cooperation. Visits by reserve managers took place, management experiences were exchanged, and a joint vision evolved, namely to establish a wildlife corridor between two large protected areas and to coordinate conservation management in an ecoregion divided by a boundary between two states that were not yet ready to cooperate on conservation matters. At that time it seemed neither possible nor feasible to formalize this cooperation at the higher government level.

The Role of Donors in Galvanizing Grassroots Conservation

Significant sustainable development challenges face governments in sub-Saharan Africa. The international donors who choose to support development and conservation in these countries need effective avenues to further the work of addressing these challenges.[2] Local politics affect the acceptance, success, and implementation of community-based wildlife management, but this detail is often overlooked by international donors and foreign development agencies.[3] Decentralization of program control is not automatically successful and is often fraught with problems, but the examples of transboundary community-based wildlife management initiatives presented in this chapter bring some hope.

Because the need for conservation was defined from the grassroots level, the donor agencies did not have to direct establishment of objectives as they would in a more top-down approach. In this case government played the role of endorser and consultant, but not the role of director. Since most examples of community-based conservation in Africa are spearheaded by donor, development, and government agencies, grassroots conservation initiatives have not been highlighted on an international level. However, community-driven initiatives can provide a refreshing alternative model with more effective roles for donors.

In Tanzania, grassroots innovation and creative approaches to community conservation provide valuable case studies and models that may be of interest to donors, development agencies, and governments in their quest to achieve the Millennium Development Goals of the United Nations. The need for steadier, more robust flow of resources from international donors to Africa is an important requirement for the local institutional development. At a time when donors and multilateral agencies are likely to be on the lookout for new ideas,[4] projects like those presented here are especially timely. These projects can help reveal useful future directions for the current deliberation and research on relationships between capacity building, donor funding, and expatriate advisory roles.[5]

From War to Peace—But Poverty Remains

The coastal areas of both Tanzania and Mozambique felt the impact of foreign influence as early as the eighth century, when Arab traders arrived at the southeast coast of Africa. By 1498 the Portuguese had explored the "Swaheli coast" and claimed control over it while destroying highly developed cities and trading states built by the Arabs.

The Portuguese control of present-day Tanzania was restricted to a few coastal settlements. Assisted by Omani Arabs, the indigenous people succeeded in driving the Portuguese from the area north of the Ruvuma River by the early eighteenth century. After the Anglo-German agreements were negotiated that delineated the British and German spheres of influence in the interior of East Africa and along the coastal strip previously claimed by the Omani Sultan of Zanzibar, the German government took over direct administration of the territory in 1891. Germany's colonial rule ended after World War I, when control of most of the territory passed to the United Kingdom under a League of Nations mandate. After World War II, Tanganyika became a UN trust territory under British control. Subsequent years witnessed the country moving gradually toward self-government until it reached its independence peacefully in 1961. In 1964, the mainland united with the island of Zanzibar and was thereafter named the United Republic of Tanzania.

Mozambique became a Portuguese colony in the seventeenth century. The Portuguese government ruled with an autocratic hand, and Mozambique gained independence only in 1975 after years of bloody guerrilla warfare. Independence did not bring peace. Instead, one of Africa's cruelest civil wars continued with foreign involvement until a UN-negotiated peace agreement ended the fighting in 1992. It left destroyed infrastructure, 900,000 deaths, and 1.3 million refugees in the neighboring countries.

The Mozambiquan liberation war and the subsequent civil war contributed to the destabilization of the whole of southern Africa. The president of Tanzania, Julius Nyerere, known for his pan-African and anti-colonial sentiment, supported Mozambique even at the expense of the welfare of his own country. Consequently the south of Tanzania suffered from the ongoing war in the northern provinces of Mozambique. Many thousands of war refugees were hosted in numerous refugee camps in the south. Spurred by the permanent fear of an invasion, the Tanzanian government closed the southern border area for many years and moved villages along the Ruvuma River farther north, thus negatively affecting the infrastructure and economy.

After 1992, the war refugees of Mozambique were peacefully repatriated, but still the common border of 756 kilometers in length remains a battle zone and one of the most underdeveloped zones of both countries. There is not even a single bridge crossing the Ruvuma River, which forms the boundary, thus preventing transport and communication.

The end of the civil war in Mozambique changed the regional political scenery. The peace settlement led to the establishment of majority governments in Zimbabwe and South Africa, and enabled both countries to redefine the contours of their policies. Multiparty elections are provided for as well as development toward a free market economy. Mozambique and Tanzania are members of the South African Development Community (SADC), which holds the vision of a common, more prosperous and peaceful future within a regional community. An initiative by the governments of Tanzania, Mozambique, Malawi, and Zambia—and supported by South Africa—aims to develop the Mtwara Development Corridor, including the Ruvuma interface. This plan focuses on the regional integration of infrastructure networks and all kinds of development initiatives. Its aim is to reduce poverty by stimulating broad-based economic growth. Wildlife and forests are still abundant, however threatened, and their sustainable use including tourism could develop into an important regional asset.

The border between Tanzania and Mozambique is more a zone of underdevelopment than a zone of hostility or border conflicts. For the peace to establish itself, it needs the development of infrastructure, communication, and economic growth, all only possible if the two countries join forces. The Selous–Niassa Wildlife Corridor, which is discussed in this chapter, will be an integral and important component of this peace-building process.

Two Reserves and Wildlife Management Areas

With an area of around 48,000 square kilometers, representing 6 percent of Tanzania's land surface, the World Heritage Site of the "Selous Game Reserve" is the largest single protected area in Africa. It is also the oldest dating back to 1896. The reserve contains some of the largest and most important populations of elephants, buffalos, antelopes, wild dogs, and other predators in Africa. With its extensive area of natural miombo forests, the Selous is one of the largest forest areas under protection on the continent.

During the 1980s the rapid increase in poaching for ivory and rhino horn led to a steep decline of the elephant and rhino populations and threatened the reserve's ecological integrity and survival. With assistance from the Federal Republic of Germany and through the Tanzanian–German Selous Conservation Program (SCP) it was successfully rehabilitated to a self-financing conservation area[6] between 1987 and 2003.

With a surface of 42,000 square kilometers, the Niassa Reserve[7] is the largest conservation area of Mozambique, and it contains the greatest concentration of wildlife in this country. Located in northern Mozambique, the Niassa Reserve core area of 22,000 square kilometers is bordered by the Ruvuma River in the north (along the Tanzanian border), the Lugenda River in the southeast, the Luatize River in the southwest, and the Lussanhando River in the west. A buffer area of 20,000 square kilometers, which was divided into six management concessions, is also part of the Niassa reserve. The total area is twice the size of Kruger National Park in South Africa, or comparable in size to Wales or Denmark.

Despite many years of liberation, civil wars, and conflict, the reserve supports a rich and diverse collection of wildlife. Of particular interest are the endangered African wild dog and three endemic subspecies that exist in Niassa but are rare elsewhere: namely the Niassa wildebeest, Boehm's zebra, and Johnston's impala. Over 100 elephants are known to be carrying heavy ivory of over 80 pounds on each side, where poaching used to be common but is rare nowadays. The reserve has a rich bird life and an abundant raptor population. The Ruvuma River is an important bird area and over 370 bird species have already been identified.

The reserve is managed by the Sociedade de Gestao e Desenvolvimento da Reserva do Niassa in the form of an innovative partnership between the public and the private sector, where the government of Mozambique retains ownership of the land and wildlife, and assists in the management. This group has received support from a number of private and governmental donors, the major one presently being Fauna and Flora International.

Conservation efforts in the Selous Game Reserve did not end at the reserve's boundaries. A specific Tanzanian concept for community based natural resource management was developed for the integration of local communities and for securing buffer zones outside the reserve. Village wildlife management areas (WMA) are its major component. They contribute not only to the achievement of conservation targets but, more important, to development and poverty alleviation in the rural areas.

The idea of government recognized community based conservation in Tanzania originated with the Selous Conservation Programme (SCP), which was a joint initiative begun by the Tanzanian and German governments at the end of 1987. SCP was run by the Wildlife Division of the Tanzanian Ministry of Natural Resources and Tourism with the technical support of GTZ until 2003 when it was converted to a self-financing conservation area.[8] A government advisor (the lead author of this chapter) added by GTZ was to assist until 2005 in further developing the concept of community-based conservation and incorporating it fully into the work of the Tanzanian government's Wildlife Division, and to coordinate the GTZ's Conservation and Sustainable Use of Natural Resources in Forestry and Wildlife project sector throughout Tanzania. The concept of WMA has now been incorporated into the wildlife policy of Tanzania[9] and is also implemented in other parts of the country.

The establishment of WMA around the Selous, including the southern buffer zone[10] and the area south of the Selous, has resulted in the intentions of 26 villages in three districts to register three WMA shortly, with a total area of approximately 9,000 km^2.

The Selous–Niassa Wildlife Corridor

Communities in villages situated farther south toward the Ruvuma River approached the Tanzanian authorities and the SCP to ask for inclusion into the program, since their village land still supports wildlife populations. Unfortunately, much wildlife was already lost due to poaching, but because the land is not settled yet, these populations can recover provided that proper conservation is introduced. According to the local people, traditional elephant migration routes exist between the Selous and the Niassa Game Reserve in Mozambique. But they warn that elephant poaching and an agricultural establishment bloc threaten these corridors. These activities might have adverse effects on the wildlife in the protected areas. Some villagers have already started their own informal community organizations and conservation activities without SCP support, which have subsequently caught the attention of the authorities. The creation and protection of a Selous–Niassa wildlife corridor[11] would close a critical gap separating the two reserves. It would stretch from 120 to 160 km between the southern border of the Selous Game Reserve and the Ruvuma River, which constitutes the northern boundary of the Niassa Game Reserve.

The Selous–Niassa Corridor Research

To obtain a sound scientific foundation for the conservation of the wildlife corridor the Institute for Zoo Biology and Wildlife Research of Berlin/Germany, the Sokoine

University of Morogoro/Tanzania and GTZ carried out a research project from 2000 to 2003.[12] While motivation for the project came from the Wildlife Division, the district administrations, and a number of community members, most of the necessary scientific muscle came from the Institute of Zoo Biology and Wildlife Research and Tanzania Wildlife Research Institute. The project combined advanced satellite-based observation techniques with the application of local indigenous knowledge in order to achieve a better understanding of geographic, physical, and ecological conditions contributing to the corridor's landscape. The project was financed by the German government in cooperation with the Wildlife Division.

Ten elephants were radio-collared[13] in 2000 and 2001 and their movements observed via satellite. The breeding status of most of them was known due to ultra-sound analysis applied in the field. They were decollared in 2003 under extremely difficult conditions, as they were spread over large, inaccessible areas. One elephant had meanwhile been killed by poachers. One (possibly two) more was severely injured by shots. A Tanzanian scientist collected further data working together with Wildlife Division staff, traditional hunters, and local people.[14] During the dry season of 2000, coordinated aerial censes were carried out in both countries.[15] The total elephant population of the entire ecosystem was estimated at approximately 65,000, with the majority (85 percent) in Tanzania. Globally significant populations of Lichtenstein's hartebeest (*Alcelaphalus buselaphus lichtensteinii*), African buffalo (*Syncerus caffer*), Niassa wildebeest (*Connochaetes taurinus cooksoni*), eland (*Taurotragus oryx*), greater kudu (*Tragelaphus strepsiceros*), common waterbuck (*Kobus ellipsiprymnus*), bushbuck (*Tragelaphus scriptus*), common reedbuck (*Redunca arundinum*), and different duiker species are linked by the corridor. Their distributions and occurrences vary substantially depending on the season (rainy or dry) and location in the corridor. Large numbers of Roosevelt's sable antelope (*Hippotragus niger roosevelti*) are widespread throughout the corridor. Besides these species the corridor is home to a variety of large carnivores including African wild dog (*Lyacon pictus*), lion (*Panthera leo*), leopard (*Panthera pardus*), and spotted hyena (*Crocuta crocuta*). Other wildlife includes crocodile (*Crocodilus niloticus*), hippopotamus (*Hippopotamus amphibius*), smaller mammals, other rare Tanzanian fauna, and a diversity of bird life.

The area studied encompasses a wide variety of wildlife habitats. Those are wooded grassland, substantial areas of open savannah, granite kopjes, seasonal and permanent wetland, and riverine forests along numerous streams draining either toward the Rufiji River or the Ruvuma River. The area supports a high number of globally significant, threatened, and CITES-listed large mammal species. The dimensions of the corridor allow even the largest herbivore, the African elephant, to migrate between the two largest elephant ranges on the continent, the Selous and Niassa Game Reserves.

The research proved[16] that the planned wildlife corridor provides a significant biological landscape linkage between the two largest game reserves of Africa and consequently preserving the unfragmented miombo woodland ecosystem, which covers about 150,000 square kilometers of the whole Selous–Niassa complex. The corridor is not only a transit route for elephants between the two game reserves in the north and south, it also sustains its own sizable resident population. At least 2,400 elephants are residents or use the area part-time. This population appears to be expanding, with a healthy calf: female ratio and outstanding value in terms of the reproductive quality of semen of breeding bulls.

As indicated by the details of radio-tracked movements of individuals, particularly in the center of the corridor, the biological linkages stretch further in an east-westerly direction than initially expected. Some elephants make use of large sections of the corridor by virtue of maintaining very large home ranges. They travel across the central and southern sections, with extensive movements between Tanzania and Mozambique and within the Niassa Game Reserve. The fact that there are conspicuous and well-established major elephant movement routes that cross the entire corridor also suggests that some elephants may be entirely transient and move between the adjacent game reserves on a regular basis. Hence any fragmentation of elephant habitat would be a grave disadvantage.

The movement of the elephants depends on the season and correlates with the availability of food, water, and shelter and the chance to cross the major rivers without risk. Large breeding bulls frequently move between the southern sections of the corridor in Tanzania and the Niassa Game Reserve. Not only does this emphasize the status of the area as a true transboundary ecosystem, it also pinpoints the value of the corridor as a link between the Selous and the Niassa elephant populations in terms of breeding and genetic exchange.

The key results of this research, like ecosystem description, identification of migration routes, wildlife population size, and human–wildlife conflict areas, have been used to make a strong case for the long-term conservation of this corridor.

Threats to Biodiversity

There are severe threats to the continued existence of the corridor that, if left unattended, will block this important link. The immediate threats are the bush-meat trade supplying the local markets, and ivory poaching, which is a transboundary problem. Already as early as 1989, our surveys[17] showed that the poaching of elephants for meat in southern Tanzania was a grave problem. It could not be solved by international trade bans. An agreement on cross-border law enforcement between the Tanzanian and Mozambiquan governments does not exist. A further threat is habitat degradation due to uncontrolled wildfires caused by the local population. In the long run the high population growth and associated agricultural expansion (e.g., tobacco farming) will increasingly convert this still biologically intact corridor to cultivation.

The loss and fragmentation of natural habitat will form a genetic blockade between the world's largest protected miombo forest ecosystems and elephant habitats. The obstruction of the movements of large herbivores such as the African elephant will ultimately result in increased human–wildlife conflicts. Crop damage by elephants and man-eating by lions[18] are already major problems.

According to experience, people will intensify farming and settlement along roads. There are still gaps between villages along the major routes through which the animals move, but it can be expected that in the future, in particular, along the Songea–Tunduru highway, these gaps will be blocked by intensified human activity. This road is due to be upgraded under the Mtwara Development Corridor program. Much speculation is already taking place in expectation of rising land values and quick action is required to secure and protect these areas.

Dense human population and agricultural activities to the east along the Ruvuma River in Tunduru district already prevent wildlife movements between the Niassa Game Reserve and Tanzania. Within the corridor, concentrated fishing activities and extensive snare lines from poachers along the river and some of its tributaries disturb the wildlife movements there. This effect is greatest in the dry season, when fishing activities peak and the animals are most dependent on water from the river. Furthermore destructive and uncontrolled fishing methods (use of poison) deplete the fish stocks of the river and disturb the aquatic fauna.

Uncontrolled commercial logging for valuable and marketable hardwoods was also observed and will increase with the growth of the major towns and the improvement of the road system. If not controlled or prevented, it will soon lead to a genetic depletion of some valuable species. Massive logging of all hardwood species for local consumption, and increasingly for export, has depleted the natural forests in all of eastern Tanzania in the last fifteen years. It was openly done despite being mostly illegal. As forests became empty, this logging has moved slowly south, passing Rufiji and Kilwa districts, and is now reaching the corridor area. Another concern is uncontrolled artisanal exploration for precious stones and gold that results in major ecological damage.

The 380 square kilometer Sasawara Forest Reserve, located almost in the center of the corridor, is a core area for the protection of biodiversity amid increasing human activity. However, reports backed up by satellite images show that heavy encroachment and destruction from activities such as farming and settlement take place in the eastern part of the forest reserve. With the high human population growth rate and its impact on the mostly still intact natural habitat, predictions are that the corridor will be fragmented and destroyed soon unless adequately managed.

The Conservation of the Selous–Niassa Wildlife Corridor (SNWC)

The Development and Management of the Selous–Niassa Wildlife Corridor project is to be financed by the Global Environment Facility, a joint program of the UNDP, UNEP, and World Bank. It has a budget of US$ one million for four years and was started in 2005 after four years of negotiations and preparation.

The objective of this conservation project is to extend the concept of community-based natural resources management from the southern buffer zone of Selous Game Reserve toward the border with Mozambique. Suitable villages in Namtumbo and Tunduru Districts will be able to establish a network of village WMA, eventually linking the Selous and Niassa game reserves. Specifically local competence in sustainable resources management within the corridor will be developed. These activities aim at empowering people at the local village level and involving them in all aspects of conservation in compliance with the wildlife policy of Tanzania and its relevant laws and regulations. Benefits from natural resources management and the development of eco-tourism and sustainable tourist hunting in the area will enhance the livelihood security of the villages, maintain biodiversity, and promote the long-term conservation of the corridor. GTZ-International Services implements this cooperation project as executing agency on behalf of and under a contractual agreement with the Wildlife Division and the United Nations Development Program (UNDP). In 2004 the "Mtwara Development Corridor," GTZ, and the Ministry of Natural Resources and Tourism jointly conducted a major study into wildlife, forestry, tourism, and infrastructure with a view to attract investment.[19]

In response to proposals of the GTZ Wildlife Program, the German Ministry for Economic Cooperation and Development provided a grant of 5 million euros for the corridor as financial assistance to Tanzania in early 2005. The funds will be made available through the German Development Bank (KfW), which has meanwhile appraised the project. KfW's aim is to overcome structural obstacles and to initiate sustainable development. The overall objective is the conservation of biodiversity. Special emphasis will be given to community-based natural resources management in the corridor. On the basis of a participatory planning process of all stakeholders involved, financing operations will be directed at (1) the infrastructure necessary for the development of the corridor, the existing protected areas, and WMA; (2) the extension of the corridor area while establishing additional village WMA; and (3) improvement of the livelihoods of local communities by sustainable use of natural resources to combat poverty.

The Mtwara Development Corridor

While the workings of the Selous–Niassa Wildlife Corridor were prepared with the assistance of GTZ, political cooperation between the two countries at governmental level also proceeded independently of them. The respective heads of state signed a multilateral agreement between Malawi, Mozambique, Tanzania, and Zambia on the Mtwara Development Corridor (MtDC) on December 15, 2004, in Lilongwe, Malawi. The objectives of the MtDC are, inter alia, to (1) achieve socioeconomic development and integration by unlocking the underdeveloped potential within the corridor area; (2) develop reliable and efficient transport, communication, and energy systems; (3) foster and catalyse growth in the agro-industrial, tourism, and services sectors through the involvement of the private sector and the strengthening of public–private partnerships; (4) optimize the sustainable utilization of natural resources for development in the region; and (5) re-orient the corridor trade and investment through the promotion of human settlement, minimize rural–urban migration and eradicate poverty. The Selous–Niassa Wildlife Corridor was identified at a technical level as one of the anchor projects to unlock tourism potential in the Mtwara Development Corridor.

In September 2004 a Memorandum of Understanding between the regional government of Mtwara in Tanzania and the provincial government of Cabo Delgado Province in Mozambique was signed in Pemba, Mozambique. In this agreement the government of the Mtwara region and Cabo Delgado province (1) expressed through local level transboundary cooperation their desire to promote regional economic growth and development, (2) recognized the need to establish adequate and efficient trade routes for regional integration, and (3) noted the great economic potential of the Mtwara region and Cabo Delgado province, given the rich agricultural, game park, tourism, transport, and communications infrastructure. In July 2005 preparatory meetings for the regional agreements of cooperation between the Niassa province and Ruvuma region were organized by the Mtwara Development Corridor and by regional administrations from both countries. A hierarchy of regional and local level cooperation agreements are now in place or in the process of being finalized. These can act as entry points for local transboundary dialogue that renders a practical dimension to regional-level integration. Initial dialogues on scoping and analysis of transboundary issues took place already and close cooperation between the Niassa Reserve management and the SNWC project regarding the sharing of information and cooperation on studies along the common border has been established as a guiding principle.

Capacity Building in Transboundary and Community Conservation

A number of conservation activities have been started in the corridor on the basis of the research and joint planning and cooperation between the Wildlife Division, the regional and district authorities, and the German Wildlife Program. The role of the GTZ and other branches of the German government in the establishment of the corridor has been to provide support for technical, financial and communication capacity building where the Wildlife Division and local authorities have determined that stronger capacity is needed to pave the way to success. German advisors, who are not subject to the internal bureaucracy and inherent constraints of the public sector in Tanzania and other sub-Saharan African nations, can act efficiently when called upon by local officials to promote the innovation that spurs the way to adaptive learning. The main activities of the GTZ government advisor in the Wildlife Division with regard to community-based conservation (CBC) are summarized as follows: (1) facilitation of law amendments conducive to CBC; (2) advice and training for relevant staff at various levels; (3) approval and technical advice for management plans in WMA; (4) provision of technical advice to NGOs at their request; (5) provision of technical advice to authorized agencies on development of contracts with tour operators and other private sector organizations; (6) development of financial systems to facilitate equitable handling of fees and other benefits related to wildlife utilization by communities; (7) follow-up, implementation assistance, and promotion of CBC policy at various levels of Tanzanian government; and

(8) identification of new project areas and their implementation needs followed by assistance seeking to fill these needs. In addition to these on the ground capacity-building services from GTZ, the German government has provided additional resources through Capacity-Building International, Germany (InWEnt), a collaborative capacity-building agency concerned with organizational development and institutional strengthening.

The GTZ Wildlife Program in association with InWEnt and the Environmental Law Center of the IUCN sponsored a series of workshops in Tanzania from 2002 to 2004 entitled "Transboundary Protected Areas: Guiding Transboundary Boundary Protected Areas (TBPA) approaches and processes in East Africa" at the Mweka College of African Wildlife Management. Specific objectives of these workshops aimed at (1) deepening of planning and management capacity for TBPA in East Africa; (2) stimulation of policy dialogue amongst the key role-players in TBPA development for the region; (3) amplification of knowledge and skills for policy, participation, and monitoring.[20] Several key lessons that will foster continued progress and adaptive learning on the issue of transboundary conservation in Tanzania and the greater region came out of the workshop. The principles of these selected lessons are as follows:[21]

- Develop an effective institutional framework that engages all relevant stakeholders.
- Ensure that joint management of the TBPA is achieved.
- Involve local communities early in the planning process and sustain participation. TBPA is successful if clear socioeconomic benefits are associated with its development.
- Lobby for and secure sufficient funding for implementation and management of TBPA.
- Implement processes and measures to achieve sustained good relations between main partners.

The demands of creating an arena for empowered, community-backed conservation and development are difficult to meet. The need for stronger government capacity in Southern Africa would likely go unmet without the involvement of the international community and donor organizations. The "fortress" style conservation of the past was in many ways easier for states to implement.[22] With the advent of formalized shared management between communities and government agencies, and the added dimension of cross-border cooperation, the role of outside technical assistance is increasingly important. Outside financial support will also continue to be necessary to advance innovative biodiversity conservation. As demonstrated in the Selous–Niassa Wildlife Corridor efforts, it is possible for donors and governments to play a positive and necessary role in transboundary wildlife protection when they are backing rather than directing the work of community-led conservation. Considering the complicated, barrier-laden labyrinth of capacity building in Southern Africa,[23] these creative efforts from the communities involved local authorities, and the supporting German aid agencies are difficult, exciting, and necessary.

The Future: Toward Transboundary Cooperation

Tanzania and Mozambique share the same ecosystem and hence the natural resources along the Ruvuma River, which is the longest river of eastern Africa. Not only the common species, ecosystem services, landscape features, but also the common culture of the local communities are of great significance. The planned corridor will, however, be completely on Tanzanian territory. The described conservation activities are all exclusively under Tanzanian management. While preparing, the programs' informal cooperation, first at local levels, slowly evolved. There is still a strong bond between the people on both sides of the river. During the long civil war, southern Tanzania hosted many refugees from Mozambique. Most of them returned after the war. The others remained and have settled. This all has increased the transboundary traffic. As there is no bridge yet over the Ruvuma, all crossing is by dug out canoe. The informal cooperation consisted of communication, visits by reserve managers, and exchange of information on conservation issues including on poaching and security matters. Much support by Niassa Game Reserve was needed during the research project.

Whereas many transboundary projects in southern Africa have started from above with the signing of protocols by the heads of state; in this case both sides were convinced that a bottom-up approach was more appropriate. Cooperation should grow from the grassroots, and if at a later stage all involved parties are convinced that strengthened formal relations are necessary, this can be induced. These first steps constitute a move toward transboundary cooperation, and the future will show whether this will indeed end in a kind of transboundary protected area.

Selous and Niassa game reserves together with the corridor have the potential to become such a protected area and would constitute the largest one in Africa. Southern Tanzania and northern Mozambique have been a zone of instability during the independence war and the consecutive civil war. Transboundary conservation will not only add to increased biodiversity preservation and increased rural incomes but can also stabilize and strengthen peace in a once troubled part of Africa.

Notes

1. Addison et al. (2005).
2. Virtanen (2005).
3. Kayizzi-Mugerwa (2005).
4. Wubneh (2003).
5. Baldus et al. (2003a).
6. ⟨www.niassareserve.com⟩.
7. Baldus et al. (2003a).
8. Wildlife Division (1998).
9. Baldus et al. (2001).
10. Baldus et al. (2003b).
11. Hofer et al. (2004).
12. Mpanduji et al. (2003).
13. Mpanduji (2004).
14. Tanzania Wildlife Research Institute (2001).
15. Mpanduji et al. (2002).
16. Ndunguru (1989).
17. Baldus (2004a).
18. Smith and Baldus (2005) and Hahn et al. (2005).
19. Baldus (2004b).
20. InWEnt (2003).
21. Adams and Hulme (2001) and Barrow et al. (2001).
22. Ramsbotham (2005).

7

The "W" International Peace Park: Transforming Conservation and Conflict in West Africa

Aissetou Dramé-Yayé, Diallo Daouda Boubacar, and Juliette Koudénoukpo Biao

Many conservation zones in Africa were delineated during colonial times for political expediency, leisure activities for colonialists or selective resource exploitation, rather than for environmental protection. This was particularly the case in West Africa where French colonial control extended over vast territories that were often defined by necessity on grounds of ecosystem boundaries and topographic ranges. However, when colonial control began to crumble, there was a tendency to demarcate borders based on narrow definitions of ethnicity. When different colonial powers controlled territory, borders were also formed in spite of ethnic similarities, which often resulted in violence and a quest for identity in later years.[1]

Postcolonial Africa has often been characterized by theories of state weakness and failure that might be considered by some analysts as a rationale for why transboundary initiatives might be easier to establish. However, in reality, such internal weakness of the state usually results in a divisive paranoia about borders. Often in such cases the borders are likely to be even more guarded and difficult to traverse. Visa restrictions within Africa and prohibitions of even photographing border areas by tourists are emblematic of such challenges to transboundary conservation. At the same time environmental conservation zones in tropical climates may also have forests that provide a natural cover for armed militias. Some researchers have therefore proposed that conservation zones can themselves be security threats and havens for criminal activity, calling for "peace park threat assessments," particularly for parts of Africa.[2] With scant empirical evidence of particular threats, such concerns appear to be overblown and are more a critique of ineffective management than of the concept of peace parks themselves.

In West Africa the French colonialists developed some wildlife reserves that have surprisingly endured the vagaries of colonial borders being set and may indeed provide a counter argument to security concerns over peace park establishment. Thus they can be a means of bringing states that were fractured by the colonialists during struggles for independence to come together and cooperate regionally. A natural reserve along an ecologically important part of the Niger valley exemplifies this phenomenon. Indeed this is a region that has been able to avoid many of the transboundary conflicts and clashes that have beset many other parts of Africa such as Rwanda and Congo. In a recent study the World Bank also profiled the Niger River basin countries for their relative cooperation in environmental management.[3]

Institutional Evolution of the W Park

The W Park is a vast international complex of protected areas, with a surface coverage of about 1 million hectares distributed between three countries: Benin, Burkina Faso, and Niger. The park owes its name "W" to the geometrical figure made by the Niger River between the mouth of the Tapoa River in Burkina Faso and that of the Mékrou River in Benin (figure 7.1).

This chapter intends to present and analyze the collective strategy, based on community conservation of the W complex and its peripheral area, aiming at transforming the W Park into a real peace park.

The institutional evolution of the W Park has also been described in several recent works.[4] Before the establishment of any modern administrative structures, local populations, composed first of hunters and gatherers and then of stockbreeders and farmers, managed the park's resources in an empirical way and in a customary context.[5]

The W Park was delimited as a wildlife reserve (Refuge Park) in 1926 by the colonial federal administration located in Dakar (Senegal). For this reason it received specific attention from the International Conference on the Conservation of African Fauna held in London in 1933. This conference resulted in the Convention on the Safeguarding of Natural Fauna and Flora. The Convention recommended the classification of the refuge W Park as a National Park; this was satisfied on August 4, 1954. During the colonial period the legal basis of land regulation was the 1932 Decree on the reorganization of landed property Regime. Colonial laws were dedicated to ensuring natural plant and animal regeneration, and avoiding the abusive destruction of wild fauna. This was the case of the 1947 Decree regulating hunting in the overseas African territories, and the 1954 Decree on Integral Natural Reserves and National Parks police in the French West African territories.

After the independence movements of the 1960s, the centralized management of the W Park gave way to national management through different and uncoordinated

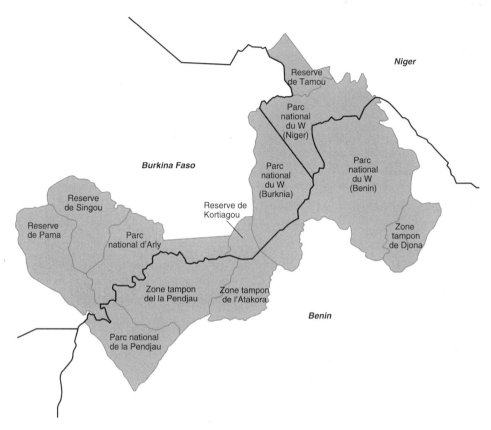

Figure 7.1
The W International Peace Park

political and legislative bodies. The postindependence laws were sector-related and repressive, with restrictions to protect the resources from being destroyed by the populations who were considered as "predators."

It was only in 1986 that the idea of transboundary collaboration for the management of the park (and especially for the fight against poaching) materialized. With the elaboration of laws related to environment management, Benin, Burkina Faso, and Niger turned toward a regional, concerted, and sustainable management of their natural resources. In Benin, the Outline Law on the Environment was defined through the February 12, 1999, law number 98-030. In Burkina Faso, the December 30, 1997, law number 005-97/ADP on the Environment Code was enacted. In Niger, the relevant law was the December 29, 1998, law number 98-56 titled the

Outline Law on Environmental Management. As their names suggest, the Outline laws on environment merely provide a general framework for environmental protection. They are reinforced by fundamental laws contained in the three countries' constitutions, which give obligation to each state to "Protect the environment" because "every person has the right to have a safe environment."

The Economic Community of West African States (ECOWAS) was established in 1975 to provide "collective self-sufficiency" for the member states by means of economic and monetary union creating a single large trading bloc."[6] However, it has also been a forum for exchange of ideas on environmental management. An ECOWAS decision established the international transhumance certificate, which replaced the vaccination book. Unfortunately, a lack of administrative diffusion of these agreements at the local level makes their application inadequate. Nevertheless, ECOWAS has transcended the tendency for postcolonial states to adhere to the linguistic colonial lineage which in the case of West Africa has been dominated by Francophone states. Nigeria, as the most populous state in the region and indeed in all of Africa, is an integral part of ECOWAS despite being Anglophone, as are Sierra Leone, Gambia, and Liberia. Guinea Bissau and Cape Verde are two Lusophone members of the organization.

Natural areas and Transboundary zones are also protected by some regional regulations: the Agreement Protocol for Transhumance between member countries of the Council of Understanding composed of Burkina Faso, Niger, Mali, and Ivory Coast, signed in 1989; the Transhumance Agreement between the countries of the Interstate Committee against Drought in the Sahel (in French, CILSS), signed in 1991; and the Decision A/DEC of October 5, 1998. The Transhumance Agreement regulates animal circulation throughout pre-identified axes in the protected transboundary areas of the CILSS states.

The W Park has been recognized by various international conventions on the environment over the past three decades. The Ramsar Convention of 1971 registered the W Park as a Wetlands of International Importance in 1987. The Convention of the World Heritage (1972) registered the park as a World Heritage site of UNESCO in 1996, and the Man and Biosphere Program (MAB) of 1974 added the park to its list of Biosphere Reserves in 2000, before listing it in 2002 as a Transboundary Biosphere Reserve. All these conventions required that Benin, Burkina Faso, and Niger be engaged in the management of the park at national, regional, and international levels.

The Park has achieved this status largely due to its extraordinary biodiversity. Many species of mammals are represented including the lion (*Leo leo*), elephant

(*Loxodonta africana*), buffalo (*Syncerus caffer*), hypotrague (*Hypotragus aquinus*), warthog (*Phacocoerus aethiopicum*), and cob Defassa (*Kobus defassa*). There are also hundreds of fish and bird species in the park as well. Public access as well as agroforestry and pastoral activities are regulated by a series of laws and regulations that are enforced by all three countries. However, the peripheral zone of the park is occupied by many villages that are impacted by immigration due to various regional conflicts.[7] The subsequent increase in population has led to an intensity of activities that are increasingly incompatible with the objectives of sustainable management. These include illegal practices such as poaching, divagation of cattle, and bush fires.[8] The competition for resource use has also become apparent resulting in conflict between farmers and stockbreeders. The latter have been forced to seek pastures beyond the borders of their countries; thus migration to those areas and ecosystems seem less affected.

Conscious of the conflicts between quasi-permanent settlers that negatively impact the conservation of Park resources and the necessity for reconciling the objectives of protection with the needs of bordering populations, the governments of Benin, Burkina Faso, and Niger decided to harmonize their strategies of conservation by applying principles from the Convention on Desertification, the Convention on Biological Diversity (CBD), the UNESCO MAB Program, and the Ramsar Convention, of which all three countries are signatories. The objective was to make the W Park a tripartite site of world interest, managed through a single regional structure called ECOPAS (the Protected Ecosystems of Sahelian Africa).[9]

This chapter evaluates the extent to which the ECOPAS Project is an instrument that can transform the W Park into a real peace park by involving outlying village populations in managing the protected areas.

Comparative Political Dimensions

The adoption of environmental and rural sector development policies has coincided with periods of sociopolitical and economic changes in Benin, Burkina Faso, and Niger. In colonial times there was no clear natural resources management policy regulating landed property in French West Africa. For example, livestock production policy was centered on measures to prevent and treat contagious diseases and to protect livestock from wild animals.

In the postindependence period (1960–1970), the young West African states tried, through regulatory measures, to modify land ownership regimes. During the exceptional administrations (1970–1990), the governments in the three countries have taken some isolated actions in favor of the environment. One of these actions is the elaboration of laws regulating hunting and of the Forest Code.

During the democratization period (1990–2000), previous environmental policies, strategies, and programs were called into questions through national debates and technical workshops, which resulted in the adoption of judicial laws on natural resources management.

Benin

The management of wildlife reserves is placed under the direct supervision of the National Center for Management of Wildlife Reserves (CENAGREF) created in 1996 under the Ministry of Agriculture, Livestock, and Fisheries (MAEP).[10] Within the framework of the Program of Conservation and Management of the National Parks (PCGPN), the CENAGREF has several aims:

- To create a network of protected areas.
- To reinforce national capacities in the management of wildlife reserves.
- To reinforce national capacities in the conservation of biodiversity.
- To develop natural resources.
- To promote participatory management of wildlife reserves.

The CENAGREF is supported in its activities by local partners such as the Village Associations for Management of Wildlife Reserves (AVIGREF). These associations are in charge of not only helping official governmental services monitor hunting areas of the W Park but also for sensitizing bordering populations to the conservation of Park resources. Initiatives for developing the bordering areas of the W Park are supported by international institutions such as the International Union for Nature Conservation (IUCN). The IUCN promotes activities that contribute to the reduction of human threats to biological diversity.

Burkina Faso

The National Council of the Environment for a Sustainable Development (CONEDD), created in 2000, and a part of the Ministry of the Environment and Living Surroundings, coordinates environmental policies and legislation. However, the design and execution of national policy related to forests and fauna are ensured by the directorate-general of the National Forestry Commission. This directorate constitutes the administrative authority necessary to supervise regional and provincial directorates.

At the province level, wildlife conservation units (WCU) manage protected areas. These units have three important aims:

- Ensuring ecological monitoring and functionality of the protected areas.
- Supervising and training actors.
- Organizing the monitoring and enforcement of poaching regulations.

At the village level, local development activities and the management of hunting areas are carried out by the Village Commissions for Local Management (CVGT). This organization was created in 2000 as a result of a law regarding agrarian organization and land tenure, and they are supported by the Village Wildlife Management Committee, which was created in 1996 according to a framework of legal reforms on wildlife management.

Niger

The directorate of Wildlife, Fisheries and Aquaculture within the Ministry of Hydraulics, the Environment, and the Control of Desertification manage the administration of the W Park in Niger. The Ministry of Hydraulics, the Environment, and the Control of Desertification has regional and departmental directorates as well as district and communal divisions that carry out activities related to the protection and management of the environment.

Park management, coordinated by a ranger, is conducted at the following levels or sections:

- The administrative section, responsible for eco-tourism and administrative affairs.
- The protection and monitoring section, which defends the integrity of the park by restricting illegal activities.
- The study, management, and development section, whose goal is to preserve the biodiversity of the park and to improve the condition of vision tourism and research activities.

Other public institutions that implement policies aimed at the sustainable management of Park's resources include the National Council of the Environment for Sustainable Development (CNEDD), the National Committee of the Permanent Inter-state Committee for the Drought Control in the Sahel (CILSS), and the National Rural Code Committee. At the local level, several partners such as the Association of the Tourist Guides of the W Park of Niger, the Bee-keepers Cooperative of the W

Park, and the Association support the state for Re-enhancing Livestock Breeding in Niger. Among international institutions, the UICN, UNESCO, and European Union all work with Niger to sustainable manage the Park's resources.

Effectiveness of the Institutional Framework of the Three Countries

The process of sustainable development in the three countries seems contingent on a modification of the role and involvement local populations have in the management of their lands. It is through this perspective that decentralization was considered an institutional response to the problem of local participation in the development of these countries. According to their constitutional laws, decentralization, at least at the national level, refers to a form of organization, which is based on the concept of local interest.

In Benin, the Policy Paper on Regional Planning (DEPONAT), adopted in October 2002, instituted policies based on the decentralization and devolution of state services. It created an opportunity for local populations to address the dynamics of development in predetermined, administrative areas.

In Burkina Faso, the law modifying the Texts on the Orientation of Decentralization (TOD), dated July 2, 2001, considered regions and provinces as administrative areas in addition to local communities. Decentralization in this case does not concern villages, which remain administrative units.

In Niger, the Law of June 11, 2002, defines decentralization as potential competences of the communities and those related to the transfer of competences and resources. Thus each region became local authorities with a legal status and financial autonomy. This means that the local authorities involved should take initiatives and have the vocation of promoting economic, social, and cultural development with their own resources.

The Management Methods of the Park and Their Constraints

The current management methods of the W Park are based on factors that ultimately limit the development of wildlife and their habitat. This study seeks a better scientific conciliation between the protection and valorization of the protected area through the sustainable management of park resources. Any management should concern both the W Park itself as well as peripheral multipurpose zones. It is on this level and within this geographical framework that the physical actions and constraints of management will be examined.

Protection Measures

Protection measures rely primarily on the organization and coordination of monitoring networks between the park and hunting zones, and the construction of surveillance tracks and infrastructures. The Management of the Forests and Protection of the Park's forestry team handle Park monitoring and protection. This team includes a ranger, heads of station, an assistant, and monitoring agents. They regulate several illegal activities including poaching, illegal fishing, illegal herding by transhumant shepherds, bush fires, illicit cutting of trees and other forest products, clearing of vegetation, and illicit human settlements in the Park.

Monitoring infrastructures include rehabilitated buildings that serve as offices and watchtowers. They also consist of watering stations for domestic cattle in order to reduce pressure on Park resources. Despite a lack of permanent monitoring tracks, the control actions for the territory give satisfactory results (e.g., in Benin the number of arrests increased by 400 percent in 2002[11]).

Valorization of the Park

The Park is enhanced primarily through vision tourism and sports hunting. These two activities are carried out in connection with campaigns against poverty and in defense of natural resource conservation. However, these initiatives remain modest, since the Park is still ecologically fragile and in the early stages of wildlife restoration.

In order to make a profession of sports hunting, an international tender was launched for the leasing of shooting zones. These zones are subject to contracts of concession signed by the Ministry in charge of the Park. One of the major elements of the agreement is the establishment of specifications, which define the rights, and duties of the leaseholder and his associates with respect to the bordering populations. This favors a participatory process of park management.

Hunting of birds is authorized in the conceded hunting zones. Traditional hunting is authorized and does not require a license, but it is limited to the administrative area where the hunter resides and can be practiced only in conceded zones.

A significant aspect of sustainable park management is the effective involvement of bordering populations in park operations. Tough measures of conservation (up to 1990) have been replaced with participatory approaches, contributing to a positive change in mentality and behavior regarding the Park. Bordering populations no longer consider the Park to be government property devoted to foreigners and civil servants, but rather as an integral part of their culture through the obligations and rights inherent in a process of comanagement.

Research and Ecological Follow-Up

Research and ecological follow-up aim to better understand ecosystem functioning and require the evaluation of management activities as well as the role of local populations in park management. As a result the Park intends to fix the annual interest generated through wildlife capital in an attempt to stave off conflict between the protection and valorisation of the Park. The conciliation of conservation with sports hunting is extremely urgent if the Park desires to uphold sustainability requirements, namely the integration of economic, social, and ecological dimensions in park management.

Research and ecological follow-up are all the more justified, since the biological equilibriums of the Park have, for the most part, not been studied, and because serious threats to the ecosystems exist.[12]

With the exception of specific research by Szaniawsky on the elephants of Alphakoara in Benin, on the human local populations that border the Park, and on a census organized in the hunting zone of Djona, little is known regarding fauna, flora, and the ecosystems that subsist in the W complex.[13]

Physical Constraints

Some of the park's physical constraints include the insufficiency (quantitative and qualitative) of access tracks, which make park monitoring and the surveillance of illegal hunters very difficult; the insufficient number of water points; the concentration of wild animals at these water points, which causes soil degradation and the destruction of flora; and the heterogeneous inhomogeneous distribution of a hydrographical network within the Park. To these constraints we could also add the non-materialization of park limits and the absence of signal devices. Indeed the lack of clarification regarding park boundaries is among the root causes of conflict between forest officers and local populations.

Constraints of Human-Origin: The Transborder Transhumance

Livestock breeding still remains an extensive communitarian activity in Benin, Burkina Faso, and Niger. In these three states thousands of Fulani stockbreeders relocate from one zone to another, depending on the rainy and dry seasons. Transhumance is a very old practice, and its origins are rooted in both culture and psychology. For the shepherd, crossing vast herbaceous areas conjure feelings of intense happiness. But today, transhumance can be explained primarily by ecological constraints. As the environment continues to degrade due to increasing human and livestock populations as well as climatic conditions, like the abominable droughts that occurred during 1913, 1954, 1973, 1974, and 1984 in the states of the Sahel, transhumance becomes a necessity rather than a choice.

The tradition of transhumance is passed down from elders and parents to young people, as are the risks. When crossing W Park borders, they face both the harshness of the forest officers and the wild beasts that overrun the area. The creation and current management of the Park generates conflicts between stockbreeders and the three states. The shepherds of the Sahel do not resist temptation to let their herds pasture in the Park. They still cannot understand that the access to this pastoral El Dorado, "this gift of god to their ancestors," is refused to them. Each year many of them are fined large sums of money by forest guards for entering nonpermitted zones with livestock. This repressive management is unlikely to support the development of the Park's periphery. Rather, the Park's management framework should consider devolving responsibility of safeguarding the environment to these populations. Measures should be taken to involve stockbreeders in the management of the W Park. This, in turn, might transform it into a peace park where peripheral populations can benefit, instead of a wild reserve where they cannot.

The equitable management of natural resources undoubtedly appears to be an important peace factor in Africa as well as the rest of the world. The proximity of local populations to the periphery of the Park, and the reliance they have on the Park's natural resources, are conditions sine qua non for the W complex to transform from a conflict zone to the status of a peace park.

The advent of centralized colonial administrations had weakened the customary laws and the authority of traditional African leaders who acted as trustees of these laws. Colonial administrations were unable to establish efficient structures for wildlife conservation, which led to a rush of illegal practices in the protected areas. Moreover the application of legal and judicial texts pertaining to protected areas was weak and marred by a lax attitude toward enforcement. This favored the disorderly occupation of reserves and parks by farmers and herders who often had, and continue to have, the blessing of local authorities.

The Project ECOPAS

The W Park's ECOPAS Project is financed by the European Union, within the framework of the 7th and 8th European Development Funds (EDF), for the conservation and rational use of contiguous surfaces protected by Benin, Burkina Faso, Niger, and their territories. The objectives of the ECOPAS are to halt and then reverse the degradation of the Park's natural resources in order to preserve frontier biodiversity for the benefit of the governments and local populations. The specific objectives are (1) the development and implementation of conservation measures and the sustainable management of natural resources (in the reserve), (2) the promotion of dialogue and regional coordination of policies and actions, (3) an effective plan to involve local communities in natural resource management, and (4) the sustainable valorization of natural resources in order to generate benefits for the governments and populations concerned.

The most important expected results of implementing this program are as follows:

• Operational management structures that are functional and effective, with objectives, work programs, and means clearly defined.

• Regional coordination structures and mechanisms connecting the three countries motivated by natural resource management policies and actions (migration of fauna, control of poaching, transhumance and divagation of the cattle, training of specialized personnel, management of bush fires, ecotourism and tourism hunting, valorization of wildlife capital, and research regarding ecosystem function).

• Participation of local communities in natural resource management (in the periphery zones of the program) on the basis of local contracts that clearly define the rights, roles, and responsibilities of all stakeholders.

• A master plan for local planning that allows for a concerted and coherent policy of regional cross-border development.

• Modes and mechanisms of natural resource exploitation to ensure real benefits for resident communities and local authorities.

• Better knowledge of scientific realities and ecological and socioeconomic dynamics of the area through a coherent and effective system of ecological monitoring.

• Enhancement and harmonization of environmental legislation that secures the integrity of the protected areas as well as the interests of the local populations.

Both the overall and specific objectives of ECOPAS are to be executed by each country and are based on methods of sustainable management. For example, park officials would assess hunting quotas for each hunting zone, implement strategies to control for bush fires, and apply Geographical Information Systems (GIS), Global Positioning Systems (GPS), and *Cyber tracker* to the three national components of the ECOPAS (ECOPAS Benin, ECOPAS Burkina Faso, and ECOPAS Niger).

Conclusion: Lessons to Be Learned

Campaigns against poverty and the decentralization processes in Benin, Burkina, and Niger provide opportunities to transform a wild reserve into a peace park. Decentralization depends on the principle that the elected local governments are capable of managing their own internal affairs. However, the application of this principle in developing countries remains highly questionable. The situation of extreme human poverty in the W Park's periphery requires a confluence of economic and environmental forces to emerge. Thus transforming this natural reserve into a true peace park is a solution that could resolve conflicts engendered by prior management. Hence this chapter exemplifies what Duffy argues in chapter 3 of this volume, that economic developments cannot be divorced from peace park formation. Indeed, if proper care is not taken, states and local elite can subvert such intentions. The ECOPAS program provides a possible model for achieving this win-win outcome. This program is likely to succeed in developing ecosystem safeguards and satisfying the vital needs of the local populations. More specifically, ECOPAS should bring several improvements:

• The creation of an atmosphere of confidence and trust between bordering populations and forest officers through multiple sensitizing missions.

• An increase in numbers and effectiveness of wildlife monitoring through the augmentation and recruitment of guards, particularly from border villages where former poachers and fishermen reside.

• The protection of bordering populations through the creation of natural resource management committees and associations of tourist guides at the village level.

• A change of attitude among bordering populations toward the protected areas. The park would be regarded as a good that belongs not to the state, expatriates, or national senior officials, but to the bordering populations as an integrated part of their heritage, reinforced through their duties and rights of park management.

With the implementation of measures such as these, the W international peace park has the potential of being a beacon of stability and sustainability in West Africa. The commitment of governments in all the participating countries appears to be fairly secured, since the joint management project of the W Park is reinforced by several declarations made by the three concerned countries.[14] This is expressed through the tripartite agreement on the fight against poaching, and the master plan for the W management.

Also the ECOPAS program has at its disposal a platform of institutional relations among international organizations such as the European Union and UNESCO, regional organizations such as the Economic and Monetary West African Union (UEMOA), the Interstate Committee against Drought in the Sahel, and technical institutions like UICN. The combined efforts of all these interlocutors contribute in accelerating and securing the involved countries commitment and the realization of the program's objectives.

Notes

1. Broch-Due, eds. (2004).
2. Langholz (2006).
3. Dione et al. (2005).
4. IUCN (2004), Oumarou (2004), and Ecole Nationale de Génie Rural et Eaux et Forêts (1992).
5. Seyni (1998).
6. ECOWAS Web site ⟨http://www.ecowas.int⟩.
7. Le Berre (1995).
8. ENGREF (1992).
9. Fonds Européen de Développement (2002).
10. Centre National de Gestion des Réserves de Faune (CENAGREF) (1999).
11. ECOPAS (2002).
12. Koudenoukpo (2001, 2002) and Programme Parc Régional W/ECOPAS (2004).
13. Szaniawsky (1982).
14. These declarations resulted from the Ministerial Meeting of Kompienga (Benin) of June 2, 1997; the Consultative Meeting for the beginning of the program ECOPAS, organized by the West African Economic and Monetary Union (ECOWAS) in Ouagadougou (Burkina Faso) in March 2000; and the Ministerial Meeting of Tapoa (Niger) in May 2000.

8

The Emerald Triangle Protected Forests Complex: An Opportunity for Regional Collaboration on Transboundary Biodiversity Conservation in Indochina

Yongyut Trisurat

Regional hegemonic powers can greatly influence environmental cooperation and the potential for cooptation of smaller states or communities. This chapter explores how a dominant economic power can be "managed," while providing constructive engagement in the formation of collaborative environmental projects. The last fifty years have seen the rise of Thailand as a regional power in Indochina because of its economically diverse economy and the growth of foreign direct investment while its neighbors were embroiled in civil conflict between communist and capitalist forces. Remarkably the ecology of the region has continued to remain resilient to major political convulsions and conflicts. International forest conservation organizations have thus given this region particular importance. The role played by the International Tropical Timber Organization in facilitating peace park formation will also be analyzed in this chapter and provides a useful comparison with analysis in chapter 6 where a single country donor, the GTZ played a facilitating role. Furthermore this study reveals how technology, namely Geographic Information Systems, can help to understand borders between communities and build trust between parties by providing a means for dependable monitoring of environmental change.

The Southeastern Indochina Dry Evergreen Forests ecoregion occurs in a broad band across northern and central Thailand into Laos, Cambodia, and Vietnam, covering approximately 124,300 square kilometers. The ecoregion is globally outstanding for the large vertebrate fauna it harbors within large intact landscapes. Among the impressive large vertebrates are the Indo-Pacific region's largest herbivore, the Asian elephant (*Elephas maximus*), and largest carnivore, the tiger (*Panthera tigris*). The list includes the second known population of the critically endangered Javan rhinoceros (*Rhinoceros sondaicus*), comprising a handful of animals in Vietnam's Cat Loc reserve—Eld's deer (*Cervus eldi*), banteng (*Bos javanicus*), gaur (*Bos gaurus*), clouded leopard (*Pardofelis nebulosa*), and common leopard (*Panthera pardus*).

About two-thirds of the original forest in this ecoregion has been cleared or seriously degraded, especially in Vietnam and Thailand, but the habitat is relatively intact in Cambodia.[1] A few large forest blocks also remain in Thailand and Laos. Currently there are more than thirty protected areas in this ecoregion covering nearly 24,000 square kilometers (19 percent) of the ecoregion.

Recent years have seen an increasing interest in the creation of transboundary protected areas, for a variety of environmental, economic, and political reasons, including the need for more effective management of politically fragmented ecosystems. With the financial assistance from the International Tropical Timber Organization (ITTO),[2] the Royal Forest Department (RFD) of Thailand has initiated a program of managing transboundary biodiversity conservation areas (TBCA). The RFD selected the Pha Taem Protected Forests Complex (PPFC), as a pilot project for one of four protected forests complexes because there is an increasing pressure on biodiversity from trade in plants and animals across the border with Cambodia and Laos. In addition to the PPFC, the Western Forest, Kaeng Kachan Forest, and Hala Bala Forest complexes also have potential to promote transboundary conservation. The "Management of the Pha Taem Protected Forests Complex to Promote Cooperation for Transboundary Biodiversity Conservation between Thailand, Cambodia, and Laos (phase I)," or ITTO Project PD 15/00 (F), had been implemented during 2001 to 2004 by the RFD to strengthen the management of the PPFC and to initiate cooperation in transboundary biodiversity conservation by the three countries. Under its first phase, the project had initiated a management planning process for the PPFC in a framework of transboundary biodiversity conservation. This involved establishing an effective organization and management system and a geographic information system (GIS) database, and the commencement of a cooperative process involving the three countries. It is noted that most activities in phase I were largely undertaken in Thailand during this period.

The PPFC Landscape

The PPFC is located in Ubon Ratchathani Province in northeast Thailand and covers an area of 1,737 square kilometers (figure 8.1). It comprises four protected areas: Pha Taem, Kaeng Tana, and Phu Jong–Na Yoi National Parks (IUCN category II),[3] Yot Dom Wildlife Sanctuary (IUCN Category Ia), and Bun Thrik–Yot Mon proposed Wildlife Sanctuary (table 8.1). The area slopes gently toward the southeast and is drained by the Mekong River, which forms the border between Thailand and Laos. The PPFC's buffer zone contains 82 villages populated by

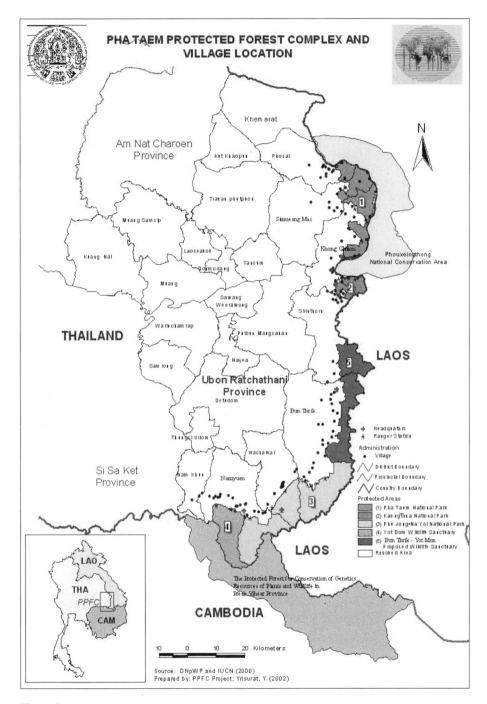

Figure 8.1
Location of the PPFC and adjoining protected areas

Table 8.1
Key features of the Pha Taem protected forest complex

Name	Establishment date[a]	Area (km²)[b]	Perimeter (km²)[c]	Country boundary km (%)[d]	Number of ranger stations	Officials[e]
Pha Taem National Park	December 31, 1991	353.2	242.7	63.3 (27%)	5	3/100
Kaeng Tana National Park	July 13, 1991	84.6	62.5	30.0 (48%)	4	2/90
Phu Jong–Na Yoi National Parks	June 1, 1987	697.4	215.9	93.9 (43%)	4	1/90
Yot Dom Wildlife Sanctuary	October 11, 1977	235.9	88.2	33.2 (37%)	4	1/60
Bun Thrik–Yot Mon Wildlife Sanctuary	Proposed	365.9	186.2	96.4 (52%)	1	1/15
Total		1,737.0	795.53	316.8 (43%)	18	8/355

a. *Royal Gazette*.
b. Calculated by GIS.
c. Excluding shared border.
d. Length of country boundary.
e. Government official/employee.

89,000 people.[4] The major occupations of the residents are agriculture, livestock, and fisheries. On the Lao side, the 1,200 square kilometers Phouxeingthong National Biodiversity Conservation Area (NCBA) is located adjacent to the northern part of the PPFC, while the 1,900 square kilometers Protected Forest for Conservation of Genetic Resources of Plants and Wildlife abuts the border on the Cambodian side. The tripartite border area has been dubbed, in Thailand at least, the Emerald Triangle because of its extensive tracts of monsoon forests.

Biological Features

Three main vegetation types were described based on the interpretation of *Landsat* satellite images in 2002: dry evergreen forest, mixed deciduous forest, and dry dipterocarp forest. More than 288 tree species are identified and at least 49 mammal, 145 bird, 30 reptile, and 13 amphibian species are recorded, but large wildlife species such as the wild elephant, banteng, freshwater crocodile, and tiger are observed only along the tri-national borders.[5] Biological features of protected areas in Cambodia and Laos were not assessed during the project's phase I. However, the ongoing UNDP/GEF[6] medium-size project for the northern plain "Establishing Conservation Areas through Landscape Management" (CALM) reveals an abundance of the populations of Eld's deer, surus crane, and giant ibis inhabiting and breeding in Preah Vihear and areas adjoining Laos.[7]

Eight focal wildlife species and two domestic species were selected to monitor their distributions and habitats using rapid ecological assessment (REA).[8] These species are Asian elephant, leopard, sambar (*Cervus unicolor*), southern serow (*Naemorhedus sumatraensis*), banteng, Siamese crocodile (*Crocodylia siamensis*), pig-tailed macaque (*Macaca nemestrina*), Siamese fireback pheasant (*Lophura diardi*), domestic cattle (*Bos taurus*), and domestic water buffalo (*Bubalus bubalis*).[9]

In addition, the *deductive model* and geographic information system (GIS) were employed to extrapolate known requirements to the spatial distributions of wildlife habitat factors. Four physical factors and three anthropogenic factors were selected to form the basis of natural wildlife habitats. These factors include (1) land use/land cover, (2) accessibility to permanent water, (3) elevation, (4) slope, (5) distance to road, (6) distance to ranger station, and (7) distance to village. Basically forest type is a significant factor in providing cover and forage for herbivore species, while water is a resource necessary for animals to survive, especially in the dry season and for reproduction (Patton 1992). Elevation and slope serve as physical barriers for wildlife migration because most species prefer to inhabit lowland area rather than rugged terrain. On the other hand, road and proximity to human settlements are negative factors for wildlife distribution.

The results indicate that Phu Jong–Na Yoi and Yot Dom are relatively high to highly suitable for eight wildlife species while Pha Taem and Kaeng Tana have relatively low suitability. For instance, relatively high to highly suitable habitats of Asian elephant extend along the national borders and in the southern portion of Phouxeingthong NBCA. In these areas paved roads are few, human settlements are distant, and dry evergreen forest is dominant. Habitat suitability of the other wildlife species is similar to that of the Asian elephant. On the other hand, the distributions of domestic cattle and domestic water buffalo are opposite to those of selected wildlife species. Highly suitable habitats are located in landscape traversed by humans, but the remote areas, especially the rugged forests of southern Phu Jong–Na Yoi and Yot Dom, are of relatively low to low suitability for cattle and buffalo.

The overlying suitable habitats of the eight focal wildlife species show that high concentrations of all species or "hot spots" are found along the tri-national borders and clustered in three places. The highest and largest area is located along the western border of Phu Jong–Na Yoi adjoining Laos. The second region is found in Phouxiengthong NBCA in Laos, and the third area extends along the northern border of Bun Thrik–Yot Mon. In the plan for transboundary biodiversity conservation among three countries, the first concentration area is recognized as the most important critical habitat because protected areas in Thailand alone cannot support the survival of these migratory species (figure 8.2).

Achievement of the PPFC Phase I

A long-term management plan constitutes the framework for the conservation objectives. It covers a period of twenty years from 2004 to 2023. The medium-term working plan spans a period of three years from 2004 to 2006, and both were documented through participation of many stakeholders.[10] Basically the working plan outlines specific action with the emphasis on transboundary conservation. It describes the activities to be carried out in each protected area. The period 2004 to 2006 corresponds to phase 2 of the project.

Management Objectives

For the transboundary protected areas with PPFC and the surrounding landscape, the overall management objectives are (1) to support long-term cooperative conservation of biodiversity, ecosystem services, and natural and cultural values across boundaries; (2) to share biodiversity and cultural resource management skills and experience, including cooperative research and information management; and (3) to enhance the benefits of conservation and promote benefit-sharing in transboundary protected areas. Consistent with the overall management objectives, the specific objectives for area management within a period of three years are as follows:

• To initiate and set up the foundation for long-term cooperative conservation of biodiversity, ecosystem services, and natural and cultural values, and to strengthen the protection of natural resources across boundaries.

• To strengthen the technical information base, organization and human resources, public relations and participation mechanism at each individual protected area within the PPFC.

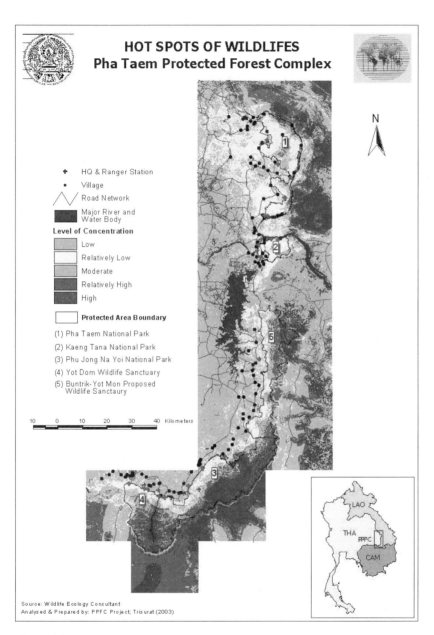

Figure 8.2
Wildlife hot spots in the PPFC landscape

• To share biodiversity and cultural resource management skills and experience, cooperative research and scientific studies, including ecotourism promotion and the generation of tourism income to the local communities.

Zoning

The management zone of the PPFC attends to the border landscape encompassing the protected areas as well as the surrounding areas. The zoning concept was adapted from the zoning scheme of UNESCO Biosphere Reserves.[11] This approach divides human-altered landscapes into four zones: core zone, buffer zone, corridor, and matrix. This way biological diversity can be maintained across the entire landscape without compromising the people's needs.

The objectives of the zoning scheme for the PPFC include protection of critical habitats, facilitation of wildlife migration across the borders, and permission of reasonable human uses. Four ecological quality factors were selected to define management zone: (1) critical habitat, (2) environmental service, (3) natural quality, and (4) remoteness for developing ecological management of the PPFC. The grid-based modeling of GIS ArcView 3.2a was used for all spatial analyses because it is powerful for large datasets and user-friendly (ESRI 1995). Spatial criteria and zoning scheme of each management zone are presented in table 8.2. The descriptions of each ecological management zones follows:

1. *Core zone* This zone covers an area of 2,945.71 square kilometers or 27.53 percent of the total PPFC landscape. It is found in Phouxiengthong NBCA and along the southern border joining Thailand, Laos, and Cambodia. The Cambodian side contains a significant proportion of core area. Of the five protected areas in Thailand, the core area is largest in the Phu Jong–Na Yoi NP, and moderate in Yot Dom WS and the Bun Thrik–Yot Mon proposed WS. The core area is smallest in the Pha Taem and Kaeng Tana NPs because these two areas are disturbed either by human presence or by grazing cattle. An integrated joint task force should be established among the three countries to combat encroachment, poaching, and illegal logging. In remote areas where accessibility is limited due to landmines, cooperation with military and border patrol police is essential.

2. *The buffer zone* Buffer zone covers the remaining forest area in the PPFC and degraded forest, as well as a few agricultural areas inside the five protected areas. The total area of buffer zone is 3,912.59 square kilometers or 36.57 percent of the PPFC area. The large patch is situated in the west of Pha Taem NP and to the west-

Table 8.2
Zoning scheme for the PPFC landscape

Character/criteria	Core zone	Buffer zone	Corridor	Matrix
Biophysical condition	Critical ecosystem(s) that supports viable population of focal species and environmental services, normally remote from disturbances and human settlements	Natural area situated around the core area to manage unfavorable impacts that flow between the core area and its surrounding landscape	Linear assemblage of mainly continuous vegetation connecting fragmented critical ecosystems to encourage and facilitate migration and dispersal	Extensive cover and connectivity in the landscape where human settlements and intensive development are conducted
Spatial criteria				
Critical habitat[a]	Critical	Moderate	Critical–moderate	Low
Environmental service[b]	Sensitive	Moderate	Sensitive–moderate	Low
Naturalness[c]	Undisturbed vegetation	Undisturbed vegetation	Remnant vegetation, slightly disturbed	Settlement, reservoir, agriculture
Remoteness[d]	>3 km from settlement >2 km from main road >1 km from large agricultural activities	Disturbed areas inside protected areas	Preferably remote from settlement	—
Physical setting	>1 km^2	Not overlapping with core zone and matrix	May overlap with core area and buffer zone	Extensive and connected

a. Represented by the concentrations of eight wildlife species; habitat suitability maps were superimposed and reclassified into three classes.
b. A significant environmental service of the PPFC is monitoring the integrity of a watershed, represented by the erosion sensitivity index.
c. The degree to which a site is either free from disturbance caused by modern technology and humans or still remains in natural condition.
d. This factor indicates how remote a site is from established settlements and the main road.

ward of Bun Thrik–Yot Mon WS in Laos. Approximately 75 percent of Pha Taem NP and 92 percent of Kaeng Tana NP are classified as a buffer zone of the PPFC because these two protected areas contain low suitability habitats for wildlife and most areas are accessed by the local people and used for cattle grazing. In addition the forested area in Laos situated to the west of Bun Thrik–Yot Mon proposed WS, where commercial logging is practiced, is defined as a buffer zone. Management of the buffer zone surrounding the core area inside a protected area boundary (*primary buffer zone*) aims at promoting training, education, and ecotourism. If local communities can benefit from tourism activities, pressure on encroachment and poaching will be greatly reduced. Activities in the remaining buffer zone outside protected areas (*secondary buffer zone*) nearby the communities might include agroforestry, community forestry, collection of mushrooms, bamboo shoots, as well as domestic animal grazing.

3. *Corridor* The ecological zone map shows that the boundary of the Kaeng Tana NP does not adjoin Pha Taem NP and Bun Thrik–Yot Mon WS. Therefore it is proposed that both Thai and Laotian wildlife scientists with local assistant should conduct a rapid wildlife survey and rehabilitate this gap as a *wildlife corridor* to link the fragmented protected areas. This corridor covers vegetation remnants along an approximately 17 kilometer length. Besides, the park rangers of the Kaeng Tana NP must regularly patrol the area in the north and raise local awareness on conservation, as well as promote alternative occupations to the local people in order to reduce dependent on forest resources.

4. *Landscape matrix* The remaining areas in the PPFC landscape dominated by extensive agriculture and human settlement are classified as a landscape matrix. The legal status of this land is either national reserved forest or privately owned land. Park officials should promote community-based conservation activities that lead to biodiversity-friendly land and water uses such as activities related to ecotourism and para rubber plantation instead of cashcrop production. In addition it is essential to enhance community awareness and their participation in the TBCA.

Management Programs

Management programs under the long-term management plan consist of six programs. These include (1) the Natural Resource Conservation and Management Program, (2) the Recreation and Ecotourism Management Program, (3) the Integrated

Community Development Program, (4) the Organization and Human Resource Development Program, (5) the Research Program, and (6) the Transboundary Cooperation Program. The annual operational plans outlined in the working plan proposed of the first phase are presently awaiting implementation in the second phase by the various concerned agencies.

Threats and Opportunities

After completion of the PPFC project's phase I, the project faced a number of threats and challenges to its effectiveness. These obstacles should be addressed if the project is to meet its transboundary biodiversity conservation and socioeconomic objectives in the second phase.[12]

Threats

1. *International relations* The management of cross-border reserves requires a lot of cooperation. However, uncertainties remain, so cooperation among the three countries has probably achieved only a level 2 or 3 of consultation (more frequent communication, sharing of information, notification of actions affecting the adjoining protected areas, and relative consultation in planning). Laos is reluctant to nominate the Phouxeingthong NBCA for inclusion in the TBCA in the project's second phase. This is because Laos is not an official member of ITTO, and it stands to gain limited direct benefits from the project.

2. *Encroachment* The forest in the buffer zone outside the PPFC has been encroached for agriculture. Further clearing could jeopardize the viability of rare large mammal species. The forest is also being degraded in Laos and part of Cambodia by large-scale commercial logging concessions. Unless there are serious protection measures undertaken, the results of a Markov chain model[13] predict the forest cover to decrease from 56.8 percent in 2002 to 40.8 percent in 2050, and the agricultural area to increase from 23.7 percent in 2002 to 53.3 percent in 2050.[14]

3. *Poaching* Wildlife is being poached and plants collected for trading along the borders of the three countries. Bush meat is in high demand because it is an important source of protein for rural households, particularly in Laos.[15] However, only one of the complex's eleven border crossings (between Thailand and its two neighbors) is a CITES[16] checkpoint.

4. *Cattle grazing and forest fire* The intensity of cattle grazing in the PPFC, especially in Pha Taem and Kaeng Tana NPs, is influenced by the number of cattle owned by the herders who reside in 82 communities around the PPFC. At the end of dry season, herders burn large tracts of grasslands to promote growth of new pasture for their animals. Also local people collecting edible plant burn dry diptercarp forest to stimulate young shoot rejuvenation. The impact of forest fire on tropical biodiversity, especially deciduous forest, is a controversial issue. Scientific research is essential to present positive and negative impacts of forest fire and how to effectively manage the deciduous forest.

5. *Capacity* Both Cambodia and Laos lack the capacity at all levels to manage their protected areas effectively. Their staffs have little or no access to training, budgets for management are very small, and there are very few park rangers and facilities on the ground, especially the protected area in Cambodia.[17]

6. *Landmines* Thousands of landmines were laid along the borders between Thailand, Cambodia, and Laos in the 1980s. Currently Thailand, with the assistance from Norway, has begun demining, but this task requires a large amount of money. These landmines constitute a major threat to park rangers, researchers, and also large mammals.[18]

Opportunities

1. *International significance* The PPFC is the only forest complex in Thailand that contains both terrestrial and aquatic ecosystems of regional significance. The area still features rare and endangered species such as tigers and freshwater crocodiles. In addition Eld's deer, the surus crane, and the giant ibis inhabit and breed in lowland forest in Cambodia and Laos.[19] The area therefore has globally significant biodiversity value, so it will continue to attract international support such as from the Wildlife Conservation Society (WCS) and UNDP/GEF. In addition the ITTO is willing to support phase II of the project.

2. *Existing cooperation* Multiple cooperation in this region has strengthened in recent years. Joint cabinet meetings between Thailand and Cambodia, and Thailand and Laos, were conducted in 2003, and it was agreed to strengthen natural resource conservation and tourism promotion in this region. In addition regional groups such as the Mekong River Commission (MRC), the Food and Agriculture Organization (FAO), and the Asian Development Bank (ADB) are interested in supporting this landscape conservation initiative. In early 2005 the ADB launched the Technical

Assistance on Biodiversity Conservation Corridors Initiatives in the Greater Mekong Subregion (GMS) with six countries—China (Yunnan), Myanmar, Thailand, Laos, Cambodia, and Vietnam—in an aim to identify potential sites and develop the strategic framework (2006–2015) and action plan (2006–2008), as was agreed by these six countries in the GMS.

3. *Trends of protected area management* Recent international initiatives on protected area management that advocate a landscape-approach to protected area management and look beyond the boundaries of individual protected areas are acknowledged both at the international and national levels. Lessons learned from ecosystem management in the Western Forest Complex[20]—located in western Thailand and jointly implemented by the RFD and the Danish Cooperation for Environment and Development (DANCED)—are proving valuable in strengthening management in the PPFC (Trisurat 2004).

Extending PPFC to the Emerald Triangle Protected Forests Complex

The target condition in the *Emerald Triangle* or adjoining forest areas along the trinational borders is transboundary biodiversity. This condition can be described in terms of wild animal population and secured transboundary ecosystems. Therefore it is essential to reinforce this landscape conservation initiative and implement biodiversity conservation activities through the involvement of local residents in the buffer zone because protected areas alone are not sufficient to conserve biodiversity or sustain ecosystem services that human life depends on.[21]

Thus RFD of Thailand and Forestry Administration (FA) of Cambodia jointly developed the project's phase II proposal and modified project name to "Management of the Emerald Triangle Protected Forests Complex to Promote Conservation for Transboundary Biodiversity Conservation between Thailand, Cambodia, and Laos (phase II)" because it has extended the implementation to cover protected forests in Cambodia and possibly Laos. It should be mentioned that Laos took less part in the proposal formulation because it is not a member of the ITTO, but Thailand and Cambodia are encouraging other donors such as ADB, MRC, and UNDP/GEF to fill this gap. The project's phase II is being spread over two years (2005–2006), and it will build on the achievements of phase I as well as resolve pending issues (table 8.3). It was approved by ITTO in December 2004 and is being co-financed through that organization by Japan, Switzerland, and United States. The overall objective is to conserve transboundary biodiversity in the Emerald Triangle. The specific

Table 8.3
Summary of project objectives and pending issues in phase I and objectives of phase II

Phase I (2000–2004) objectives	Pending issues	Phase II (2005–2006) objectives
1. Initiate cooperation among Thailand, Cambodia, and Laos for transboundary biodiversity conservation 2. Start a management planning process for Pha Taem Protected Forests Complex in the framework of TBCA	Cambodia and Laos took less part in project's phase I and Laos has been reluctant to nominate Phouxeingthong to joint the TBCA Waiting for implementation of phase II	1. Strengthen cooperation among Thailand, Cambodia, and Laos for biodiversity conservation in respective TBCA 2. Enhance protection and monitoring of biological resources along tri-national borders 3. Strengthen the involvement of local communities and stakeholders in sustainable uses of natural resources

Source: ITTO/RFD/FA (2004)

objectives are (1) to strengthen existing cooperation among the three countries, (2) to enhance protection of biological resources along the tri-national borders, and (3) to strengthen the involvement of local communities and stakeholders in sustainable use and management of natural resources in buffer zones.

Taking into account the accomplishments of phase I of the project, and following an assessment of outstanding issues that still need to be addressed, the objectives of phase II will be achieved by the further development and implementation of a number of key strategies:

1. International agreements already ratified by the three countries and support from the international community, including ITTO, IUCN, MRC, GMS, and ADB, which should further develop, stimulate, and support cooperation for transboundary conservation interventions.

2. Because Lao PDR is, at the present time, reluctant to participate, Thailand and Cambodia will initiate a range of *soft collaborative activities*, including training, consultative meetings, nature-based tourism development with Laos, and strengthening of cooperation at the provincial levels as agreed by the cabinets of three countries.

3. Through the progressive development and implementation of sustainable economic development activities, local community networks are to be established or strengthened, and livelihood opportunities increased for local people. Nature-based tourism activities as well as buffer zone development are the main areas of economic interest.

4. Phase II results will include (a) conservation of Cambodia's protected forest for genetic resources of plants and wildlife in the TBCA, (b) strengthened TBCA cooperation among the three countries, (c) increased human resource capacity, and (d) integration of conservation and development programs in buffer zones and nature-based tourism interventions as a part of the efforts to increase livelihood opportunities for local residents.

Can Environmental Cooperation Foster Regional Stability?

The objectives and concepts of peace parks and transboundary protected areas are not only the protection and maintenance of biological diversity but also the promotion of peace and cooperation in the region. The prime example is of course the establishment of cross-border protected areas in the Middle East between Israel and Jordan.

The Emerald Triangle Protected Forests Complex will require completion of wildlife distribution maps and the ecological management zones which were derived during phase I. These can be used as a contribution to the establishment of a broad collaborative framework for transboundary biodiversity conservation among the three counties. Results from habitat modeling reveal that landscape species (e.g., elephant and tiger) seasonally migrate across the frontier and that the long-term future of these species is not viable without cooperation between Thailand, Lao PDR, and Cambodia.

Basically the concentrations of landscape species is defined within the core zone where strict protection measures are an essential element as part of the effort to maintain ecological integrity and to enhance ecological processes. It is recommended that an integrated joint task force of protected areas staff from the three countries be established to develop a united approach to major conservation problems such as poaching, illegal trade, and removing existing barriers to wildlife movement along tri-national borders.

Another important strategy will be to reinforce on-the-ground collaborative research projects with a particular focus on species distribution and habitat uses, as well as the reintroduction of Eld's deer in this area. In recent years the Smithsonian Institute and the Zoological Authority of Thailand have undertaken regional workshops on the reintroduction of Eld's deer. Likewise scientists from Thailand, Lao PDR, and Cambodia have agreed to implement a reintroduction program within the Emerald Triangle.

It needs to be pointed out that the formulation of the phase II strategies and objectives was transparent and fair. In fact this was the first ITTO TBCA project where participating countries worked intensively and cohesively to submit *a single proposal*. Within this environment of trust and a sound understanding of shared objectives, six countries in the GMS are now working closely to safeguard biodiversity and improve the livelihood of local people in this region. For instance, Cambodia has officially supported Thailand's efforts to prepare a proposal for the Dong Phayayen–Khao Yai Biodiversity Conservation Corridor situated in the Phanom Dong Rak Range along the border between Thailand and Cambodia (west of the Emerald Triangle). This proposal received endorsement by the Environmental Ministers of the six GMS countries in mid-2005. The ongoing programs and activities clearly demonstrate that environmental cooperation is regarded as a viable mechanism to reduce diplomatic tension and promote regional cooperation in what is regarded as a fragmented political and socioeconomic context among countries that share ecological and political boundaries.

The Emerald Triangle for Peace and Stability

The past and current socioeconomic development and the distribution of existing tropical forests reflect contrasting realities of three countries in the Emerald Triangle region. The disparities across the borderland are environmental, economic, political, and social. The evidence, as shown in this chapter, is that transboundary cooperation and constructive dialogue can combine biodiversity conservation, peace, stability, and the needs of local residents in the Emerald Triangle landscape to some extent.

Biodiversity Conservation

Conservation and development challenges in the protected areas along the border of Thailand are due to the limited suitable habitat left to maintain viable populations of rare large mammal species, and to the encroachment of agriculture that in the buffer zone has significantly reduced forest cover. The situations in Laos and Cambodia are very different. Although forests are being degraded in both countries, relative large forest areas still remain, and this allows the habitat fragmentation to be experienced as less severe and fauna assemblages to stay more intact. Nevertheless, both countries face diverse threats from poorly protected area management, wildlife poaching, and trade of animal and plant products. A TBCA could help all three countries minimize these stresses by coordinating a joint task force to combat illegal activities. Already local development initiatives, including the provision of alternative sources of income by domesticating and commercializing local products in the buffer zone for subsistence living and the reduction of dependency on natural resources, are being implemented. The impetus is the proved success of such activities in Lanjak Entimau in Borneo. In addition protected area staffs from Laos and Cambodia are being provided opportunities to participate in a series of training workshops in Thailand, so that they can learn to effectively manage protected areas to meet TBCA standards.

Reconciliation of Country Border Politics and Minefield Defusing
The political and administrative systems of Thailand, Laos, and Cambodia are incompatible. Thailand is a democratic monarchy. Cambodia has a similar administrative system but has been reconstructing its civic structure over several decades following a long series of armed conflicts and genocide. Laos is called the People's Democratic Republic but has its own history of turmoil. In the 1980s thousands of landmines were laid along the tri-national borders, and each country deployed their military to impose national security in border areas.

Today armed conflicts and disparities of country border politics have significantly reduced. The prime ministers of all three countries have signed bilateral and multilateral agreements to promote peace, cooperation, and tourism development, as well as to strengthen natural resources conservation in the Emerald Triangle. This strategy is being supported by a number of international organizations, as discussed in the previous section. With the assistance from Norway and Japan, each country has begun removing landmines. It is believed that greater cross-border cooperation in TBCA could bring greater international interest in clearing the region of landmines, which is estimated to cost US$70 million. The by-product of this activity is the added potential for ecotourism and also the improved safety for residents. In fact the transboundary conservation for peace involves the demilitarization and the identification of all threats emerging from any one country and directed toward neighboring countries within a transboundary conservation area. Where such threats require the use of an integrated task force, and accessibility is limited due to landmines, a cooperative military staff from these three countries will deploy the appropriate task force modalities to combat poaching and illegal logging.

Economic and Human Communities

Because Thailand is a capitalist country, its economic development has increased with rapid modernization. The gross domestic product (GDP) in 2003 was US$111 billion, and the gross national product (GNP) per capita was US$2,160. In contrast, over one-third of Laotians and Cambodians live in poverty. The average GNP per capita in Laos and in Cambodia was US$260 and US$320, respectively, or one-ninth that of Thai citizens. Thus the economic prosperity in Thailand has brought about income disparities between Thai nationals and nationals of neighboring countries. These contrasting economic conditions have resulted in an influx of foreign migrant workers and asylum seekers into Thailand. In addition there is no physical border to demarcate the territory of each country, so movements around the border areas are unrestricted.

The Thai immigration office has estimated that there were approximately 525,000 illegal immigrant workers and 20,000 Indochinese refugees staying in Thailand in 1994. Of this number the largest group came from Burma, followed by China and Indochina. The government of Thailand is concerned about the influx of illegal foreign migrants and consequent negative impacts, such as the spread of infectious diseases, human trafficking, and the legal problems concerning the status and rights of immigrant workers. A number of technical and development programs have been provided to neighboring countries to improve health care, education, and income generation. A TBCA may be a further mechanism to use in generating foreign income and eradicating poverty in rural areas. A TBCA not only offers excellent opportunities for the appreciation of nature, it can also provide a multinational and multicultural experience. The Emerald Triangle and the Mekong River are mysterious places that attract international tourists and greatly enrich their visits to the Thai, Laotian, and Cambodian communities. It is claimed that ecotourism will only work if local communities can benefit and become fully involved in its management. However, the tourism benefits must be fair and equitably shared among the participating countries. If the Emerald Triangle Complex succeeds, it will hold promise for a similar Western Forest Complex in Thailand, along the border of Burma where there is need to reduce Burmese immigration into Thailand. This intact forest block is listed as the next TBCA to be implemented soon.

Conclusions

In 2001, with financial support from ITTO, Thailand initiated a transboundary biodiversity conservation plan with Cambodia and Laos and selected the Pha Taem Protected Forests Complex as the pilot project. Two important outputs derived from the phase I included (1) a long-term management plan of the PPFC within the framework of transboundary biodiversity conservation and (2) cooperation among Thailand, Cambodia, and Lao PDR (a tripartite commission) and the protected area staff (a joint taskforce). During phase I the three countries achieved a certain level of cooperation. Results of technical research indicate that large wildlife species such as the wild elephant, banteng, and tiger seasonally migrate across the tri-national borders. Forest in the buffer zones of Thailand's protected areas has been encroached for agricultural practices. The PPFC project developed ecological management zones by modifying the zoning scheme of biosphere reserves. The core zone, which covers about 27.53 percent of the total PPFC landscape, is found in Phouxiengthong NBCA and along the southern border with Thailand, Laos, and Cambodia. The buffer zone covers approximately 36.57 percent and is dominant in the Pha Taem NP and Kaeng Tana NP. Presently both Thai and Laotian wildlife scientists, and their local assistants, still need to conduct wildlife surveys and assess the possibility for establishing and/or rehabilitating a *wildlife corridor* to link fragmented habitats between the Kang Tana NP and Bun Thrik–Yot Mon proposed WS.

The PPFC project faces a number of threats and impediments to its effectiveness such as the reluctance of the Laos government to join the TBCA, encroachment in the buffer zones, poaching, landmines, and low education of the field staff. Nevertheless, many opportunities to promote trans-boundary biodiversity conservation are present in this region. In order to achieve transboundary conservation, it is essential that efforts are made to reinforce the landscape conservation concept through the involvement of decision-makers from the three countries and local residents in the buffer zones.

To this end the RFD of Thailand and FA of Cambodia jointly developed a proposal on the "Management of the Emerald Triangle Protected Forests Complex to Promote Cooperation for Transboundary Biodiversity Conservation among Thailand, Cambodia, and Laos (phase II)," which was offered in 2004 to ITTO for funding support. This proposal extends the scope of the five protected forests complexes in Thailand to cover adjoining protected forests in the three countries that are named as the Emerald Triangle.

The project's phase II proposal was approved by ITTO in late 2004 for a period of two years (2005–2006). The development objective of the project is to conserve transboundary biodiversity in the Emerald Triangle Protected Forests Complex situated between Thailand, Cambodia, and Lao PDR in a framework of transboundary biodiversity conservation area (TBCA). Basically phase II will extend the lessons learned from the project's phase I. The activities are to be implemented in Cambodia, while Thailand will focus on implementation of planned biodiversity conservation activities.

The key strategies meet the desired objectives of cooperation among the participating countries in TBCA and the poverty issues concerning local residents in the buffer zone, which are to be reduced through alternative income sources and less dependency on natural resources exploitation. Given that Laos is reluctant to participate in phase II, Thailand and Cambodia will provide incentives for Laos to become involved in several soft collaborative activities, especially human resource development and research in order to build trust among the three countries. The hope is that the establishment of cross-border conservation areas in the Emerald Triangle will encourage the three countries to recognize the benefits of safeguarding the viable biodiversity in Indochina and thus reduce the tension in these politically and socioeconomically fragmented environments. In addition this experience holds the promise that positive international cooperation for biodiversity conservation can bring peace and friendship through constructive dialogue and strengthen mutual arrangements for sustainable development and stability in the border regions. The lessons learned from this project can be spread and repeated in peace parks created elsewhere in this region.

Notes

1. Wikramanayake et al. (2000).
2. The ITTO is an organization of countries that produce and consume tropical timber.
3. IUCN-The World Conservation Union (1984) classifies management protected areas into six categories: category Ia, strict nature reserve; category Ib, wilderness area; category II, national park; category III, natural monument; category IV, habitat/species management area; category V, protected landscape/seascape; and category VI, managed resource areas for sustainable use.
4. Tanakarn (2003).
5. Marod (2003) and Phumpakphan (2003).

6. The Global Environment Facility (GEF), established in 1991, helps developing countries fund projects and programs that protect the global environment.

7. Sonnon (2003) and Chheng (2004).

8. Rapid ecological assessment (REA) combines remote sensing and GIS with selected field survey inventories to provide a baseline of information on the biodiversity values of a region (Say et al. 2000).

9. Bhumpakphan (2003).

10. RFD (2004).

11. Phillips (1998) and Miller and Hamilton (1999). Biosphere reserves are vital centers of biodiversity where research and monitoring activities are conducted, with the participation of local communities, to protect and preserve healthy natural systems threatened by development established under UNESCO's man and the biosphere (MAB) program.

12. Trisurat (2003a).

13. The Markov chain model is a mathematical simulation used to examine potential long-term trends in land use/land cover changes.

14. Trisurat (2003b).

15. Galt et al. (2000).

16. CITES stands for Convention on International Trade in Endangered Species of Wild Fauna and Flora.

17. D. Chheng (November 2005), personnel communication.

18. Trisurat (2003c).

19. Sonnon (2003) and Chheng (2004).

20. The Western Forest Complex or WEFCOM includes contiguous six wildlife sanctuaries and eleven national parks.

21. Sandwith et al. (2001) and Gasana et al. (2004).

9

Conflict Avoidance and Environmental Protection: The Antarctic Paradigm

Michele Zebich-Knos

Harmony through Denial

The Antarctic regime exemplifies how conflict avoidance and environmental management can be constructed on an entire continent devoid of nation-state boundaries. It represents one of the world's most interesting examples of how many of the world's nation-states join forces in harmony to manage Antarctica's environmental well-being in the broadest sense. At the same time this harmonious management system engages in conscious denial of the claimant versus nonclaimant reality, which—if broached—could provoke considerable conflict over territorial sovereignty on the continent. Unlike the deep Arctic seabed—which is often compared to Antarctica in that it belongs to no one state—this continent has experienced claims staked *to the ice* and, more important from a legal standpoint, to the land beneath it. The denial factor that facilitates harmony is contained in Article IV of the Antarctic Treaty, which entered into force in 1963. It states that no acts or activities taking place while the present Treaty is in force will constitute a basis for asserting, supporting, or denying a claim to territorial sovereignty in Antarctica. No new claim, or enlargement of an existing claim, to territorial sovereignty can be asserted while the present Treaty is in force.[1]

The modus operandi for Antarctic management consists in denying the existing claimant reality, and freezing the ability for states to make future claims or enlarge existing ones. To date, this system based on denial works well.

The Antarctic example also fits conceptually into what Oran Young calls "governance without government"[2] in which governance is established through the creation of mechanisms that can resolve conflict, facilitate cooperation, and tackle collective-action problems among interdependent actors minus the sovereign and formal authority vested in institutions that are normally associated with state-level governments. Young distinguishes such governance from actual government frameworks and posits this as a successful way to manage the environment.[3] If any region on earth best fits the *governance without government model*, then it must be Antarctica.

However, Young's model also applies to other cases in this book. Kim's coverage in chapter 13 of the demilitarized zone (DMZ) between North and South Korea is a good example of a thriving ecosystem that belongs to neither North nor South Korea. Kim's vision of a future peace park created in the DMZ would be administered by a joint commission of North and South Korean representatives in addition to scientists from around the world. Much as Antarctica is a sovereignless continent governed through what is essentially a joint commission approach, albeit on a much larger scale, the DMZ area is perhaps the closest parallel to Antarctica because of its no-man's land status.

The continent is currently devoted to the pursuit of science which, as is claimed by the authors in this book, forms a cooperative basis upon which peace can be built. At the same time the tourism sector continues to expand. While the number of scientists residing in Antarctica ranges from 1,000 (winter) to 4,000 (austral summer),[4] approximately 19,370 tourists visited Antarctica by ship for the period from November 2003 through March 2004. This is a 46 percent increase from the 2002–2003 austral summer season.[5]

Both scientific and tourist endeavors are cooperatively managed on a global scale by their host governments and nongovernmental organizations (NGOs). The private sector tourist industry is also actively involved in supporting compliance of national guidelines that have been refined by the International Association of Antarctic Tour Operators (IAATO), which was founded in 1991. Overall management of Antarctica remains in the hands of Antarctic Treaty Consultative Party (ATCP) national institutions such as the US National Science Foundation's (NSF) Office of Polar Programs (OPP), the Australian Antarctic Division, and the British Antarctic Survey to name a few. Scientific research in Antarctica functions on a cooperative basis and serves as a common bond much like the establishment of peace parks in border areas elsewhere in the world might do.[6] It also utilizes select NGOs to assist in Antarctic scientific management, which complement national efforts. The Scientific Committee on Antarctic Research (SCAR), for example, is part of the International Council for Science and facilitates Antarctic research, identifies new issues that are subsequently brought to the attention of policy makers, and provides advice to the Antarctic Treaty Consultative Meetings.[7]

The system admittedly operates under less than optimal conditions, and efforts to resolve claimant versus nonclaimant and commercial resource extraction issues were postponed for future discussion. Can stakeholders from both the North and South find common ground should claimant or resource extraction issues reemerge? What impact will the growing ecotourism industry have on the Antarctic environment? This chapter will examine the Antarctic regime within the context of such questions, and will approach the Antarctic regime within a conceptual framework for high-stakes "environmental peacemaking" and conflict avoidance refined by Conca and Dabelko in their book *Environmental Peacemaking (2002)*.

Conca and Dabelko's general framework encourages scholars of potential, or actual, conflict areas to view environmental cooperation—in Antarctica or in any area of the globe, not as low-stakes politics, but instead as an important means capable of leading to conflict resolution or avoidance. As such, the framework for environmental peacemaking connects mundane conservation management activities, such as overseeing Antarctica's trash maintenance in its scientific stations, to much grander accomplishments such as the Madrid Protocol, which placed a fifty-year moratorium on Antarctic mining. Conservation management activities conducted by many countries in Antarctica may appear to lack relative power to influence outcomes, but they contribute to institutionalizing peaceful means of working out the continent's problems. Routine conservation management activities contribute to a body of high politics precisely because they represent cooperative building blocks upon which conflict avoidance is based. This corresponds to Lejano's theoretical assertion in chapter 2 in which relationship-building leads to cooperation.

Thus the Antarctic paradigm fits into a high-stakes environmental peacemaking framework through its mundane and grand activities alike. In the Introduction to this book Ali writes that environmental concerns can be used as a trust-building tool. I take this a step further and assert that this is already being accomplished in Antarctica and that such trust-building contributes to future conflict avoidance on the continent. When the time comes to deal with a country that actually decides to mine on the continent, the environmental peacemaking framework currently in play should be secure enough to contribute to a resolution of the problem.

After years of framing issues around environmental security and conflict, many scholars now expand their research to emphasize working together in a cooperative, cross-border manner on environmental issues that can be managed, for example, by the creation of border parks and reserves.[8] The chapters in this book endorse such an approach because it holds promise for transboundary environmental cooperation among states and may not only help resolve environmental problems but also ease the tension of previous armed conflicts or minimize future conflicts.

The main goal of such efforts is to prevent conflict and encourage confidence building, or, as Conca explains, "environmental cooperation can generate synergies for peace."[9] The causes of these synergies derive from shared norms developed as a result of carrying out Antarctic conservation activities within a peaceful discourse based on mutual concern for protecting the continent. Conca posits that such activity leads to the creation of what he calls a "shared collective identity" that makes conflict among the participants an unlikely proposition. In Conca's environmental peacemaking framework, trust, interdependence and transparency develop among stakeholders and form solid transnational linkages that prove useful in avoiding conflict. This is reminiscent of democratic peace theory in international relations in which democratic states do not wage war against each other. If Conca's approach is valid, then the components of a shared collective environmental identity which the Antarctic regime helped develop should temper future Antarctic conflicts.

Examples of existing transboundary-protected areas abound and range from the Waterton–Glacier International Peace Park along the US–Canadian border, to the Komoé–Leraba Forest–Warigué Forest along the Burkina Faso–Côte d'Ivoire border.[10]

Nowhere is there a better laboratory for environmental cooperation than the *frozen continent* of Antarctica. A key example of this cooperation is the continued acceptance for the prohibition of (1) military activities other than for a peaceful nature (i.e., logistical support for scientific research), (2) nuclear explosions, and, (3) the disposal of radioactive waste south of 60° South latitude. The term Antarctica includes all environs south of 60° South latitude excluding the high seas.[11] Devoid of the Westphalian sovereign-state divisions that exist on the other six continents, Antarctica is a model for environmental peacemaking minus state borders. Blum calls Antarctica "the last sovereignless continent" and an international anomaly in which there is no domestic legal system.[12] Its track record is excellent for no wars were fought over Antarctica—or on the continent itself.

While cross-border, or transboundary, peace parks and other protected areas offer a veritable ray of hope for environmental peacemaking and general conservation efforts in Africa, the Americas, Asia, and Europe, Antarctica displays its own specific version of peace management goals and mechanisms that are contained in the various treaties, laws, and regulations that comprise the Antarctic Treaty System

(ATS). Before examining the goals and mechanisms that are integrated into this regime, it is worthwhile to briefly consider what makes Antarctica a unique area in which to apply an environmental peacemaking and conservation framework.

Antarctica: A Polar Desert

Mention of Antarctica evokes visions of an icy continent, the last great wilderness devoid of native human inhabitants, and populated albeit sparsely by scientists from around the world.[13] It is indeed the world's last region of untamed and oft-uncharted areas that attracted adventurers like Ernest Shackleton in 1915, whose ship, the *Endurance*, was crushed by pack ice in a failed attempt to explore the continent and reach the South Pole. Today it still attracts the daring few such as New Zealander Graham Charles and his two-man team who, in 2001, kayaked Antarctica's near-frozen waters for a three-month trek along the western peninsula around the Lemaire Channel south to the Antarctic Circle. The near mystical attraction that Antarctica holds for adventurers, coupled with recent concerns over global warming's impact on polar ice, keeps the region from falling off the media's agenda. For the most part, Antarctica rarely emerges in the minds of the general public or most environmental scholars, for that matter. Generally, Antarctica takes a backseat to broader environmental issues such as global climate change, tropical deforestation, or hazardous waste disposal. Only when Antarctica is connected to global issues like climate change or stratospheric ozone depletion, do we read about it. Not surprisingly, this author frequently encountered the same response when mentioning a 2004 trip to the Antarctic Peninsula: You went *where*? It is indeed an out of the way place for most of us.

Yet Antarctica deserves greater attention in light of its unique natural characteristics. It forms about 10 percent of the earth's land surface and is the largest ice sheet on earth. Ninety-eight percent of Antarctica is covered by ice, which averages 7,086 feet thick (2,160 meters).[14] At its thickest points the ice is estimated to reach depths of nearly three miles (4,776 meters). This ice comprises 90 percent of all the world's ice, and more important, it accounts for 70 percent of the world's fresh water. The immense weight of Antarctica's ice serves to depress the eastern ice sheet so that one-third of Antarctic land lies below sea level.[15] Continental ice extends beyond the shores to form ice shelves that float atop the seas surrounding Antarctica and can reach a thickness of up to 3,900 feet (1,188 meters). The most significant of these is the Ross Ice Shelf, which is twice the size of New Zealand.

Antarctica's spectacular ice shelves are responsible for the tabular icebergs that are readily visible to ecotourists who sail around the continent. Despite this massive amount of ice, Antarctica is also one of the driest spots on earth—hence the term polar desert, and receives a mere four inches (10 centimeters) annual precipitation.[16]

Not only is Antarctica one of the driest places on earth, but it is also one of the coldest and windiest. Its climate earns it the reputation of being the most extreme on earth. The coldest temperatures on earth were thought recorded in June 1982 at the US South Pole Station when the thermometer reached −117.4°F (−47° centigrade), but were later surpassed in July 1983 when Russia's Vostok Base registered a low of −128.6°F (−53.6°C). Joyner and Theis note that at those temperatures "fuel oil turns to jelly and boiling water thrown into the air will burst into ice crystals before reaching the ground."[17] As if temperature extremes are not sufficiently harsh, strong gusts called katabatic winds whip across the continent and can attain speeds up to 100 miles (160 kilometers) per hour.

Such chilling scenarios remind us that this is a continent worthy of great respect despite its lack of abundant fauna and flora. In fact the continent has no trees, or vertebrates, and only two flowering plant species are found at the northern tip of the Antarctic Peninsula. Lichens, algae, and moss are the principal flora and live primarily along the coastal areas. An incredible example of the hardiness of Antarctic life forms is the fact that according to Joyner, some indigenous flora was detected near the South Pole.[18] Mites, lice, and springtails are its main fauna. Seals, penguins, and seabirds are considered part of the marine—not continental—ecosystem.

Just how living things can survive such harsh environmental conditions is of interest to scientists who study the area. The adaptive mechanisms of the Antarctic fauna and flora manifest themselves in the creation of antifreeze-like compounds that keep fish, for example, from freezing.[19] The Antarctic toothfish, for instance, is observed clinging beneath ice shelves in near freezing water.[20] Research on why this fish does not freeze is but one example of the type of scientific exploration undertaken under the auspices of the Antarctic Treaty System.

While Antarctica has a general paucity of indigenous fauna and flora, the surrounding Southern Ocean forms an important marine ecosystem best known for its small zooplankton inhabitants known as krill. This two-inch long shrimp-like crustacean forms the bottom of the food chain for many higher life forms. Most krill are found around the Antarctic Peninsula, which explains the area's rich marine life where whales, seals, birds, and penguins abound. Because of the *Antarctic Convergence*—or the circle surrounding the continent where Antarctic waters meet (converge) with sub-Antarctic waters—the Antarctic ecosystem is kept relatively distinct and separate, although some fishing grounds, whale zones, and krill areas do cut across the Convergence. The rough seas and gale force winds in the 100 mile-wide Antarctic Convergence served as a natural deterrent to early exploration. It helped keep Antarctica pristine.

With the post–World War II era improvement in ice-rated and ice-breaking ship construction, crossing the Antarctic Convergence no longer dissuaded scientists and the scant few adventurers who took an interest in the continent. Aircraft capable of withstanding very cold temperatures and icy conditions also began to transport scientists and supplies to various bases on the continent. Scientific exploration thus began on a larger scale than ever before.

Humans became increasingly interested in the continent, could travel and reside there more safely, and were indeed arriving in growing numbers by the mid-1950s. This raised the concern that humans might just be capable of doing damage to this pristine polar environment; hence the need for regulatory mechanisms became imminent. The mid-1950s marked the beginning of the need for recognized environmental peacemaking efforts.

Following the breakup of the Soviet Union, Russia's fleet of icebreakers and ice-rated research ships were frequently converted for dual use—research and ecotourist transport. Crossing the Drake Passage is done weekly and the Antarctic Convergence's rough seas no longer deter visitors in the least. As more ecotourists continue to visit each year, the notion of regulation and management of the area becomes ever more important.

Antarctic Treaty System: Unique Global Cooperation

The very comprehensive Antarctic Treaty System sets the groundwork for maintaining peaceful coexistence in Antarctica and is one of the most successful environmental regimes to date. An international regime is considered to be a set of rules, norms, and procedures around which the expectations of actors converge in a certain issue area; hence, we speak of the Antarctic regime.[21] Rothwell maintains "Antarctica has effectively been subject to international control under a regime which places strict limitations on the exercise of national sovereignty and jurisdiction."[22]

What makes this regime so worthy of note is the fact that since the continent has no globally recognized sovereign borders, no state has heretofore acted in a manner that would project its sovereignty in a hostile manner. Cooperation predominates, and decisions are made as if nation-states were part of one happy family. Naturally, as families often feud, occasionally scientists do the same. This is hardly reminiscent of the global arena best known for balance of power rivalries, border conflict, world wars, and general systemic tensions.

The Antarctic Treaty (1959) forms the basis for what we term the Antarctic Treaty System (ATS) and is primarily supported at the international level by the Agreed Measures for the Conservation of Antarctic Fauna and Flora (1964), Convention for the Conservation of Antarctic Seals (1972), the Convention for the Conservation of Antarctic Marine Living Resources (1980) (CCAMLR), and the Protocol on Environmental Protection to the Antarctic Treaty (1991)—the latter is often called the Madrid Protocol.

States that are a *party* to the Antarctic Treaty have also adopted laws and regulatory mechanisms designed to harmonize with this international regime. The United States, for example, enacted the Antarctic Conservation Act of 1978 and the Antarctic Science, Tourism, and Conservation Act of 1996. Regulations and permit issuance remain under the authority of the National Science Foundation's (NSF) Office of Polar Programs. In addition NGOs contribute greatly to Antarctic management and play a vital role especially in monitoring scientific and ecotourist endeavors. Regime formation, however, is not the object of this research; instead, our concern is with the ATS's actual operation once it was recognized as a *real* (viable) regime. Successful regime operation is instrumental in contributing to environmental management and ultimately to conflict avoidance. Regimes are not static and should be thought of as a framework that evolves over time. The Antarctic Treaty is the main anchor to this regime around which other treaties, regulations, and national laws are attached as supporting mechanisms.

Attempts in the 1980s to create a mining structure in the form of the Convention on the Regulation of Antarctic Mineral Resources (CRAMRA) appeared successful only to fail at the last minute when Australia and France withdrew their support. CRAMRA was a treaty designed to recognize that future commercial mineral extraction would likely occur and that a procedural framework set forth in advance is the best solution for avoiding potential conflicts down the road. According to Article 3, mineral resource activities outside the convention would be prohibited and an Antarctic Mineral Resources Commission would be created. Article 29 also provided for a "Mineral Resources Regulatory Committee."[23]

Although CRAMRA died in 1988, the Antarctic regime persisted, and member states went on to produce the highly regarded Madrid Protocol, which postponed the mining issue at least for fifty years. Article 7 calls for the prohibition of commercial mineral activity and declares, "Any activity relating to mineral resources, other than scientific research, shall be prohibited."[24] To buy time for reassessing the minerals issue, the protocol does not expire until fifty years after its entry into force in 1998.[25] Some refer to this as a *moratorium on mining*, but—based on Antarctic Treaty procedures—the protocol can potentially be amended or modified at any time. Modification of this moratorium to regulate commercial mineral exploration and/or extraction is likely to provoke a critical juncture, which would be the true test for the ATS. Fortunately, no mineral concentrations of any significance are known at this time south of 60° South latitude, although oil reserves are thought to exist within the Southern Ocean—neither could they be extracted in a cost-effective manner on the continent.

Instead of focusing on major, or high-politics issues such as commercial mineral exploration/extraction, ATS members (referred to as Antarctic Treaty Consultative Parties, ATCPs) concentrate on low-politics issues such as scientific research or tourism. CRAMRA was the only exception and, while it failed, the regime remained intact. Hence ATS formal activity today centers upon the continuation of a regime aimed at protecting the Antarctic environment, and keeping the area open for managed science and tourism. (See table 9.1.)

The most contentious Antarctic high-politics issue is that made by the seven countries that staked claims to parts of Antarctica starting in 1907.[26] Claimants include Argentina, Australia, Chile, France, New Zealand, Norway, and the United Kingdom. Avoiding a sovereignty fight over claims through denial behavior is much like adopting a mutual assured destruction (MAD) policy during the cold war.[27] The

Table 9.1
Status of major Antarctic issues

Issue	Low politics	High politics	Deferred attention	Receives attention
Tourism	Yes	No	No	Yes
Scientific research	Yes	No	No	Yes
Nuclear/military	No	Yes	No	Yes
Commercial mineral exploration or extraction	No	Yes	Yes	No
Claims	No	Yes	Yes	No

unthinkable—nuclear war—would have such devastating consequences that it was to be avoided at all costs. A disruption in the status quo would prove catastrophic according to MAD. Similarly, if claims were forcefully asserted, then such sovereignty-linked turmoil would surely turn into a force capable of regime meltdown. Instead, states continue about their business while pretending that they do not see the claims, even though Argentina, for example, appends the word Antarctica to its southernmost province (Tierra del Fuego, Antarctica, and the South Atlantic Islands) and makes reference to it as Antarctica, Argentina, as street signs openly reveal. It is considered to be part of Argentina's national territory according to Argentina's National Antarctic Institute.[28] While not incorporating its claim as national territory, Australia does refer to the claim as the Australian Antarctic Territory as does the United Kingdom with its British Antarctic Territory. France also uses the term French Antarctica. France's Polar Institute Web site explains how the Terres Australes et Antarctiques Françaises (TAAF), or French Southern and Antarctic Lands, are responsible for maintaining French sovereignty.[29] TAAF has official French status as an Overseas Territory and even has a préfet (prefect) responsible for administering the area much like a governor would do the United States, although that person does not actually live in Antarctica.[30]

Dealing with Incidents in Antarctica

Despite efforts expended by claimant states to keep their claims active, all states essentially work around the sovereignty issue, relegate it to a stage-set backdrop, and normal scientific research or tourism proceeds—no matter that one may be working in someone else's claim area. It is business as usual.

Sovereignty over claims remains an unyielding force that is so cumbersome no one dares move it to front stage. To do so would insert state-centric conflict into the regime and possibly lead to war in what is now a relatively pristine environment. To date, disputes have been relatively minor. States caught up in Antarctic disputes have not used a great degree of sovereign power to win the dispute. In fact winning is not even considered an option. Instead, a group mentality prevails in which everyone is acutely aware that asserting true power could let the genie out of the bottle.

ATCPs cannot let the unthinkable happen, so they cooperate, or de-escalate the tension. This was exactly what happened in 1985 when France built an offshore airstrip at Point Géologie, in Adélie Land. During construction French engineers destroyed several penguin rookeries and, according to Joyner, there was sufficient evidence that the Agreed Measures for the Conservation of Antarctic Fauna and Flora treaty was breached.[31] No admonishment or sanctions were imposed by ATCPs. Joyner notes that cooperation prevailed in the end:

> If the French said that they were in compliance with the Agreed Measures, then that explanation was sufficient for other ATCP governments. No punitive measures, formal or informal, were taken against the French. Environmentalists branded this incident a failure on the part of the ATCPs to protect treaty norms and Antarctic values. The ATCPs, however, regarded their nonactions as a small environmental price for long-term regime cohesion.[32]

In a similar apparent breach of the Antarctic Treaty, the Chinese bore the brunt of muted concerns when, in 1981, they inaugurated their new Great Wall Research Station. To celebrate this occasion the Chinese brought in doves, which are not an indigenous species to Antarctica—and hence prohibited under the ATS. The doves were released and undoubtedly died shortly thereafter.[33] Adding more injury to insult, some Chinese delegates brought their dogs with them to the station—another prohibited action—and were seen kicking and generally abusing penguins. This sorry state of affairs was another apparent breach of the Agreed Measures, which could have resulted in ATCP objections. Instead, "low-level, informal consultations" were used to inform the Chinese station chief about the misconduct[34] while compliance with conservation norms is treated as an educative process. Norms and rules in the ATS regime are not viewed by member states as black-and-white situations that give cause for punitive sanctions.[35]

While this response may sound weak, it reiterates the obvious—that strong diplomatic actions such as sanctions and harsh words are not the methods employed by the ATCPs to induce regime compliance. Not only is understatement the preferred means of rebuke, it has also "tended to produce more impressive results."[36] Antarctica is a very dangerous place—especially for ships in its surrounding ice-laden waters. Accidents are always a concern as is liability for causing harm to the environment. Article 16 of the Madrid Protocol places the liability burden on the parties to create "rules and procedures relating to liability for damage arising from activities taking place in the Antarctic Treaty area."[37] Oil spills from ships can pose a possible disaster, and in 1989, the Argentine Navy supply ship *Bahia Paraiso* ran aground near Arthur Harbor.[38] It was doing double duty carrying eighty-one tourists and supplies. Approximately 250,000 gallons (946,000 liters) of diesel fuel were spilled into the Southern Ocean to the detriment of penguins and krill that were killed during breeding season as a result of the accidental spill. Assistance to the *Bahia Paraiso* came from the US Palmer Station.[39] This is in keeping with Annex

IV, Article 12, of the Madrid Protocol that addresses emergency response issues and encourages parties operating in the Antarctic Treaty area to create contingency plans for ships in distress within the Southern Ocean. Article 12 states that parties shall "cooperate in the formulation and implementation of such plans..." and also "establish procedures for cooperative response to pollution emergencies...."[40]

Cooperation in this harsh environment is essential where immediate help is available only from the closest scientific station or nearest vessel in the area, which are few and far between. This explains why ice-rated tourist vessels are frequently called upon to assist in search and rescue operations, as did the Russian ship *Akademik Sergei Vavilov* when it aided eight Korean scientists at King Sejong Station during difficult weather in the 2003/2004 travel season.[41]

Alas, there are no guarantees of immediate rescue if pack ice prevents vessel mobility. It is perhaps this most imposing of natural environments that encourages—indeed requires—all humans in the area to cooperate for the common good and is akin to Garrett Hardin's "tragedy of the commons"[42] in reverse. Instead of ruining (overgrazing) the commons for one's personal gain (herd), Antarctic visitors—and we are all visitors to this continent—be they scientists, adventurers, tourists, or logistical support personnel, develop a sense of camaraderie amid this vast wilderness so far from civilization and lacking in indigenous inhabitants. Out of the need to survive in this harsh environment a sense of dependency and closeness develops regardless of nationality. Such dependency, both real and affective, is quantified by the number of inter-country rescues and the incidence of cross-national cooperative ventures.

The 2005 Russian–US icebreaking effort into McMurdo Sound is one such example in which the Russian icebreaker, *Krasin*, was chartered to clear a path for delivery of supplies to the US-operated McMurdo Station. Because the US icebreaker, *Polar Sea*, was in dry dock, this turned into a good financial venture for Russia's Far East Shipping Company that owned the *Krasin*.[43] Yet other instances, such as the one involving the Russian ship *Akademik Sergei Vavilov* mentioned above, involved no financial gain. In only a few places worldwide can this feeling of dependency and closeness be duplicated. It is possible that these harsh conditions facilitate a more forgiving application of the ATS in which punishment is eschewed in favor of cooperation, and forgiveness for one's mistakes prevails. Somehow the cooperative Antarctic behavior percolates back to Washington, Paris, London, Moscow and other ATCP capitals where political leaders and policy makers set the tone, and administrators apply regulations.

The most recent example of this low-keyed style is how the ATCPs, and the United States, in particular, handle disagreement over core drilling into Lake Vostok, a subglacial lake the size of Lake Ontario. At present there is a lack of consensus and polarization among scientists, but the controversy remains at a scientific, or low, level. Russian and French scientists believe the 2.5 mile deep lake is sterile and devoid of any life. US scientists maintain that as yet undiscovered life forms might exist and that Russian drilling into Lake Vostok would contaminate it.[44] A scientific controversy has developed over some microbes discovered in an ice core obtained from drilling above the lake. The Russians and French maintain that the microbes are contaminants from the drilling process in which sterile techniques were not employed. In addition they say the lake's high oxygen levels make life unsustainable. US scientists assert that there could be microbes in the lake. The lake also sparked interest because it has not been exposed to sunlight in 15 million years and some believe it may have been part of Gondwanaland, the continent that once included South America, India, Australia, Africa, and Antarctica.

At issue is the Russian drilling method, which the Americans believe will surely destroy any microbes or life forms. Scientific controversy not withstanding, the lake lies below the Russian research station, and it is the Russians who want to drill an ice core using their own technique. That technique uses kerosene to prevent the hole from refreezing as the drill punctures the ice on Lake Vostok's surface. The fear of contamination fuels this debate, and the Russians insist that they are within Antarctic international parameters governing research and environmental protection.

Compromise solutions include a proposal from "an international panel to drill a new, cleaner hole through the ice and enter the lake using a self-sterilising 'cryobot,' [while] the Russians plan to enter the lake from the existing, highly contaminated hole."[45] At present the Russians appear intent on drilling into Lake Vostok, and the alternate drilling hole may very well offer a face-saving way to resolve this dispute. However, the kerosene process still remains a contentious point.

In addition the United States intends to be involved in any international drilling effort. Let us recall that the existence of one country's scientific base (Russia's Vostok Base) does not exclude other scientific teams from exploring and/or drilling at the lake's site. No border guards physically prevent people from entering Antarctic areas; consequently scientific exploration benefits from this freedom of movement. As manager of the US Antarctic Program, the National Science Foundation's (NSF) Office of Polar Programs (OPP) would coordinate US efforts. In the meantime OPP has created a steering committee to study subglacial lake exploration.[46]

While these examples illustrate how human activity operates under the existing ATS, let us now examine concrete environmental structures that facilitate cooperative environmental management.

Specially Protected Areas and Specially Managed Areas

The closest Antarctic parallel to transboundary protected areas (TBPAs),[47] commonly called peace parks, are the specially protected areas (SPAs) set forth in Annex V of the Madrid Protocol. Annex V came into force in May 2002. While Article 2 of the protocol designates Antarctica as a "natural reserve, devoted to peace and science," Annex V details the SPAs objectives and goes a step further by creating specially managed areas (SMAs) as well.[48]

SPAs are intended to protect a continental or marine area that holds environmental, scientific, historic, aesthetic, or wilderness value and must be kept free from human interference. SPA designation may be for species protection, to protect geological or glaciological or other geomorphological features, or to protect sites and monuments. Entry into SPAs is prohibited without a special permit granted by one's country of origin. Exceptions are made for valid scientific research projects.[49]

As a corollary to this permit process, it behooves us to mention something about the general permit issuance procedure. An appropriate national authority specified by each Party issues permits. Each Party's Antarctic Division, NSF/OPP, or equivalent, is tasked with granting permits overall, and the SPA permit falls under this system. Permit issuance for travel to, and around, areas of Antarctica is akin to being granted a visa except that one's own government is the issuer.[50]

There is, of course, no receiving government's Ministry of Foreign Affairs to complete this task, since the visitor is not actually "setting foot in a foreign country." In fact, when this author arrived at Argentina's Esperanza Base, the entry procedure is to step out of the boat, or raft in my case, and climb up a ladder to shore. Despite the fact that Argentina considers this Argentine *soil*, no customs or immigration procedures were in place, and the entire visit was very informal. Passports were not required for entry and would be stamped only if requested by visitors. By downplaying (denying) the exercise of Argentine state sovereignty, harmony is maintained in its claim that also overlaps the British, French, and Chilean claims.

Specially Managed Areas

SMAs are similar to SPAs except that they include areas where "activities are being conducted or may in the future be conducted ... to assist in the planning and coordination of activities, avoid possible conflicts, improve cooperation between Parties or minimize environmental impacts."[51] Entry into SMAs does not require a special permit. A SMA may also contain within its boundaries one or more SPAs. Zones within the SMA can be placed "off-limits" and hence become *prohibited* areas in order to achieve, for example, a scientific research objective.

Article 4, Annex V, specifies that any Party or ATS organization—as well as the Scientific Committee on Antarctic Research (SCAR)—may propose an area intended to become a SPA or SMA.[52] Thus we see that an NGO is actually named in the Madrid Protocol and is treated much like an international governmental organization (INGO) such as the Committee for Environmental Protection, which was created in Article 11 of the protocol.[53]

Comparing TBPAs to Antarctic Specially Protected Areas

If we examine TBPAs traits and compare them to Antarctic SPAs, we see many similarities. Both strive for environmental conservation, and species and ecosystem protection. Both also encourage increased cooperation among member countries. Some TBPAs may seek to more effectively manage shared natural resources such as the Dead Sea, thereby enhancing environmental security. Antarctica's SPAs seek to manage and preserve shared environmental or historic patrimony that forms part of an Antarctic common heritage of mankind. While TBPAs may also seek to maximize economic gain through jobs creation and development of communities that abut the peace parks, this is not a concern in Antarctica.

Antarctica's SPAs and SMAs are akin to La Amistad International Park and World Heritage Site, the jointly managed and one of the oldest TBPAs along Panama and Costa Rica's border. Cooperation goes back to the 1970s when the Ministry of Planning and Economic Cooperation from both countries initially agreed to integrate border management.[54] Other TBPAs demonstrate similar cross-border cooperation, which is akin to what takes place in Antarctica except for one big difference—the continent has claimant and nonclaimant states in which SPAs are located. This means that there is no sovereign territorial boundary that divides the SPA. On the contrary, borders distinguish the Costa Rican side from the Panamanian side of La Amistad.

We can never ignore the presence of the border, despite the cooperation that transcends it. Presently in Antarctica, borders are a non-issue. Figures 9.1 and 9.2 create a conceptual rendition of both the traditional TBPA—or peace park—and the Antarctic equivalent. The Antarctic example illustrates how an SPA can be located within an overlay zone of different claims.

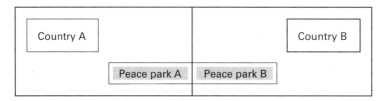

Figure 9.1
Traditional peace park concepts within the nation-state system

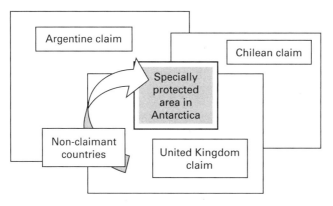

Figure 9.2
Example of overlapping relationship between claimant/nonclaimant areas in which a specially protected area may be located

Applying Lessons Learned in Antarctica: Backseat Sovereignty

Specially protected areas are created mainly for environmental conservation reasons and are jointly administered. If an area in need just so happens to fall within overlapping claims, then so be it—it can still be categorized as a SPA. That is the current Antarctic philosophy, and ATS member participants accept this as a working formula. In fact several scientific bases are located in the claims of other nation-states. For example, the US Palmer Station on the Antarctic Peninsula is located within the Argentine, Chilean, and British overlapping claim area. The United States, however, is solely responsible for managing the station, but cooperative assistance from other ATCPs will likely be forthcoming at Palmer Station, if requested.

Perhaps this approach can hold promise for replication elsewhere in the world. Is it not possible to take cross-border cooperation a step further in parts of Africa or

Asia and create a park truly governed by a joint commission—somewhat akin to an autonomous region in which sovereignty either takes a backseat, or is jointly shared by nation-states that border the park? Let us recall that, in 1979, Costa Rica and Panama signed the first Border Cooperation Agreement for joint development and technical assistance projects in La Amistad.[55]

Over time more agreements were signed to cover joint projects and administration of roads, forest conservation, and municipal development—to name a few examples. While these agreements enable administration of joint projects, this is not the same as having a bi-national commission whose sole purpose is to govern the area.

Joint sovereignty—or condominium—is *not* legally what occurs in Antarctica. Joyner describes a condominium as "an international arrangement of shared rights and duties, with several states capable of exercising rights of sovereignty [jointly]."[56] Shaw explains this process simply as one in which "two or more states equally exercise sovereignty with respect to a territory and its inhabitants."[57] States would combine governance, responsibilities and enforcement under the auspices of a new international institution. While an Antarctic Public Heritage Agency and an Antarctic Environmental Protection Agency were discussed in the past and would have led to a condominium arrangement, nothing came to fruition.

Yet ATCPs do act as if sovereignty were less of an issue in the management of the continent and of its SPAs. Much as the claims issue takes a back seat to administration of Antarctica's environmental good, so might individual state sovereignty be given a backseat within a TBPA, or peace park. Bordering park states could also proceed one step further to create a structure of joint sovereignty (condominium) for the park. Primary nation-state sovereignty would then take a backseat much like Antarctic claims do today.

While there is no guarantee that de facto backseat sovereignty or de jure joint sovereignty will make park administration any easier, it could serve to defuse border conflicts by giving both parties a greater stake in the disputed area. Environmental conservation might provide a *border buffer*, or a way to confront political conflict that has no other apparent solution. The way to confront the conflict is not head on, but through the common ground called environmental conservation. While war rages in the Congo–Rwanda–Uganda border area and violence prevails, no one dares speak ill against the mountain gorilla that roams (albeit now in decreased numbers) the area. It is the gorilla that represents the environmental common bond to that part of Africa much as penguins, albatross, and seals do in the Antarctic.

While SPAs do prohibit human entry into the area without special permit, this would not be the best solution for a populated border area turned into a peace park. Since this chapter focuses on Antarctica, developing details for parks that adopt backseat or joint sovereignty is a task best reserved for future research. It is, however, an applicable lesson learned from the Antarctic Treaty System's environmental management system.

Other lessons learned from Antarctica relate to the harmony and denial approach. Too often states try to achieve harmony but fail to forget ongoing disputes. While Antarctic overlapping claims, nonclaimant concerns, and commercial mineral extraction are very real potential issues, they are forgotten or denied, the CRAMRA attempt not withstanding. It is in this manner that scientists and tour operators alike go about their business in Antarctica. Harmony is thus institutionalized in the ATS. When the commercial mineral exploration and/or extraction issue again raises its head, we can only hope that, by then, the ATS's harmony-seeking code of conduct will have become deeply institutionalized, that it will prevail, and that a solution will ensue. This harmony is, after all, what makes the ATS unique and a role model for other regimes and peace parks.

Notes

1. United Nations (1959: Art. IV).
2. Young (1994: 17–18).
3. Young (1994: 18).
4. *CIA World Factbook* (2004).
5. International Association of Antarctic Tour Operators (2004: 2–3).
6. The IUCN/World Conservation Union defines peace park as "transboundary protected areas that are formally dedicated to the protection and maintenance of biological diversity, and of natural and associated cultural resources, and to the promotion of peace and co-operation." The term peace park is frequently used interchangeably with the term transboundary protected area. See Sandwith et al. (2001: 3).
7. Scientific Committee on Antarctic Research (SCAR) (2004a: 3–5).
8. Conca and Dabelko (2002).
9. Environment, Development and Sustainable Peace: Finding Paths to Environmental Peacemaking, *Conference Report* (2004: 2); see also Conca (2002: 4).
10. For an extensive list of such areas see "Parks for Peace, International Conference on Transboundary Protected Areas as a Vehicle for International Co-operation," held on September 16–18, 1997, in Somerset West (near Cape Town), South Africa.

11. United Nations (1959: Arts. I and V).
12. Blum (1994: 667).
13. Joyner and Theis (1997); Joyner (1998).
14. US National Science Foundation (1994).
15. US National Science Foundation (1994).
16. Joyner and Theis (1997: 10–13).
17. Joyner and Theis (1997: 12).
18. Joyner (1998: 5).
19. US National Science Foundation (1994).
20. US National Science Foundation (2005).
21. Krasner (1983) and Young and Levy (1999: 1–32).
22. Rothwell (2000).
23. United Nations (1988: Arts. 3 and 29).
24. United Nations (1991: Art. 7). The Madrid Protocol entered into force in 1998.
25. United Nations (1991: Art. 25, sec. 2).
26. For a history of these claims, see Joyner (1998: 14–19).
27. Schelling (1960).
28. Argentina's National Antarctic Institute (2005).
29. French Polar Institute (2005).
30. French Southern and Antarctic Lands (2005). See also TAAF's official French government Web site for more detailed information (French only) ⟨http://www.taaf.fr/index.htm⟩.
31. Joyner (1998: 109).
32. Joyner (1998: 109).
33. Introduction of nonnative species carries with it the possible introduction of diseases, parasites, and other health issues that can endanger native Antarctic species.
34. Joyner (1998: 110).
35. Joyner (1998: 110).
36. Joyner (1998: 110).
37. United Nations (1991: Art. 16).
38. Mari Skåre (2000: 166).
39. Joyner (1998: 208–209).
40. United Nations (1991: Ann. IV, Art. 12).
41. International Association of Antarctic Tour Operators (2004: 10).
42. Hardin (1968).
43. "Icebreakers Clear Channel into McMurdo Station," Press Release 05-13.
44. Lake Vostok DEEP ICE Project (2005).

45. George (2004: 6–8) and Scientific Committee on Antarctic Research (SCAR) (2004b: 22).
46. US National Science Foundation (2002).
47. The IUCN/World Conservation Union defines a TBPA as "an area of land and/or sea that straddles one or more boundaries between states, subnational units such as provinces and regions, autonomous areas, and/or areas beyond the limits of national sovereignty or jurisdiction, whose constituent parts are especially dedicated to the protection and maintenance of biological diversity, and of natural and associated cultural resources." Sandwith et al. (2001: 3).
48. United Nations (1991: Art. 2 and Ann. V).
49. United Nations (1991: Ann. V).
50. In case of arrival by ship, the country of registry would grant a permit to the ship and a travel company would obtain a general permit to transport tourists to Antarctica.
51. United Nations (1991: Ann. V, Art. 4).
52. United Nations (1991: Ann. V, Art. 4).
53. United Nations (1991: Art. 11).
54. Castro-Chamberlain (1997).
55. Castro-Chamberlain (1997: 49–55).
56. Joyner (1998: 50).
57. Shaw (2003: 206).

10

The Waterton–Glacier International Peace Park: Conservation amid Border Security

Randy Tanner, Wayne Freimund, Brace Hayden, and Bill Dolan

In 1932 Glacier National Park of the United States and Waterton Lakes National Park of Canada combined their efforts to establish the Waterton–Glacier International Peace Park (Waterton–Glacier). As the world's first peace park, Waterton–Glacier was the forerunner of what would become a significant movement in conservation worldwide. However, unlike many peace parks or transboundary protected areas that have been established since 1932, Waterton–Glacier was formally created as a symbol of peace and goodwill between the two nations rather than as a mechanism for conflict resolution. While the two parks have a long history of peace and cooperation, the events of September 11, 2001, and the US subsequent border security initiatives illustrate the challenges that peace parks face when they must be responsive to the demands of both border security and conservation. Recognizing that border security and transboundary conservation may often be competing interests, both countries have engaged in collaborative efforts to ensure that the resilience of the peace park is maintained while implementing the requisite security measures. The purpose of this chapter is to explore how the recent escalation of border security initiatives along the United States–Canadian border has affected the cooperative efforts of the Waterton–Glacier International Peace Park.

A Brief History of Waterton–Glacier

The Rocky Mountains of North America are shared by the nations of Canada and the United States. At the convergence of this mountainous divide and the international boundary lay two protected natural areas: Waterton Lakes National Park in the Canadian province of Alberta and Glacier National Park in state of Montana (see figure 10.1). The parks encompass striking mountain landscapes, deep eutrophic lakes, dramatic evidence of glaciation, and a rich assemblage of flora and

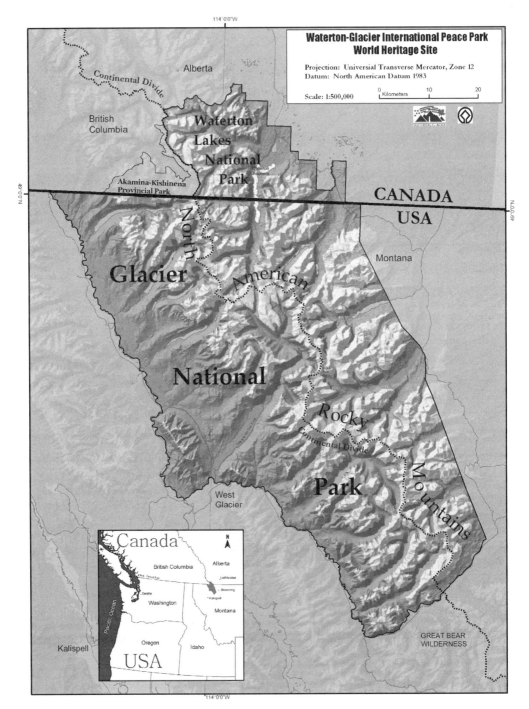

Figure 10.1
Waterton–Glacier International Peace Park

fauna. The parks are also situated on the western margin of the Interior Great Plains of North America and contain important elements of Great Plains biota.

Waterton Lakes and Glacier National Parks were established by the national governments of Canada and the United States in 1895 and 1910 respectively. Both countries have long enjoyed a close, peaceful relationship and the international boundary between the parks goes largely unseen as it traverses the wilderness. Close cooperation among the staffs of the two parks has occurred since their establishment. Like many peace parks and transboundary protected areas around the world, the creation of Waterton–Glacier was catalyzed by a nongovernmental organization. In July 1931 Rotary Chapters from Alberta and Montana met in the town of Waterton and unanimously endorsed the idea that the two Parks be joined as a symbol of friendship between the United States and Canada.[1] In 1932 the Canadian Parliament and United States Congress acted on the Rotarians lobbying efforts by designating Waterton Lakes and Glacier National Parks as units of the Waterton–Glacier International Peace Park. In doing so, they created the world's first peace park. The designation commemorates the peace and goodwill that exists between the United States and Canada, two nations that share the largest undefended border in the world.

In 1976 and 1979 respectively, Glacier and Waterton were each designated as separate Biosphere Reserves under the Man and Biosphere Program of the United Nations Educational, Scientific, and Cultural Organization (UNESCO).[2] The two main tenets of the Man and Biosphere Program are preservation of core natural values and encouraging a sustainable area economy that will protect those values. The two parks strive to build close relationships with their neighbors to help meet the goals of the Program. In 1995, a further recognition was bestowed when the Waterton–Glacier International Peace Park was designated as a World Heritage site by the Convention Concerning the Protection of the World Cultural and Natural Heritage, a part of UNESCO.[3] Waterton–Glacier met UNESCO's world heritage site criteria for natural area nominations. The World Heritage site designation signifies that the Waterton–Glacier International Peace Park possesses outstanding, universal value to the world's peoples and it requires that the United States and Canada refrain from actions that might damage the values of this world heritage site.

While the peace park designation does not impact national sovereignty, fiscal autonomy, or management responsibilities of both parks, the managers of Waterton and Glacier acknowledge that the effective management of shared natural resources requires coordinated and collaborative cooperation. Such cooperation has led to improved research related to both natural and cultural resources, more expedient search and rescues, enhanced visitor services, and partnerships that extend beyond the peace park.

The Nature of Cooperation within Waterton–Glacier

In order to better understand the nature of transboundary cooperation in Waterton–Glacier and how the peace park has responded to the pressures incurred by border-security initiatives, it is useful to envisage Waterton–Glacier as an "international regime." As defined by Krasner, *international regimes* are "principles, norms, rules, and decision-making procedures" that govern issue-specific relations among states.[4] International regimes, then, may be thought of as the framework by which an international issue is governed. In the case of Waterton–Glacier the regime framework consists of guiding principles and norms for cooperative research, search and rescue, environmental education and interpretation, wildfire management, visitor access, and wildlife management. While Waterton–Glacier is formally established, these guiding principles, for the most part, are informal. In other words, cooperation is not governed by written treaties or binding agreements, nor are there sanctions imposed for a failure to cooperate.

Envisaging Waterton–Glacier as an international regime facilitates the evaluation of cooperation in terms of its "effectiveness" and "resilience." For instance, Waterton–Glacier would be considered *effective* if—within the context of its regime framework—it achieves the objectives and purposes for which it was established.[5] Evaluating the effectiveness of informally governed peace parks, such as Waterton–Glacier can be difficult, since specific objectives and purposes are not always formally defined.

Nevertheless, informal cooperation may be evaluated if one recognizes that most informal cooperation in the peace park is "functional." *Functional cooperation*, for the purposes of this chapter, may be defined as cooperation that is driven by bottom-up technical and situational demands.[6] In Waterton–Glacier, examples of functional cooperation include transboundary wildfire management, search and rescue, and wildlife management. Effectiveness, then, for the Waterton–Glacier regime is largely measured by the extent to which the objectives and goals of specific transboundary projects and programs are achieved. This is not to say that functional cooperation does not lead to the realization of broader goals such as peace and enhanced regional conservation, but if these goals are achieved, they are achieved through a variety of technical and situational projects and programs.[7]

Despite the effectiveness of a peace park regime at a given moment, it may experience pressures (either internal or external) to dissolve.[8] For instance, the regimes might initially be effective, but for a variety of reasons—including decreased funding, high staff turnover, or elevated conflict among participating states—effectiveness may not be sustainable.[9] In addition to effectiveness, then, another factor critical to the success of a regime is *resilience*, or the ability to maintain effectiveness despite pressures to dissolve.[10] For Waterton–Glacier, resiliency has historically not been difficult to maintain—both Canada and the United States have a long history of peaceful cooperation, both parks are relatively well-funded, and there is a high degree of institutional memory among the park staff that facilitates continued cooperation.

Cooperation in the Waterton–Glacier Regime: A Functional Approach

The effectiveness of cooperation within the Waterton–Glacier regime is rooted in personal relationships among the staff of both parks that can be traced back to John George "Kootenai" Brown—Waterton Lakes' first Park official—and Glacier National Park Ranger Henry "Death on the Trail" Reynolds. These relationships have persisted since as early as 1911 and have provided a foundation for the collaborative management regime. While much of the cooperation in the peace park is informal, formal arrangements have been made to support the objectives of the regime. These range from administrative agreements for operational support to science and planning programs.

In recent years a Memorandum of Understanding between Parks Canada and the United States National Park Service (1998) has, in part, provided a strong endorsement of this relationship. Specific agreements to implement a number of joint projects and provide operational support are covered by this Memorandum and other administrative documents.

Waterton–Glacier's effectiveness as a regime is most apparent in the functional dimensions of cooperation. Four examples of such cooperation, which have both informal and formal components, are vegetation restoration, education and interpretation, wildfire and public safety, and the Crown Managers Partnership. Below we briefly describe each area of cooperation. In each of these areas, collaboration has been the common factor contributing to cooperative effectiveness. And, despite recent challenges imposed by border security initiatives, all four areas of cooperation have remained resilient.

Vegetation Restoration

Over the past several years the peace park has implemented an aggressive cooperative program for invasive plant species. Waterton Lakes initiated their program in the late 1970s and Glacier followed in the 1980s. Both parks regularly share knowledge and experience across the border, as evidenced by harmonized research and management activities. In recent years Waterton Lakes has benefited from the native plant restoration program in Glacier. This program, which was originally established for highway restoration activities, includes the collection and propagation of native seed from a park greenhouse.

Plant material is subsequently used in restoration of disturbed sites such as backcountry campgrounds and high visitor use areas. These activities also help reduce the population of invasive species in both parks. In 2002 the two parks initiated a cooperative project where Glacier staff have supported the collection of native seed in Waterton Lakes. This seed is taken back to the Glacier greenhouse and propagated for the next growing season. The plants are then returned to Waterton Lakes for restoration of disturbed sites and in support of an educational native plant garden in the Park.

Education and Interpretation

Over the past several years there has been considerable cooperation and integration in the area of education and interpretation. A common publication, the *Waterton–Glacier Guide*, is produced annually and provides basic visitor information to the Peace Park as well as articles on current management issues and challenges. There is also a common brochure for the Waterton–Glacier International Peace Park that was revised in 2000 and is made available to all park visitors.

Cooperation is also critical to the regime's interpretive programming during the summer months. Park interpreters from both parks provide a weekly theater program in the other park. Every Friday evening, a park interpreter from Glacier travels to Waterton Lakes and presents a program to park visitors in Canada. At the same time an interpreter from Waterton Lakes travels to Glacier to present a theater program to park visitors in the United States. A joint interpretive hike is also led by interpreters from each park. These popular hikes down the Waterton Valley begin in Waterton Lakes, cross the international boundary, and end at the Goat Haunt Ranger Station in Glacier. Hikers return to Waterton Lakes on a tour boat on Upper Waterton Lake.

In 1985 the superintendents of both parks held a "Superintendents Hike" to celebrate the centennial of the establishment of Canada's first national park (Banff) and the seventy-fifth anniversary of the establishment of Glacier National Park. The three day Hike was very popular and became an annual event. Each superintendent invites up to 10 participants from their respective countries. Participants come from a diverse background, including ranchers, local business persons, local governments, politicians, and First Nations and other government agencies. The Hike has been supported logistically by Glacier park rangers and culminates with a barbeque hosted by the Waterton Lakes superintendent. The Hike offers opportunities for participants to share their knowledge, experience and values with regard to managing the peace park.

Wildfire and Public Safety

Both Waterton Lakes and Glacier support one another in matters related to emergency response, such as wildfire suppression and search and rescue. During the Sofa (Waterton Lakes) and Flattop Fires (Glacier) in 1998, both parks provided staff to work with the respective Incident Command Teams to ensure effective communication and cooperation between the parks. In other years, Waterton Lakes has supported Glacier with initial response to fires in the Upper Waterton Valley. Similar cooperation has been demonstrated in the peace park's public safety program. The two parks have frequently worked together on backcountry searches when visitors are reported missing. In some cases these incidents involve other governmental agencies in both countries when there are concerns around border security. Although this concern has heightened in recent years, increased interest in border security has not substantially impacted the effectiveness of the cooperative public safety program. Given the rugged terrain and high visitor use, both parks respond to a number of mountain rescues each year. These rescues may be technical or

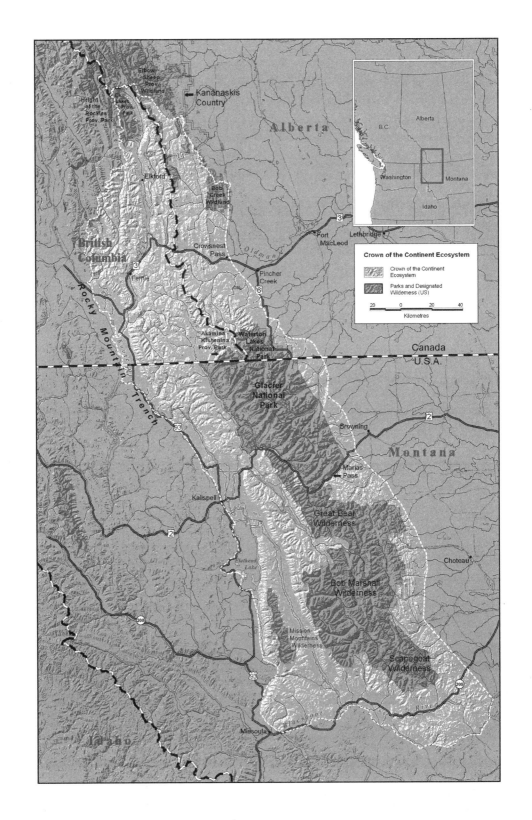

nontechnical in nature and may also be located in steep terrain with difficult access to the victim. In many cases the injuries are significant and/or may involve a fatality. Waterton Lakes maintains a Heli-sling rescue capability that provides for quick access to remote backcountry locations to recover injured hikers, scramblers, or climbers. Glacier does not have this capability and has called on park wardens from Waterton Lakes to provide support for an average of one to two rescues a year.

Crown Managers Partnership

The Waterton–Glacier International Peace Park is at the center of an ecosystem that extends well beyond the park's boundary. This ecosystem is commonly referred to as the crown of the continent ecosystem (see figure 10.2). The term was originally coined by George Grinnell, who was instrumental in the establishment of Glacier National Park, as a descriptor for the only area in North America where watersheds flow to three distinct oceans: the Pacific, Atlantic and Arctic oceans.

In 2001 the peace park played a leadership role in developing a workshop for 15 to 20 land and resource management agencies from Montana, Alberta, and British Columbia. The purpose of the workshop was threefold:

• To build awareness of common interests and issues in the Crown of the Continent Ecosystem.

• To build relationships and opportunities for collaboration across mandates and borders.

• To report on collaborative work already underway and identify opportunities for further cooperation.

This workshop was the initial step in building the "Crown Managers Partnership"—an annual forum hosted by the superintendents of the Waterton–Glacier International Peace Park. The Partnership includes national, provincial, state, and local governments as well leaders from surrounding indigenous communities. Participation is encouraged at senior and middle manager levels, as well professional and technical staff whose responsibilities and projects are pertinent at the ecosystem scale. A "work-plan project" is presented and approved annually at the

Figure 10.2
The Crown of the Continent Ecosystem

forum. As a rule, the purpose of the work-plan is to develop management tools, data management, and research programs at the ecosystem level in cooperation with academic institutions. The design and implementation of the work-plan is coordinated by a multi-agency steering committee that is representative of the broader membership.

The four functional areas of cooperation highlighted above serve as an illustration of Waterton–Glacier's effectiveness as a peace park regime. Since the creation of the peace park, there have been relatively few impediments to the implementation of these four areas of cooperation and a host of others. This resiliency can be attributed to the peaceful relationship between Canada and the United States, a high degree of institutional memory in both parks, and strong working relationships among the staff. Despite its resilience, Waterton–Glacier has recently experienced external pressures—catalyzed by the US post–September 11, 2001, border policy—that have challenged cooperative efforts in the regime.

Border Security and Visitor Access within Waterton–Glacier

If anything has threatened to impair the effectiveness of cooperation within the Waterton–Glacier regime, it has been the pressures incurred from an increasing interest in ensuring that the US–Canadian border is secure. Particularly on the US side of the border, increased border security has equated to a slight, but noticeable, change in the park staffs' ability to collaborate across the border. Glacier National Park's staff, for instance, has been forced to re-allocate funds to train park rangers as border patrol officers, instead of using that funding for cooperative programs with Waterton Lakes. Organizing meetings where staff from both parks are present is also increasingly difficult, due to the restrictive nature of the US post–September 11, 2001, border policy. Despite these relatively minor impediments, both parks have found ways to work alongside border security agencies to ensure that the peace park maintains its resiliency in a dynamic security environment. One area of cooperation where these collaborative efforts are most apparent is the management of cross-border visitor access.

With the establishment of the Waterton–Glacier regime, the governments of the United States and Canada made a conscious effort to attenuate the existence of the international boundary in the name of peace and ecosystem conservation. While not necessary for the existence of a peace park, transboundary visitor access has been an important consideration for both Glacier and Waterton Lakes. Allowing visitors to travel across the border with minimum restrictions has facilitated cross-border tourism, raised awareness of conservation efforts at an ecosystem level, and fortified the cooperative relationship between the two parks. The events of September 11, 2001, though, reframed the parks' relationship and required the peace park regime to be responsive to the demands of the evolving border policy in addition to those of visitor access. While the increased interest in border security could have threatened the effectiveness of the regime, the collaborative efforts of both parks with a host of border security agencies demonstrates that it is possible to secure borders while simultaneously practicing transboundary conservation and fostering continued peace.

Border Security and Visitor Access: A Collaborative Resolution

Border security may be simply defined as "the effort to restrict territorial access."[11] Initiatives to restrict border access, therefore, can be classified according to what type of access is being restricted. Broadly speaking, borders may be classified as military, economic, or police (or a combination thereof).[12] The purpose of economic borders, for instance, is to collect revenue and taxes related to interstate commerce and to protect domestic producers.[13] Military and police borders, on the other hand, are intended to deter interstate military threats, illegal immigration, and access by individuals who might threaten national security.[14] While most borders could certainly be characterized in terms of these three categories, peace parks are the manifestation of a different type of border—namely a conservation border. Prior to September 11, the border between Waterton Lakes and Glacier was the epitome of a conservation border—one of the guiding principles of the regime was that the presence of the border was minimized to the extent that conservation initiatives could be effectively implemented at an international level while simultaneously offering liberal access to transboundary visitors.

Within the Waterton–Glacier Peace Park there are two designated border crossings: the Goat Haunt Ranger Station located at the southern end of Waterton Lake and the Chief Mountain Border crossing located near the eastern boundary of the Peace Park (see figure 10.1). Unlike the Chief Mountain border crossing, which is a "frontcountry" crossing with vehicle access, the Goat Haunt Ranger Station in Glacier—as a "backcountry" crossing—was not equipped with the technology and infrastructure (computers, satellite equipment, metal detectors, etc.) required for detailed monitoring of transboundary visitation prior to September 11, 2001.[15] As a result any visitor, upon crossing the international border at Goat Haunt, was not required to immediately register with US or Canadian customs agencies or pass through any security screening. If, in the park the visitor was traveling to, they were considered a foreign national, they were merely instructed to register with the appropriate security agency at "an appropriate time." In effect visitors were essentially free to travel through the Goat Haunt crossing as long as they departed in the same country they entered through. This approach to border security was based on a number of assumptions:

- An overwhelming majority of those who crossed the US–Canadian border were citizens of either country as opposed to countries that posed a military/police threat to either the United States or Canada.
- Economic border security and conservation are compatible and perhaps symbiotic (e.g., cross-border access in Waterton–Glacier leads to increased tourism revenue for both countries).
- The backcountry border between Waterton and Glacier is a very rugged and inopportune location for illegal immigration (less strenuous crossings could be made elsewhere).

Consequently the effectiveness of the regime with respect to visitor access management was defined by allowing as much visitor access across the border as legally possible and maintaining the environmental integrity of both parks.

Following the events of September 11, however, the United States reprioritized its border policy in such a way that the military/police dimensions of security were emphasized to a greater degree.[16] The US increased interest in tracking and identifying who was crossing the border led to uncertainty as to whether or not the border could be secured, particularly in the backcountry, while continuing to permit liberal transboundary visitor access also increased. Although no substantial changes were recommended for the Chief Mountain crossing, the US government initially proposed that the Goat Haunt crossing be closed unless the US National Park Service invested in the infrastructure needed to operate a more secure crossing. Developing the necessary infrastructure would have most likely required road construction, building and maintaining electricity power lines, increasing surveillance along the border, and investing in expensive equipment to monitor transboundary visitation. These changes would have been inconsistent with the conservation and wilderness goals of the peace park.

Conversely, closing the backcountry border would have affected nearly 35,000 visitors annually and resulted in substantial financial losses for both parks. Therefore the two parks, Shoreline Cruises (a concessionaire operating boat tours on Waterton Lake), and Canadian and US security agencies—including US Customs and Immigration, US Border Patrol, and the Royal Canadian Mounted Police—collaborated in an effort to permit transboundary visitor access without investing in additional infrastructure.[17]

The collaborative efforts of the various agencies culminated in an operational agreement that respected the statutory obligations of all agencies and permitted continued access for Canadian and US citizens across the border into each respective park (albeit under greater scrutiny). Citizens of other countries, however, who comprise roughly 15 percent (5,250 annually) of backcountry visitor crossings, would no longer be able to cross through Goat Haunt into Glacier without first going through a more secure crossing such as Chief Mountain. One exception to this is when the visitor is part of a boat tour on Waterton Lake, which crosses the international boundary. In this case all visitors are allowed to dock at the Goat Haunt Ranger Station but are not permitted to leave the premises.

On the US side of the border there would also be an increased presence of border security agencies immediately outside the boundary of Glacier National Park and overhead. While the border security agencies do not routinely walk the border within the peace park, there has been a substantial increase in the use of sensor and surveillance equipment and monitoring flights. In general, park rangers report that visitors have been very cognizant of the increased border security, but that presence is not something that border security agencies (particularly in the United States) are interested in concealing. If anything, the presence is a deterrent to illegal crossings, and there may be an interest in increasing the visibility.

Despite the increased presence of United States border security agencies, the peace park continues to effectively function in much the same way it has since 1932. For the most part, border security initiatives have had little or no impact on many of the functional areas of cooperation. Arguably, the only significant change has been that visitors who are neither US nor Canadian citizens must take an extra preparatory step before crossing the border along the backcountry into the United States (i.e., visiting a more secure border crossing station such as Chief Mountain). In fact the post–September 11, 2001, cooperative efforts have perhaps strengthened the relationship between the two parks by forcing them to work through a difficult issue involving a number of agencies and interests. The successful transition to the contemporary security environment has been premised on transboundary and interagency collaboration that acknowledges if the peace park regime is to maintain its resilience, competing interests must be harmonized. Fundamentally this requires an understanding of how each agency involved defines success and what conditions are necessary for each agency to achieve their objectives.

Lessons for Collaboration in Peace Park Regimes

The basic components of any international issue are power and interests.[18] As a result resolving international issues necessarily involves negotiating agreements (both formal and informal) that satisfy interests and/or wielding power to force a state to behave in a particular way. In the context of peace park regimes, where the goal is to achieve a common interest—peace and transboundary conservation—it is both unrealistic and counterproductive to expect that cooperative efforts could be coerced either through intimidation or sanctions.[19] Perhaps the most important mechanism, then, for resolving conflicts or reconciling competing interests in peace park regimes is *collaboration*—the collective integration of interests throughout decision-making processes.[20] The efforts of the Waterton–Glacier regime to reconcile conservation with border security provide three valuable insights into how collaboration may be employed in peace park regimes.

First, before border security initiatives were implemented, all of the parties involved—the parks, border security agencies, and their respective governments—collaboratively assessed the impact that their particular interests would have upon one another. As Adler and Haas note, "between international structures and human volition lies interpretation. Before choices involving cooperation can be made, circumstances must be assessed and interests identified."[21] By proactively engaging in collaboration before the issue of border security became entrenched in conflict, the parks, agencies, and national governments developed a relationship grounded in mutual trust and a common understanding of interests. In the absence of those relationships and a common knowledge base, legitimate and transparent outcomes would have been impossible.

One of the most critical components of any collaborative process is the identification of an objective framework by which outcomes can be evaluated.[22] Peace parks exist as a regime within a framework that predetermines the acceptability of collaborative outcomes and serves as a focal point for the decision-making process. For instance, in Waterton–Glacier a collaborative outcome is acceptable to the extent that functional cooperation in areas such as transboundary visitor access or cooperative wildfire management remain effective. Having a thorough understanding of the regime's guiding principles and cooperative framework allowed Waterton–Glacier to define a set of acceptable outcomes prior to negotiating an agreement.

Finally, while Waterton–Glacier has resolved many of the issues that have resulted from the US increased interest in border security, the resilience of the peace park must be constantly attended to and not taken for granted. In February 2005 the US House of Representatives passed a bill that would exempt the Department of Homeland Security from having to comply with any environmental laws when constructing barriers and roads for the purposes of border security.[23] While it is unlikely that a barrier will be constructed along the US–Canadian border, should the bill become law, the Department of Homeland Security would theoretically be permitted to construct roads in the US portion of the peace park (including the backcountry areas) without consulting either Glacier NP or Waterton Lakes NP. It seems unlikely, though, that a road can be built along the border in the peace park—the rugged terrain presents an immediate and significant deterrent in and of itself. But, more important, the peace park and border security agencies have developed a relationship characterized by trust, and in the absence of a substantial normative change, the peace park regime will likely continue to guide activities along the border.

Conclusion

Waterton–Glacier, as the world's first peace park, can provide governments and park managers around the world with the most complete historical account of how regime resiliency evolves and is maintained. Granted, unlike many peace parks throughout the world, the Waterton–Glacier is shared by two countries with a long history of cooperation that can be traced back to long before its formation. As such, it may not experience the degree of conflict as evident in some areas of southern Africa or Southeast Asia. Nevertheless, even for countries in a state of Kantian peace, the effectiveness and resilience of cooperative efforts must be continually attended to when issues such as border security are a concern.

Peace parks will continue to be an important tool for achieving peace and practicing cross-border conservation, but they are a manifestation of only a subset of international interests among many others that are often conflicting or competing. As

Wolmer notes, "it seems wistful thinking that [peace parks] are likely to be anything other than a very low priority for governments in actual conflict situations."[24] It follows that the resilience of a peace park regime—even one shared among peaceful countries—may be largely dependent on its ability to collaboratively engage stakeholders who have interests that may run counter to or compete with those of the peace park. Peace parks may therefore be better equipped to effectively serve as instruments for conflict resolution and international conservation in a dynamic political environment.

Notes

1. Rotary International is a worldwide organization of business and professional leaders whose mission is to provide humanitarian service, encourage high ethical standards in all vocations, and help build goodwill and peace among the world's peoples.

2. UNESCO (2005).

3. UNESCO (2005).

4. Krasner (1983: 1).

5. Underdal (1992: 227–40).

6. Haggard and Simmons (1987: 491–517).

7. Pedynowski (2003: 1261–69).

8. Danby (1997: 1–14) and Fall (2003: 81–103).

9. Wolmer (2004: 137–47), Wolmer (2003: 261–79), Fall (2003: 81–103), and Zbicz (2003: 19–34).

10. Hasenclaver, Mayer, and Rittberger (1996: 177–228).

11. Andreas (2003: 78–111).

12. Ibid.

13. Ibid.

14. Ibid.

15. Frontcountry areas in the peace park are those areas that are adjacent to major roads or human-made infrastructure. Backcountry areas, on the other hand, are those areas in the peace park that are a considerable distance from roads or other human-made infrastructure.

16. For a detailed discussion of the US post–September 11th border policy, see Meyers (2003).

17. The boat tour operated by Shoreline Cruises begins in the village of Waterton in Waterton Lakes National Park and stops at the Goat Haunt Ranger Station in Glacier National Park.

18. Hasenclaver, Mayer, and Rittberger (1996: 177–228).

19. Brown and Jacobson (1999: 16–20, 37–45) and Tanner (2004: 67).
20. McKinney and Harmon (2004: 26).
21. Adler and Haas (1992: 367).
22. Conley and Moote (2003: 371–86).
23. United States House of Representatives 2005, §102 H.R. 418. The bill is currently being considered in the United States Senate.
24. Wolmer (2004: 142).

III

Peace Parks and Regional Governance Regimes:
Redefining Security and Realism

If warfare has, in fact, been based on rational behavior for much of human history, then deciding that warfare is bad and should be stopped will not solve our problem.... If we have reached a point at which we can live within Earth's carrying capacity, we can eliminate warfare in the same way we can eliminate infectious diseases: not perfectly, not immediately, but slowly and surely.
—Steven Le Blanc, *Constant Battles*, 2003

I think the environment should be put in the category of our national security. Defense of our resources is just as important as defense abroad. Otherwise, what is there to defend?
—Robert Redford, 1998

11

Bridging Conservation across *La Frontera*: An Unfinished Agenda for Peace Parks along the US–Mexico Divide

Belinda Sifford and Charles Chester

The 2,000 mile US–Mexican border—*La Frontera*—is hardly a line in the sand. Defined by diverse cultures and languages, it embodies an ancient human footprint of conquests and losses, of mass migration south and north, and more recently, of legions of patrolling security forces. With the human dimension of *La Frontera* demanding so much attention, the landscape itself rarely comes to mind. This is particularly true in terrain that is vast, arid, and seemingly monotonous—a desertscape that many believe to be a lifeless wasteland, "the big empty."

Contrary to such common sentiment, a dedicated cadre of scientists and conservationists see the desert as a source of enormous natural wealth fundamental to ancient and modern agriculture. As desert scholar Gary Paul Nabhan has noted, the dry tropics are "the most probable wellspring of seed agriculture...the funding gene pools of most food crops."[1] The two major North American deserts sweeping across the border—the Sonoran and the Chihuahuan—speak to the vitality of desert ecosystems.[2] Recognizing the value of such landscapes, both Mexico and the United States have formally designated protected areas in each desert.

International conservation efforts in these two deserts also extend to borderland rivers, albeit in piecemeal fashion. Limited base water resources as well as severe droughts have led to international accords and commissions reaching back to 1884, yet even longstanding agreements over water have often failed to quell suspicion, political vitriol, and lawsuits over both water shortages and water degradation. As recently as March 2005, for example, after several contentious years, Mexico and the United States resolved Mexico's water debt under the 1944 Utilization of Waters bilateral treaty. The settlement garnered positive media coverage, prompting suggestions to use this cooperative mood to resolve other longstanding issues—notably the growing numbers of undocumented Mexican laborers in the United States.[3] The subsequent months, however, saw no cooperative political momentum. To the contrary, the US construction of steel fences along various stretches of its border with Mexico exacerbated the political tensions between the two countries. This scenario, so dramatically different from US–Canadian relations to the north, highlights the need for a peace initiative of some kind on the US–Mexico border.

Unbeknown to many, since the 1930s there have been several public and private efforts emanating from both Mexico and the United States to create transborder protected landscapes. Prompted by the peace spirit engendered by Rotary International in 1931 that resulted in the creation of Glacier–Waterton International Peace Park on the US–Canadian border (see chapter 10 in this volume), it seemed fitting that a complimentary park should exist on the US–Mexico border. Despite this sentiment no park materialized then or later, largely due to different governmental priorities, political contretemps, and cultural missteps. Yet decades of lower level cooperative endeavors raise the question of whether an international park initiative is necessary when smaller projects are helping land management efforts on both sides of the border. A second question concerns the viability of an international park in the context of government policies so dominated by issues of border security in a post 9/11 world—namely the illicit drug trade and illegal immigration. In short, does laboring for a peace park constitute a naïve waste of resources or a window of opportunity for both countries?

A comparative look into the Chihuahuan and Sonoran Desert initiatives over the past seventy years provides clues to the answers to these questions—as well as a lens into the future. Although the outcome of future intergovernmental cooperation is never certain, a joint initiative would likely result in both improved land stewardship and long-lasting symbolic ties between the two countries. With effort and imagination, the two countries can create enduring transborder protected landscapes along *La Frontera*.

The Chihuahua Desert: Protected Lands across a River Grande and Bravo

The 500,000 square kilometer Chihuahuan desert lacks widespread name recognition in the United States. Lying mainly in Mexico, with the northern third extending into southwestern Texas, southern New Mexico, and southeast Arizona, its distance from major population centers keeps the Chihuahuan Desert a secret. One localized exception is Big Bend National Park (BBNP), which is 90 percent Chihuahuan desertscape (the Chisos Mountains and Rio Grande habitats comprising the other ecosystems within the park).[4] Given the ecotones between these ecosystems and elevations ranging from 1,350 to 7,825 feet, BBNP contains a vast array of plant and animal species, including the greatest diversity of bat, cactus, and bird species of any US national park—as well as recently reintroduced black bears.[5] Although the Chisos Mountains constitute BBNP's most noticeable feature, the area is named for the Rio Grande (called the *Rio Bravo* in Mexico[6]), which "bends" north forming the distinctive curve of southern Texas as well as the southern boundary of BBNP. Equally as impressive as the mountains are the Mariscal, Elena and Boquillas canyons of the Rio Grande.

Along with BBNP, other US protected areas in the region include Big Bend Ranch State Park (established 1988), Black Gap Wildlife Management Area (established 1948 by the state of Texas), and 196 miles of the Rio Grande Wild and Scenic River (designated 1978). Combined with two protected areas that the Mexican government created in 1994 along the Rio Grande's canyons, *Cañon de Santa Elena* and *Maderas del Carmen*, the total land protected nearly 1,000,000 hectares. In comparison, Yellowstone National Park contains 898,321 hectares and the international Waterton–Glacier International Peace Park on the Canadian–US border covers 462,799 hectares.

The Rio Grande has largely defined the region's cultural history. Crude tools crafted by ancient indigenous peoples drawn to the water date back 10,000 to 15,000 years. Recent centuries have seen a parade of other peoples pass through the region, including the Hispanos, Comanches, Apaches, Mexicans, Seminole blacks, Kickapoo—and finally the Texans, US military, and various wanderers who stayed to homestead or ranch or just continued on.

BBNP evolved in the early 1930s as cash-strapped Texas considered the establishment of a national park as a potential revenue stream. Big Bend country seemed like the logical choice. The area was remote and thus relatively undisturbed by human contact. Although not the traditional coniferous landscape of most other national parks of the time, Big Bend's vistas were wild, and advocates for a park viewed this area on the Rio Grande as a significant bequest to the country's national park heritage. Enlightened NPS representatives even saw such a park as a vehicle for building stronger "sentiments...between the Mexican and American peoples."[7] Indeed, in 1934 US representatives approached newly elected Mexican President Lazaro Cardenas with the idea of an international park that would exemplify US President Roosevelt's Good Neighbor policy and Cardenas's natural resource conservation agenda. With naturalists desiring land and species protection, business interests envisioning fanciful resorts on both sides of the border, and "native sons" seeking national and international recognition for their first federal park, forces aligned to convince Congress to authorize Big Bend National Park (BBNP) in 1935.[8] Authorization, however, meant only permission to create a park; land first had to be acquired through donation or purchase to form a contiguous territory of significant size and resource distinction.[9] With land titles cloudy and owners unwilling to sell, years of negotiation lay ahead before the park was officially created in 1944.[10]

In the meantime during these closing years of the Mexican Revolution,[11] the Mexican government had turned its attention to strictly domestic matters—primarily the tasks of building an economic base and initiating agrarian reform. Fundamental to Cardenas's agenda had been the goal of land redistribution, an issue that would take on significant consequence in later years regarding land degradation and protected landscape management. Cardenas granted land to peasants through the *ejido*, a collective management system of typically small settlements that were divided into communal and individual parcels. *Ejido* members (*ejidatarios*) were given a right to work particular plots and pass on that right, but not to purchase or sell *ejido* property. The federal government retained absolute ownership.[12]

Unfortunately, the majority of *ejido* lands designated by Cardenas were marginally productive and typically far from population centers (and the associated electricity and roads). In addition *ejido* lands in the desert north often lacked water or water rights.[13]

Cardenas had nonetheless officially promoted conservation, arguing that if natural resources were destroyed, the agrarian economy would flounder. He created the first centralized environmental agency during his presidency, and added 40 national parks to the two that existed.[14] However, conservation was but one of many competing and often contradictory demands voiced by the Mexican populace. Expropriating land for a park that entailed high infrastructure costs and no immediate economic return never became a priority for Cardenas or subsequent administrations, especially when the enterprise would be located in geographic proximity to looming US domination.

Despite the lack of movement toward an international park, the United States and Mexico conducted cooperative conservation activities over the next decade that would strengthen borderland bonds. A few examples demonstrate the breadth of this activity. In the mid-1930s representatives of both governments traveled by horse and boat to inspect at close range the natural features of the Chihuahuan Desert.[15]

In 1944 the Utilization of Waters Treaty was signed establishing water exchange flows for various rivers, including the Colorado, Rio Grande, and *Los Conchos*. During World War II a presidential exchange of letters supported an international park as conducive to strengthening the ties between the two countries. And as the war concluded, Mexico cooperated with the US Fish and Wildlife Department and other public and private entities to conduct the first ecological survey of the Sierra del Carmen in order to "gain a more satisfactory insight into the relationships of the natural resources on both sides of the international boundary."[16]

Postwar attention on economic growth relegated conservation policy, including the international park initiative, to a secondary position. With few restraints, Mexico's industrialization efforts depleted natural resources at a rapid pace.[17] In Maderas del Carmen, for example, the richly diverse coniferous forests were heavily cut and the lower elevation land strip-mined. Although the United States likewise exploited its natural resources for industrial growth, its economic strength and concomitant infrastructure allowed the government to promote National Parks for recreational tourism as a natural and cultural resource.[18] With a large middle class, automobile culture booming, and an entrepreneurial NPS director at the helm, the public flocked to the parks in droves.[19]

It was thus that success in the NPS came to be judged in terms of visitation numbers rather than resource stewardship. Nevertheless, park enthusiasts once again looked to Mexico, envisioning an international park as another "can't lose" opportunity. Rotary International, for example, continued to promote peace through parks, as did certain local and national park employees. This included the BBNP superintendent of the time, who spoke to the Rotary Club in Saltillo Mexico 1954 to discuss this plan.[20] Notably the Five State Good Neighbor Council passed international park resolutions in 1954 and 1956.[21] But other interests ruled the day, including US citizens with ranches in Coahuila and Chihuahua who feared the Mexican government would expropriate their land to create a park. Moreover the US State Department resented the ad hoc international efforts of the Park Service and private groups. Ultimately this tension resulted in the 1962 establishment of the Office of International Affairs within the NPS, but not an international park.

As ecosystem science emerged in environmental policy debates during the early 1960s, environmental groups and concerned citizens became more vocal about environmental degradation. And while politicians passed major environmental legislation in both countries during the late 1960s and early 1970s,[22] the NPS was slowly changing its traditional notions about science-infused management policies, diverse landscape values, and the consequences of no-holds-barred visitation. Particularly relevant to the borderlands was increased attention to the influence of land use practices outside a park, whereby park managers came to recognize regional ecosystems as at least deserving of consideration—if not primary attention.[23] This change in thinking was reflected in the designation of BBNP as a biosphere reserve in 1976.[24]

During the 1980s BBNP's managers reassessed the park's historical relationship with Mexico.[25] Rather than continue to ignore the Mexican villages that border BBNP—Santa Elena, San Vicente and Boquillas del Carmen—the park administration began to view them as part of cultural resource base of the area, and to recognize that Mexican land was part of the greater ecoregion. Superintendent Gil Lusk, for example, arranged meetings with Mexican state officials and local villagers to learn more about the region's *ejidos* and *colonias*,[26] where communities typically lacked electricity and running water in all homes, and even paved roads. Park management encouraged employees to speak Spanish and visitors to hire Mexican villagers to row them across the Rio Grande for a meal or beverage in Mexico. The "Good Neighbor Fiesta" became an annual event for park personnel and Mexican neighbors. As resource planning benefited from improved relations with Mexican officials, the NPS regional office encouraged cooperative attitudes in selecting its own staff.[27]

In 1991 the NPS director and the Mexican cabinet secretary of SEDUE[28] attempted to obtain funding for an international park. Their failure was but one indicator that all was not idyllic in the region's transborder relations, and tensions flared over drug trafficking and illegal migration across the border.

Rio Grande water quality was another source of friction. Both mercury runoff from abandoned mines and fecal bacteria from livestock made recreational use under the US Wild and Scenic River program problematic. Furthermore it was plain to BBNP superintendents that managing half a river—namely to the midway point in the Rio Grande—was meaningless. And from a broader perspective, public eyes in both countries had turned to the borderlands largely due to debate over the North American Free Trade Agreement (NAFTA). Both NAFTA and the supplemental North American Agreement on Environmental Cooperation (NAAEC) emerged from protracted and contentious negotiations, with environmental topics being particularly polarizing.[29]

Ultimately, when NAFTA took effect January 1, 1994, the United States, Canada, and Mexico had committed themselves to free trade with at least some minimal environmental safeguards. That same year Mexico established two "Areas de Protección de Flora y Fauna" adjacent to BBNP: *Maderas del Carmen* and *Cañon de Santa Elena*. Each was of significant size and natural resource richness.[30] While NPS officials greeted this news enthusiastically, they recognized that Mexico's protected lands posed distinct challenges. Approximately 80 to 85 percent of the two protected areas were (and remain) in private hands.[31]

Although the "protected area" designation recognized certain lands for natural distinction and mandates conservation with limited natural resource extraction, government enforcement of the "limited" resource extraction proviso has been weak—particularly in light of the far north location of these areas as well as the region's sparse population and poor roads. Private landholders have thus been entrusted with the responsibility—perhaps better described as a choice—of conserving the land or not.

In Maderas del Carmen the private sector appears to be implementing strong stewardship efforts largely due to CEMEX, a leading producer of cement, ready mix concrete, and aggregates that are marketed worldwide. With a record of supporting environmental initiatives as part of both its internal operations and corporate social responsibility strategy,[32] CEMEX turned its attention to conservation of the Chihuahuan Desert in 1992. Formally recognizing the wealth of the ecosystem, the extent of degradation, and the lack of governmental protection, the corporation began purchasing land in the state of Coahuila under the advice of Agrupación Sierra Madre (a Mexican conservation NGO), and with local rancher support.

CEMEX now owns over 75,000 hectares of land within or adjacent to the Maderas del Carmen Protected Area and has entered into long-term conservation agreements with adjacent private landowners for an additional 25,000 hectares. The company has a fulltime manager and staff for its El Carmen lands, and a number of initiatives underway such as reintroduction of desert big horn sheep, habitat restoration, and a long-term study of black bears.[33]

A 1992 change to the agrarian land code could also have broad impact in the region. *Ejidos* were given the right to "dis-incorporate" as communal entities, thus permitting individual ownership of *ejido* land. However, with many *ejidatarios* living far from the land (often in the United States), individuals having passed away, and poor land records kept, ascertaining title and rights is complicated. CEMEX has the financial and legal resources to purchase *ejido* land, but this may not be the best social or political solution. Mexican federal, state, and local officials are working with *ejidotarios* and private landowners, including CEMEX, to create viable land resource models and bring sustainable jobs to the desert north.[34]

Although the Mexican government has purchased land in neither the Maderas del Carmen nor Santa Elena Cañon Protected Areas, over time the region's status as a protected area has given clear authority for various federal and state agencies to work together on managing border landscapes—an authority that has subsequently legitimized the peace park dialogue. Mexican officials, for example, traveled to Waterton–Glacier at the invitation of BBNP's and Glacier's superintendents in 1996 for a firsthand view of cooperative practices. Subsequently Secretary of the Department of Interior Bruce Babbit signed a Letter of Intent for Adjacent Protected Areas (LOI) with his Mexican counterpart, Julia Carabias of SEMARNAT.[35] While expressly recognizing the sovereignty of the two countries, the LOI created pilot projects "in the conservation of contiguous natural protected areas" in the border zones of the northern Chihuahuan Desert and the Western Sonoran Desert.[36]

Following the signing of the LOI, Big Bend Superintendent Jose Cisneros organized a 1998 meeting with over 60 participants to jumpstart an international park effort. Enthusiasm ran high, but the backing, funding, and congressional legislation were still not forthcoming.[37] Regardless, shared natural resource initiatives continue even in a time of increased attention to border security. For example, the multi-agency BRAVO project published in 2004 combined public and private resources in both countries to inventory the sources of visibility impairment in BBNP. Although Mexico chose not to participate in the final study, the joint effort in the preliminary phase provided a richness of scientific data for both countries.[38]

The Sonoran Desert: Cooperation across the Dry Borders of El Gran Despoblado

With an approximate total land area of 310,300 km^2, the Sonoran Desert straddles the US/Mexico border in an inverted "U" around the Gulf of California. In Mexico, the Sonoran Desert covers most of the Mexican states of Sonora, Baja California, and Baja California Sur, while in the United States it covers most of southwestern Arizona with significant portions in southeastern Arizona and southeastern California.[39]

While the region's indigenous O'odham people—formerly called the Papago and Pima—had occupied the region for centuries, Mexican and "Anglo" colonization began through a network of missionaries in the eastern portion of the Sonoran

Desert in the late seventeenth century. Further settlement came soon after the establishment of "the devil's highway," a treacherous route largely established to accommodate the wave of gold rushers from the Mexican frontier to California.[40] But by the 1930s, neither colonizing culture had achieved any significant utilization of the desert's natural resources; the Cardenas administration's *ejidos* were remote and hardscrabble, and few white settlers in the region found success in wresting economic success out of the harsh landscape. A result of the desert's inhospitable environment was the US allowance of granting a sizable reservation for the O'odham. Established in 1916, the Tohono O'odham Nation comprises the second largest tribal reservation in the United Stated at 1,122,815 hectares. Notably, the Sonoran Desert is the only place where three nations—US, Mexican, and tribal—meet together along the entire US–Mexican border.[41]

Despite the size of the O'odham reservation, federal usurpation and control over much of the landscape outside the O'odham Reservation created conducive conditions for the establishment of relatively large protected areas on the US side of the border. Two of the earliest protected areas in the region were created largely out of concern over dwindling game species, particularly desert bighorn sheep, and the protection of cactus species unique to the border. On April 13, 1937, President Franklin D. Roosevelt established Organ Pipe Cactus National Monument (ORPI).[42] At 133,825 hectares, ORPI has 141.6 kilometers of international borders including its boundary with the Tohono O'odham Reservation. Two years later, Roosevelt established the Cabeza Prieta Game Range to the north and west of the national monument. In 1976 the Range was redesignated as a National Wildlife Refuge (NWR).[43] Sharing a 56 mile international border with Mexico, Cabeza Prieta is the third largest NWR in the lower 48 states at 348,034 hectares.[44]

Efforts to integrate ORPI and Cabeza Prieta date back at least to the early 1960s when the NPS proposed tying the two together as the "Sonoran Desert National Park."[45] Credit for that idea is generally attributed to former Secretary of the Interior Stewart Udall (1961–1969). Conservation organizations such as the Sierra Club and influential decision-makers such as Secretary Udall's brother, Congressman Morris Udall (D-AZ), soon took up the banner for such a park.[46] Yet former Secretary Udall would later recall that the idea "didn't receive a lot of publicity" and that he could not convince President Lyndon B. Johnson to do it.[47] Although Johnson ignored Udall's proposal, the seed of an idea had been planted.

Significantly, the proposal issued by the NPS stated that "...it can only be concluded that the entire area is eminently qualified for National—if not

International—Park status."[48] But for many of the same reasons relevant to the Chihuahuan Desert, the establishment of protected areas on the Mexican side of the border came much slower than in the United States. Consequently the idea of an International Peace Park found little traction in Mexico. Notably Secretary Udall recalled that at the time he was thinking of a Sonoran Desert National Park during the 1960s, the idea of an international park "was kind of a dream at that point. We didn't get down to practical cases."[49] Exequiel Ezcurra, a prominent Mexican biologist and high-ranking government official, further noted that the Mexican federal government "was at that time opposed to decreeing protected areas along the Mexico–US border.... Mexico perceived that the setup of national parks along the Mexico–US border, like Big Bend or Organ Pipe, were really things that the US did to define its boundaries and territories and to have control of the border—to keep the border under some sort of governmental control."

Despite such concerns, as early as 1943 the Mexican government had investigated the possibility of establishing a game refuge in the Pinacate and had collaborated with the United States in a small, research-oriented transborder conservation initiative.[50] Compared to the Big Bend region, however, little official transborder activity appears to have occurred in the region during the ensuing decades. It was only after sporadic attempts at transborder cooperative initiatives during the 1970s[51] that scientists and conservationists came to the fore in terms of advocating for international cooperation. In 1980 the Centro Ecológico de Sonora with support from the Nature Conservancy began actively investigating the possibility of biosphere reserve status for the Pinacate.

Yet despite a number of investigations through the rest of that decade, the effort to designate the Pinacate as a biosphere reserve stalled until the 1988 "Symposium on the Pinacate Ecological Area." Although few realized it at the time, this research conference would generate considerable momentum toward international cooperation in the region as well as the establishment of biosphere reserves in Mexico.

Significantly, the Symposium not only brought together scientists and conservationists from both sides of the border, but was the first time that the O'odham had been included in such a transborder forum.[52] Representatives from one group of the O'odham, the Hia Ced O'odham, expressed strong concern that the designation of the Pinacate region as an international biosphere reserve would be a significant violation of their rights.[53] Coming out of that 1988 conference, participants agreed that there was a need for "a larger public forum...to promote dialogue among residents of the Sonoran Desert."⁵⁴ Although it took several years to organize, a well-attended "Land Use Forum" took place in 1992, which in turn sparked the genesis of the International Sonoran Desert Alliance (ISDA).⁵⁵

The formation of ISDA played a critical role in the 1993 designation of *Reserva de la Biosfera El Pinacate y Gran Desierto de Altar*, situated across the border to the south of ORPI and CPNWR. Along with the concurrently designated *Reserva de la Biosfera Alto Golfo de California y Delta del Rio Colorado*, the two biosphere reserves cover over 1.6 million hectares.⁵⁶ Beyond the benefit of elevating a desert landscape as worthy for protection, the biosphere designation enabled Mexico to adhere to a land conservation plan different from that of the NPS, most notable for the 67 *ejidos* contained within the biosphere reserve (although only 20 of them are "minimally inhabited").⁵⁷ At least in theory, the biosphere reserve included community sustainable development as part of conservation strategy.⁵⁸

As a tri-national grassroots network, ISDA focused on a wide range of objectives ranging from improved border crossings to better access to health care. Although much of ISDA's agenda was focused on such social issues, its leadership and membership became intricately entwined with an effort to establish an International Sonoran Desert Biosphere Reserve.⁵⁹ Largely due to an anti-internationalist agenda within the United States that was aimed at US cooperation in international cooperative programs—particularly, the Man and the Biosphere Program and the World Heritage Areas Program—the effort died a sudden political death in 1996, after which ISDA turned to other activities.⁶⁰ Yet despite the disappearance of the biosphere reserve proposal, the energy spawned by the effort generated two somewhat disjointed initiatives, one within the government and one in the conservation community.

Within the government, high-level government officials were still interested in working across the border. In November 1996, Arizona Governor Fife Symington and Sonora Governor Manlio Fabio Beltrones signed a Memorandum of Understanding (MOU) announcing a joint endorsement of a "Binational Network of Sonoran Desert Biosphere Reserves."⁶¹ Incorporating most of the public lands in the Western Sonoran Desert, the network was created to protect cultural values, promote sustainable community and economic development in the region, and to promote "cooperation between the contiguous protected areas on both sides of the border so as to motivate collaborative resource management of the region's shared resources."⁶² A few months later, Babbitt and Carabias signed the previously mentioned LOI in Mexico City. Although not under a biosphere reserve agreement, both federal governments were now at least informally working together in the Sonoran Desert under the nominal title of "sister areas."[63]

Outside the government, a handful of activists continued to work on an international land designation—but at this point reverting from the "international biosphere reserve" concept back to the original concept of an "International Sonoran Desert Peace Park." Prominent among these activists was Bill Broyles, a retired teacher from the Tucson public schools who was the principal catalyst in establishing the organization Sonoran Desert National Park Friends (SDNPF).[64]

In promoting the idea of a peace park, Broyles and his colleagues decided first to focus on the US side of the border—not hesitating to make a critical juxtaposition between the "fragmented management" on the US side of the border against the administratively unified management on the Mexican side. In 1999 and 2000 SDNPF ran a publicity campaign that generated substantial press attention.[65] SDNPF published a "citizen's proposal" for a Sonoran Desert National Park and Preserve, a large format full-color pamphlet with photographs by Pulitzer Prize winning photographer Jack Dykinga and text by the noted author Charles Bowden.[66] The proposal not only harkened back to Udall's longstanding idea of conjoining ORPI and CPNWR but also added much of the Goldwater Range to the proposal. The proposal also noted that Air Force required only 6 percent of the Goldwater range. Although military activities could still be undertaken in an expanded park, under NPS management "the heart of the Sonoran Desert would be preserved and the responsibility for this preservation would fall into dedicated and able hands."[67] Thus the full proposal actually promoted a "Sonoran Desert National Park *and Preserve*"—the latter being the name the National Park Service uses to designate areas still open to hunting and certain military uses.

A 1999 opinion poll, found that Arizonans supported the park idea at an "astounding" rate of 84 percent, with 9 percent dissenting and 7 percent of no opinion.[68] Yet despite the idea's apparent popularity, no official movement occurred. Some positive conservation news occurred, however, with SDNPF's involvement in promoting the designation of two large BLM tracts of land to the northeast of ORPI as a "Sonoran Desert National Monument." This effort successfully culminated on January 17, 2001, when, three days before the end of his tenure, President William Clinton designated the Monument (along with the establishment or expansion of seven others).[69]

In 2003 the Arizona State Senate's Natural Resources and Transportation Committee discussed a "memorial" bill "urging the United States Congress and the Department of the Interior to take the necessary steps to establish the Sonoran Desert Peace Park."[70] Its sponsors noted that the redesignation "will be cost free for the state and federal governments," and both Broyles and a representative from the Arizona Tourism Alliance testified in favor of the bill. Notably, a representative from the city of Glendale testified in opposition to the bill, arguing that it could affect the military's ability to use the Goldwater Range.[71] Either lack of consensus or inertia prevented even a committee vote on the bill.[72]

Broyles points to three primary reasons why, despite the momentum of the 1990s, there is no Sonoran Desert Peace Park: first, chronic underfunding from the federal government for conservation, second, the diversion of funds by the Homeland Security Department away from border conservation and cooperation; and third, the increase in illegal immigration and drug smuggling on this stretch of the border. For these reasons, the campaign has at least temporarily lowered its sights to focus on redesignating ORPI alone from national monument status to national park status (a strategy based on the fact that "parks generally receive higher budgets and stronger support from the public and Congress"[73]).

Because concerns over the international border are unlikely to change significantly in the near future, other approaches will be required if the United States and Mexico are ever to achieve transborder cooperation. As we argue below, such strategies revolve around the general theme of first understanding the forces aligned against international designations, and then using those contrary forces to one's advantage.

Conclusion: Building Bridges over Homeland Security fences

Land managers in the Sonoran and Chihuahuan deserts face complex yet distinct challenges. On the Sonoran border, increased drug trafficking and illegal immigration—along with the potential risk of terrorist infiltration—has pushed the region to the top of the US border concerns. Where once a handful of border patrol agents operated, 350 are currently assigned, with 150 more available on short notice, and 2,000 new agents authorized under December 2004 legislation.[74] Be it on foot, ATV, or a Border Patrol Chevy Suburban, all of this human movement has been harsh on the landscape, diverting Park Service and Fish and Wildlife staff toward spending more of their time educating the Border Patrol on the effects of its activities. The Border Patrol officers are not unsympathetic, especially when they see the widespread debris and newly etched desert roads left by illegal trafficking. Their job, however, is to secure the border under the mandates of the Homeland Security Act.

The high drama in the Sonoran Desert has an understated but compelling counterpart in the Big Bend region, where its remoteness makes it a less desirable area for smuggling and illegal entry. For several decades prior to 9/11, visitors to Big Bend were able to take a rowboat ride across the Rio Grande to have a meal or a beer in a Mexican village. The village economy in fact depended on this informal arrangement. Likewise Mexican villagers routinely rowed across the river to buy food supplies in Big Bend, a courtesy that saved hours of driving time over dirt roads to the nearest Mexican outpost. Since 9/11, crossing in either direction has been prohibited, with violators subject to arrest and fines. Not surprisingly, the current policy has had a devastating effect on village life.

The situations in both deserts are unfortunate, with local land and human needs seemingly irreconcilable with US government policy, particularly in 2006 as proposed immigration legislation includes steel fence construction along much of, if not the entire, US–Mexican border. However, the establishment of cross-border protected areas could actually help diminish current, seemingly irreconcilable, tensions while benefiting long-term cooperation in other arenas besides that of conservation. Yet this can only occur if certain conditions are met.

First, both countries must *look beyond* their own political posturing, including vituperative allegations over Mexican border violence and US maltreatment of the undocumented. This is not an empty call for these problems to simply "go away"—an unlikely prospect at best. Rather, it is to recognize that policy makers *can* go beyond grandstanding to generate international cooperation—just as they did in creating NAFTA. Whatever one thinks of NAFTA, its passage demonstrated what can be accomplished against what once were perceived as insurmountable odds.

Second, both countries must implement and publicize specific small- to medium-scale natural resource and cross-cultural initiatives on the border so that the public understands the extent and benefit of existing cooperation. For example, the NPS Mexican Affairs Office signed a Memorandum of Understanding in 2000 for continued cross-border technical exchanges and cooperation over the next five years.

Similarly joint projects that inventory bats in the Sonoran Desert and track black bears in the Chihuahuan Desert deserve attention for their global value. Also deserving of note has been the utilization of Mexican firefighters—*los diablos*—from local Mexican villages adjacent to Big Bend to assist fight fires on US protected lands. Highlighting such cooperative projects on the Web sites of BBNP and ORPI would spread awareness of such activities.

Finally, both countries must recognize that they are inextricably dependant on each other. Next to Canada, after all, Mexico is the United State's principal trading partner.[75] For better or worse (or more likely, both), NAFTA and a myriad number of cooperative enterprises are moving the two countries ever closer. The border parks should be a symbol of this interdependence, and the interdependence should occur at two levels.

First, they must involve the local economy to demonstrate the value of having land be protected. To a small degree this is happening; for example, joint projects to remove tamarisk along the Rio Grande have involved US and Mexican biologists and natural resource managers—as well as local villagers from Santa Elena who have been paid to remove a tamarisk and replant the native cottonwood tree.[76] Similarly the purchase of several sewing machines and a generator for the local Mexican villages in spring 2005 was a gesture of friendship, but also an action with significant potential to help the local economic community.[77]

Second, interdependence needs to be recognized at a higher level, whether it be through the NPS and SEMARNAT or through the NAFTA Commission on Environmental Cooperation—which has specifically promoted several on-the-ground initiatives under its program on the Conservation of Biodiversity.[78]

Certainly, in a time of constrictive budgets both countries must locate nontraditional funding opportunities and nongovernmental partners for assistance and even leadership. Such opportunities may come from universities, corporations, civic associations such as Rotary International, environmental organizations such as the Nature Conservancy and Pronatura Noreste,[79] or simply private individuals. The World Wildlife Fund, for example, has initiated a Chihuahuan ecoregion project that brings together various sectors to develop better land use practices, including sustainable goat grazing techniques. Under a different initiative the World Bank has paid staff salaries at various sites, such as the Pinacate, without which basic on-site services could not be provided.[80] And CEMEX is creating a new public-private conservation model for the region, notably exemplified by the Mexican government's designation of the El Carmen Wilderness on CEMEX-owned land. In collaboration with CONANP and national and international conservation groups, CEMEX will manage the land as the first certified wilderness on private land in Latin America.[81]

Another tentative project signals the importance both of addressing local economic needs and of involving a range of expertise in the practice of "conservation area design" in the northern deserts. Discussion has begun on creating an additional protected area or biosphere reserve in the Chihuahuan Desert, an area that would physically link the Santa Elena area to Maderas del Carmen, ideally including the land along the Rio Bravo. Currently called *El Ocampo* (the name of the town and municipal government in Coahuila that would administer the protected area), such a designation at the federal level would bring resources to the region to improve the local infrastructure and to provide jobs. University, conservation, and tourism representatives are assisting in this initiative by analyzing proposals to see how best to utilize the land through sustainable economic practices.[82] The Mexican government's attention to *Ocampo* can only benefit transborder cooperation in the future.

Overall, peace park advocates must think big and be imaginative. Cooperation can take many forms, ranging from information sharing via email to full-fledged collaborative management. Similarly supporters should not view "peace park" as the only moniker for an international protected area, particularly since Mexico and the United States have not been at war since 1847. Fortunately, alternative names such as "sister park" of "international free zone" already exist. Indeed, although the Letter of Intent has not led to any major programmatic transborder initiatives, discussions over the "sister park" idea for both deserts continued under the tenure of former Secretary of the Interior Gale Norton.[83] Thinking "bigger" might mean accepting an international park in sequential fashion—one desert at a time—or by following the lead of former Superintendent Jose Cisneros and others, promoting a green corridor (perhaps a "peace corridor") running the entire breadth and length of the Chihuahuan (or Sonoran) Desert.

Any international protected area would play an integral role in establishing a mutually supportable border management system—or at the very least could help jump-start bilateral negotiations. No matter how it occurs or where, designating an international desertscape would be a lasting popular legacy, enabling both countries to move beyond the trauma of the current border, and across the continent's immense and diverse deserts with a vision.

Notes

1. Nabhan (1989: 26–27).

2. The Great Basin and the Mojave are the other two large North American deserts.

3. Samuels (2003); phone interview, Rick Lobello, a Rotary International member, El Paso, TX, February 2005.

4. At over 324,000 hectares, BBNP is approximately the size of the US state of Rhode Island and benefits from having the far corners of northern and tropical ecosystems overlap; see the US National Park Service Big Bend National Park visitor brochure and map at ⟨http://www.nps.gov/bibe⟩ (accessed April 30, 2005).

5. See NPCA (2003), which focuses on the resources in distress. BBNP itself provides thorough information on the park's natural resources at ⟨http://www.nps.gov/bibe⟩.

6. The river has held many names, two of Spanish origin that have stuck: Rio Grande (big river) and Rio Bravo (wild/untamed river). Due to years of overdevelopment and drought, the river volume is estimated to be six times less than in earlier periods of recorded history. See BBNP video *Land of Contrasts*, 2003.

7. Others, to be sure, saw merely a chance to capitalize on stereotyped images of Mexico. See Cisneros (1999: 4) and Welsh (2002: ch. 3, 3).

8. Roth (1992: 1).

9. In contrast to other western states, Texas has never had a large acreage or percentage of public land. In 1998 it was only 1.7 percent, compared with California's 44.9 percent and Alaska's 67.9 percent; see Prescott (2003: 42).

10. For a detailed history of park creation, see Welsh (2002) or Jameson (1996).

11. The thirty years of the Mexican Revolution (1910–1940) were largely ignited by disenfranchised peasants (*campesinos*) who had been pushed off the land by large estate holders; see Krauze (1997: 239–44) and Simon (1997: 32).

12. Roth (1992: 13–14).

13. In later years when many *ejidatarios* abandoned the land for economic survival, they left with only an in-perpetuity right to use the land—land that was typically exhausted from desperate attempts at subsistence living.

14. These parks were small (5,000 square acres) by US standards and most were near population centers in central Mexico. Cardenas also created the first nature reserves; see Simonian (1995: 228).

15. Jameson (1996: 26).

16. Welsh (2002: ch. 12, 1).

17. Simonian (1995: 111).

18. Sellers (1997: 173).

19. Conrad Wirth served as director from 1951 to 1964, and inaugurated the Mission 66 program that championed visitation. His policies encouraged other entities such as Chambers of Commerce, Lions Clubs, and the popular press to promote the parks. Just as the railroad

had done in the previous century for Yellowstone, the automobile was a major factor in increased park visitation rates, which grew from 61 millions visitors in 1956 to 211 million in 1972; see Sellars (1997: 206) and Runte (1987).

20. A related proposal for an "international free zone" in the Big Bend–Sierra del Carmen area would have moved customs and immigration back to park peripheries in each country, with visitors moving unfettered back and forth across the border.

21. The Council encompassed Texas and its four neighboring Mexican states: Tamaulipas, Nuevo Leon, Coahuila, and Chihuahua. See Roth (1992: 35).

22. During the same period that the United States was adopting its major environmental laws (notably, the Clean Air and Clean Water Acts and the Endangered Species Act), Mexico passed its Law for the Prevention and Control of Pollution in 1971.

23. Sellers (1997: ch. 6).

24. Initiated in the early 1970s, UNESCO's Man and the Biosphere Program (MAB) focused on the interrelationship of diverse adjacent landscapes and varying land-use patterns, resulting in the concept of "biosphere reserves"; see Chester (2006).

25. Welsh (2003: ch. 18).

26. Another form of restrictive land holding under Mexican agrarian laws.

27. Welsh (2003: ch. 18); interview with Gil Lusk, 2005.

28. Secretaría de Desarrollo Urbano y Ecología, Mexico's federal agency in charge of environmental issues at the time.

29. Much has been written on NAFTA and the economic benefits and environmental consequences, and it appears at this juncture that neither NAFTA's supporters nor detractors accurately assessed the treaty's full environmental consequences; see Gallagher (2004) and Deere and Esty (2002: esp. chs. 1–4).

30. Maderas del Carmen contains 514,701 acres (208,381 ha); Santa Elena contains 684,706 acres (277,209 ha); see CONANP (2003).

31. Interviews with Dan Roe, CEMEX, February 2005.

32. CEMEX was awarded the World Environmental Center Gold Medal for International Corporate Achievement for its efforts to develop state-of-the-art technology for energy efficiency and cleaner emissions; see Gallagher (2004: 55), Roe (February 2005), and the video *El Carmen: Un reto, un compromiso*, produced by CEMEX and Agrupación Sierra Madre.

33. See, for example, ⟨www.beartrust.org⟩.

34. Dan Roe of CEMEX, personal communications, July 2005; Jose Dominquez (resource manager Maderas del Carmen) in a July 2005 interview commented that with 14 *ejidos* in Maderas del Carmen in various condition of viability, the Mexican government is giving the area increased attention.

35. Secretaría de Medio Ambiente y Recursos Naturals, which oversees the Comisión Nacional de Áreas Naturales Protegidas (CONANP), the principal agency governing the main categories of protected areas including Biosphere Reserves, Special Reserves, National Parks, Natural Monuments, Protected Areas of Flora and Fauna, and Natural Protected Areas

including Forests; see ⟨http://www.semarmat.gov.mx⟩ (and link to the English language "protected areas" Web page).

36. Carabias and Babbitt (1997).

37. Cisneros (1999: 8).

38. The report concluded that various sources contributed to the visibility degradation, apportioning one-third to Mexico and two-thirds to the United States. Sulfate and carbonaceous compounds and dust comprise the majority of the haze. Two power plants in Mexico constitute the largest point sources, but total only 9 percent; Texan emissions generally contribute a greater amount, and the remaining contribution comes from far and wide. See BRAVO 2004 summary at ⟨www.nps.gov/bibe/pressreleases/2004⟩. No immediate plan exists to act on the findings. Obviously, having both countries design a study with enforcement in mind is the best approach. However, to paraphrase one participant, even if Mexico had participated and the study had found Mexico exclusively responsible, eliminating haze in a sparsely settled region on the border would simply not rank as a priority in comparison to the severe air quality issues so prevalent in the country's major population centers.

39. See Dimmitt (2000: 14–18), MacMahon (2000: 286, 297).

40. USFWS (2002) and Urrea (2004: 4).

41. Waldman (1985: 196–97).

42. Pearson (1998: 1), Felger et al. (2006), and US NPS (nd).

43. While national parks are administered by the NPS for both conservation and recreation, NWRs are administered by the US Fish and Wildlife Service primarily for wildlife conservation.

44. US FWS (nd). In total, US federal agencies currently manage approximately 3,041,302 million hectares within five protected areas and one "de facto" protected area; see Felger et al. (2006). This de facto area is controlled by the Department of Defense. In 1941 President Roosevelt withdrew a vast tract of land to the north and west of Cabeza Prieta NWR as the Luke Gunnery Range for military training purposes; see Ripley et al. 2000. After several name changes, Congress renamed the area as the Barry M. Goldwater Range (BMGR) in 1987; see Luke Air Force Base (nd). As described below, this area would come into play in later advocacy for an international peace park.

45. US National Park Service (1965).

46. Udall (1966).

47. "What I was proposing to President Johnson," Udall recalled, "was seven million acres; that would have been a record, in terms of acreage, and I just said it right out for him, 'Mr. President, if four million acres was just about right for Herbert Hoover [US president, 1929–1933], seven million is right for Lyndon Johnson.' But he didn't...." See Udall (1997: 316–18).

48. US National Park Service (1965: 29).

49. Udall (1997: 317).

50. Pearson (1998: 6).

51. Pearson (1998: 7).

52. Sonoran Institute and Alliance (nd) and Laird, Murrieta-Saldivar, and Shepard (1997).
53. Joquin (1988: 13).
54. Laird, Murrieta-Saldivar, and Shepard (1997).
55. See Chester (2006) for an in-depth case study of ISDA.
56. Felger et al. (2006).
57. Walker (nd).
58. Simon (1995: 160).
59. Chester (2003) and Nabhan (1995, 1996).
60. Chester (2006).
61. US National Committee for the Man and the Biosphere Program (1997) and Pearson (1998: 13).
62. US National Committee for the Man and the Biosphere Program (1997).
63. US–Mexico Border Field Coordinating Committee (2001).
64. According to Broyles, the group has used the alternative names of Sonoran Desert Project, Friends of the Sonoran Desert National Park, Sonoran Desert National Park Project, Sonoran Desert Friends and "maybe a few more variations."
65. Gibson and Evans (2002); for press coverage, see Allen (1999) and Landau (1999).
66. SNDPP (1999).
67. SDNPP (1999).
68. The poll was conducted by Behavior Research Center, a Phoenix-based marketing and public opinion research firm; see BRC (1999).
69. The Monument is located much closer to Phoenix than to the international border.
70. Arizona State Senate (2003b,d).
71. Arizona State Senate (2003c).
72. Arizona State Senate (2003a).
73. Broyles (2004: 35).
74. Aside from the Western Sonoran Desert's open landscape, border patrol crack downs along other stretches of border have made this area attractive for those wishing to enter the United States improperly; see Jordan (2005).
75. Mexico is the second largest export market for US products and the US imports 10 percent of all goods from Mexico; see Deere and Esty (2002: 79).
76. Interview with Vidal Davila, Big Bend Chief of Science and Resource Management (July 2005).
77. Superintendent John King personal communication February 2005 regarding efforts of Martha King and others.
78. NACEC (nd).
79. These two organizations have worked together in the Chihuahuan Desert to purchase land and establish the first conservation easements in Coahuila. See ⟨http://www.nature.org/pressroom⟩.

80. The Pinacate visitor center is small but functional with a knowledgeable—albeit again small—staff. As Maderas del Carmen and Santa Elena Cañon do not have on-site centers, this might be a project for CEMEX in the future should the company choose to open its land to visitors. See Roth (1992: 56–59) for a discussion of the World Bank's debt-for-nature programs initiated in other areas of Mexico.

81. CONANP operates all 155 federally designated protected areas. The collaboration with CEMEX, Conservation International, Agrupación, Sierra Madre, The Wild Foundation, and Birdlife International will not only establish the El Carmen Wilderness but provide a certification program for private landowners who wish to conserve their land as wilderness. See ⟨www.celb.org/xp/CELB/news-events/press_releases/10012005.xml⟩. BBNP will offer technical assistance for the El Carmen Wilderness as needed. (John King, personal communication, January 2006.)

82. Dan Roe, personal communication, July 2005.

83. Broyles (2004) and Office of International Affairs (nd).

12

Liberia: Securing the Peace through Parks

Arthur G. Blundell and Tyler Christie

Liberia is the heart of the hotspot[1] for biodiversity in West Africa. Liberia has also been the epicenter of conflict in the region, driven in large part by competition for natural resources.[2] In this chapter we briefly describe the background of the regional conflict, the importance of Liberia's ecosystems, and how peace parks can protect both biodiversity and West Africans.

In the early 1990s the rebel leader Charles Taylor captured control of rural Liberia, including almost all the logging areas and an iron ore mine, which reportedly paid him $10 million per month.[3] Taylor also supported the RUF (Revolutionary United Front) rebels in Sierra Leone to control their diamond trade, which further destabilized the region.[4] Once Taylor became president of Liberia, timber companies used private security forces—often composed of Taylor's ex-combatants—that looted and attacked civilians in both Liberia and neighboring Ivory Coast.[5] This conflict to control resources and territory has undermined rule of law, especially in rural areas. In the past, rebel groups took advantage of this vacuum, seizing control of forests using the revenue generated from natural resources to fuel insurgencies in remote border areas.

Could Peace Parks Work?

Around half of all civil wars exist in countries that return to war after less than a decade of peace,[6] generally because combatants obtain revenue from natural resources to fund the resumption of conflict. Instead, peace parks could help establish rural authority over forest areas without militarizing borders.

As outlined in UN Security Council Resolution 1647, peace and security may be within reach if:

- Liberians find employment, in particular, if ex-combatants avoid recruitment by militias;
- Rebels are denied revenue from the illegal exploitation of natural resources; and
- Countries in the region cooperate to end the violence.

Critics argue that conservation cannot achieve these objectives, especially provide the jobs and revenue necessary for the development of a country recovering from war. Such conventional wisdom is wrong. While a network of peace parks would reduce the area available for logging by approximately 20 percent, the network would provide rural Liberians with greater benefits than many alternative land uses, including logging in areas of low timber value.

While Taylor was president (1997–2003), logging grew to about $100 million annually, but most of the money did not remain in Liberia.[7] Most of the skilled labor was foreign. Tax evasion was widespread—only 14 percent of taxes were paid, and timber companies even paid arms dealers in lieu of taxes.[8] Logging companies evaded obligations to fund schools, clinics, and other community projects. Except for a few main "highways," logging companies failed to maintain roads. Logging never employed more than 0.2 percent of rural Liberians, or less than 0.0012 people per hectare of logging concessions (6,000 workers/5,000,000 ha).[9] Seasonal employment paid approximately $100/month. Despite widespread logging, rural development stagnated.

A network of peace parks has the potential to provide more. One thousand workers could be employed to manage one million hectares of protected areas. These employees—many from local communities—would be permanent rural residents paid above logging wages, thereby making a significant contribution to the rural economy. (If paid at the same rate as loggers, protected areas could easily employ more workers than logging does.)

This parks staff would be the core of a redesigned Forestry Development Authority (FDA), the parastatal responsible for managing Liberia's forests. At present, the FDA "employs" about 600 workers (at approximately $60 a month), but most do little in the way of work and their salaries are (on average) eight months in arrears. This lack of real employment has led to corruption. Taylor's administration encouraged "Operation Pay Yourself," where civil servants were instructed to replace unpaid government salaries with bribes or looting from civilians. Peace parks would help Liberia replace a system of official criminality with one able to pay legitimate wages to legitimate civil servants.

The Historical Context

To explain the context in which peace parks can function, we briefly review the history of conflict in the region. In most cases insurgencies began along uncontrolled border regions—the focus of the network of peace parks.

Liberia and Her Neighbors

In the 1950s and 1960s Ghana, Nigeria, and Sierra Leone won independence from Britain, and Guinea, Ivory Coast, Benin, and Togo from France. In contrast, Liberia has been independent since 1847, after its "founding" by eighty-six settlers from the United States who set up the new country at the barrel of a gun. Although ex-slaves themselves, the American settlers effectively enslaved the local peoples. This Americo-elite (never more than 5 percent of the population) ruled Liberia until 1980 when President William R. Tolbert was killed in a coup led by the semi-literate Sergeant Samuel Doe, an ethnic Krahn from the border region with Ivory Coast. The coup began a period of unrest that continues to affect the entire region, as various factions have launched insurgencies on Liberia from neighboring countries, and vice versa. The United Nations now has a presence in Liberia (UNMIL), Sierra Leone (UNIOSIL), and Ivory Coast (UNOCI).

Ivory Coast

President Félix Houphouët-Boigny of Ivory Coast opposed Doe, in part because his son-in-law (President Tolbert's son) had been killed in the Liberian coup, despite Doe's promise of protection. Thus Ivory Coast was sympathetic to the rebel leader Charles Taylor and his Libyan-backed insurrection. On Christmas Eve 1989 Taylor attacked Nimba County in Liberia from western Ivory Coast. Fighting raged for seven years between the government of Liberia, Taylor, several other rebel factions, and a peacekeeping force from West Africa (mainly Nigeria). In 1997, under a UN-sponsored peace accord, Taylor won the presidential election. However, Houphouët-Boigny had died in 1993 and Ivory Coast ceased to be friendly to Taylor.

Taylor tried to consolidate power cynically, by destabilizing neighboring countries. Militias from Liberia attacked, looted, and recruited soldiers in Ivory Coast. Civil war erupted in Ivory Coast in 1999. In 2003 a French-backed peace treaty essentially divided the country in two: a forested belt in the south controlled by President Laucent Gbagbo, and a savannah belt in the north controlled by the rebel

Forces Nouvelles. The Taylor-backed MPIGO rebels have repeatedly attacked Gbagbo from Danane in the north. In retaliation, Gbagbo backed the rebel group MODEL, which attacked Liberia from western Ivory Coast in 2003. "After 15 years, the wheel of forest war had come full circle."[10]

Sierra Leone

In 1991, while fighting the civil war in Liberia, Taylor backed the RUF. In 2000, the RUF was involved in an unsuccessful insurrection into neighboring Guinea to unseat President Lansana Conté, an enemy of Taylor. Sierra Leone's civil war ended in a UN-sponsored peace treaty in 2002.

Guinea

When the French offered independence, their African colonies were given the choice of remaining with France, complete independence, or a hybrid—a republic within the French community. All except Guinea chose the latter. Penalized for choosing full independence, France abandoned Guinea. In 1965, almost a decade after independence, President Sékou Touré broke off relations with France, accusing them of plotting against him. The result has been a precipitous drop in living standards.

Since 2000, rebel incursions from Liberia and Sierra Leone have claimed more than 1,000 lives and caused massive population displacement. President Conté has accused Taylor, the RUF, and army mutineers of trying to destabilize Guinea. To counter Taylor, Conté supported the rebel group LURD, which attacked Liberia through Lofa County. In mid-2003, LURD laid siege on Monrovia, the capital of Liberia, forcing a peace agreement. Taylor fled to exile in Nigeria (thus avoiding—at least temporarily—the Special Court in Sierra Leone, which has indicted him on crimes against humanity).

Future for Regional Cooperation

Given the rebellions that have formed along Liberia's borders, it may seem unlikely that the neighboring countries can find common ground. However, peace parks aimed at decreasing the likelihood of insurgencies may provide such a common cause. The main reason for hope is that although the governments have been belligerents, the people themselves on either side of the borders are linked by kinship and marriage. For example, Mano people live on both sides of the Cavaly River (eastern

Liberia and western Ivory Coast). Local commerce and social interactions are common along the footpaths that cross the unregulated border.[11] Peace parks would keep the military from interfering with this cooperative, familial spirit.

"Ironically, the use of military to impose national security in border areas has often led to a decrease in community security."[12] Locals are likely to support peace parks as a preferable alternative to the militarization of their borders.

In place of the military, parks officers could increase vigilance in frontier areas that have served as centers for rebel recruitment and the formation of insurgencies in the past.[13] It is unrealistic to expect parks staff to prevent all illegal activity, but the officers could provide an "early warning system." Staff will be embedded in communities, enabling them to access local knowledge. As locals, staff would want to stop any rebel activity before it represents a threat. This would be invaluable as the state of Liberia has almost no presence along its undefended borders.

As in the Sapo case (see the sidebar), instead of fueling conflict, the presence of parks staff could strengthen the rule of law. An effective protected area network could moreover promote good governance. Conservation officers in rural areas could enhance development if they act as extension officers to help locals, for example, in farming or managing wildlife populations as a source of protein. At least the presence of an authority would provide some security to allow locals to pursue livelihoods—activities almost impossible in the current vacuum. Parks would also protect the wildlife and ecosystems[15] that provide essential services to those

---

Peace in Sapo National Park

In 2004 Sapo, Liberia's only park, was invaded by more than 4,000 illegal gold miners and hunters. As many were ex-combatants, the invasion posed a critical security threat. Sinoe County, where Sapo is located, has only 18 police officers to cover one million hectares (about 10 percent of the country). When UNMIL (UN Mission in Liberia) decided to remove the squatters, the UN collaborated closely with FDA park guards. NGOs—including Liberia's Sustainable Development Institute, Fauna and Flora International (FFI), and Conservation International (CI)—provided assistance, such as uniforms and honorariums, that raised the guards' morale. The guards knew the local people and the trails used by the miners, so they were able to conduct an "information campaign" using their social capital with the surrounding villagers to raise awareness that squatters had to vacate the park. Fortunately this partnership of UNMIL, FDA, and local and international NGOs worked to get all the miners to leave the park peacefully.[14]

outside the protected areas, many of whom depend on forests for their lives and livelihoods (e.g., as source of bush meat,[16] medicine, and freshwater).

Increased good governance is bound to improve regional cooperation. As outlined above, Liberia and its neighbors have been belligerents, but peace parks offer the opportunity to cooperate to minimize the common threat from insurgencies. Regional park managers can also share lessons learned about conservation as well as development opportunities. Indeed, such cooperation of ecosystem management is fundamental to the conservation of regional biodiversity, because ecosystems know no borders.[17] Regional cooperation should convince the international community to support peace parks, financially and otherwise.[18] International partners should be looking for any inroads that will promote dialogue among West African countries. A small financial investment in conservation (approximately $25 million in a trust fund, or about 5 percent of Liberia's peacekeeping budget for 2005) could endow the management of Liberia's peace parks in perpetuity, yielding large dividends in regional security.

Importance of Liberia's Biodiversity

In addition to regional security, parks would yield global benefits to biodiversity. Liberia is the heart of the hotspot. While forests in West Africa have been reduced by approximately 95 percent,[19] Liberia holds about half of the remaining forest—approximately 4.52 million hectares, in two large blocks of mostly continuous forest: the evergreen lowland forests in the southeast and the semideciduous montane forests in the northwest. Together, these blocks contain more than 2,000 known flowering plant species, 240 timber species, 150 mammal, 620 bird, and 125 reptile and amphibian species.[20] They include charismatic species such as the western chimpanzee (*Pan troglodytes*), red colobus monkey (*Piliocolobus badius*), Diana monkey (*Cercopithecus diana diana*), pygmy hippopotamus (*Hexaprotodron liberiensis*), and forest elephant (*Loxodonta africana cyclotis*).

The government of Liberia has recognized the importance of Liberia's forests, including signing international conventions (e.g., the Convention on Biological Diversity) and national laws (e.g., Protected Areas Act of 2003 that requires the establishment of a network of protected areas). Quick action is needed to scale up conservation to reverse the threats to Liberia's biodiversity. Almost all (81 percent) of Liberia's forests are now within three kilometers of a road.[21]

Where Are the Peace Parks?

The FDA, CI, and FFI have collected basic ecological and socioeconomic data to design a network of protected areas. They are now consulting with rural communities to identify the appropriate interventions that will allow conservation to coincide with rural development.

Key sites for creating peace parks including the following:

Sapo-Taï

Transboundary connectivity The Sapo-Taï corridor presents the best opportunity to conserve large areas of forest in West Africa. The proposed corridor extends from coastal Liberia, at the Cestos-Senkwehn, through Sapo Park, and the proposed Grebo Park, in to Taï Park, and the proposed Haute Dodo and Cavaly protected areas in Ivory Coast. The entire corridor is under increasing threat from illegal logging, mining, hunting, and settlement. The forest areas have also been a source of conflict, serving as a source of revenue and a staging area for rebel insurgencies, and providing safe passage for arms smuggling and other illicit activities.

Biological value As the largest block of forest in West Africa, the Sapo-Taï corridor is vital for mammal conservation. Recent studies by the Liberia Forest Reassessment[22] and a Rapid Biological Assessment (RAP) carried out by CI in Liberia and Ivory Coast found evidence of several endemic and threatened species including pygmy hippos, forest elephants, leopards, and the largest known population of western chimpanzees.

Ecological value The corridor encompasses several major rivers including the Cavaly, Cestos, Senkwehn, Doube, and Sinoe. The rivers provide critical ecological services to the many agriculture-based communities, the majority of which do not have access to clean drinking water sources other than these rivers.

Greater Nimba Highlands

Transboundary connectivity The Nimba corridor connects forest blocks in three countries (Liberia, Guinea, and Ivory Coast) in one of the highest elevation forest ecosystems in West Africa. Central to the corridor is the UNESCO World Heritage Site straddling the border of Ivory Coast and Guinea along the ridge of the Nimba Mountain range, and proposed to expand in to the Nimba Nature Reserve in

Liberia. Other important forest blocks include the Diecke Massif du Ziama forests in Guinea and the proposed Wologizi and Wonegizi protected areas in Liberia. Nimba County, in Liberia, is home to the abandoned Swedish LAMCO iron ore mine. The Swedish government may choose to support a Greater Nimba Highlands Peace Park to mitigate the impacts of the mine. Likewise Mittal Steel, which recently won a concession in Nimba, should ensure that their site is "biodiversity neutral," that is, they should offset their impacts by financing conservation projects that prevent an equivalent amount of biodiversity being lost elsewhere.[23]

Biological value The highlands are known to provide habitat for more than 200 endemic and endangered species including chimpanzees, forest elephants, and the Mount Nimba toad (*Nimbaphrynoides occidentalis*), which is unique in that it gives live birth after a nine-month gestation. The area is also likely to encompass important botanical diversity unparalleled on the continent, given its elevation gradient (1,750 ms) and because it served as a forest refugium during past glacial periods when the surrounding landscape was covered in savannah.

Ecological value The corridor contains the headwaters of several important rivers including the Cavaly and the St. John.

Gola-Lofa-Mano

Transboundary connectivity The corridor is the second largest block of forest in the hotspot and extends from high altitudes down to lowlands. The Lofa and Kpelle National Forest Reserves in Liberia and the Gola Forest Reserve in Sierra Leone provide a rare opportunity for the two contries to protect a completely connected forest block. In the past this forest region has been a source of conflict, providing a porous border for illicit migration and smuggling.

Biological value Unfortunately, the past twenty years of war prevented study of the area. However, its elevational gradient suggests high biodiversity value. CI and the FDA conducted a RAP in December 2005 to fill this information gap. Preliminary results suggest the areas have high biodiversity value.

Ecological value The corridor is considered the "bread basket" of Liberia and encompasses several important rivers including the Lofa and Mano. Thus the forests provide critical environmental services for agriculture.

Economic value In addition to the direct economic benefits to locals hired as parks staff, the ecosystem services mentioned above likely have considerable value. There has not been an economic evaluation of such services in Liberia, but in Madagascar, Carret (2004) estimated that similar ecosystem services provided a net benefit of at least $10 per hectare per annum, especially through erosion control and clean water supply.

Civilian Conservation Corps

Ideally the ecosystem-service benefits would be supplemented with development assistance provided to establish and manage the parks network. Conservation International (CI), the UN Office for Project Services (UNOPS), the US Agency for International Development (USAID), and Liberia's FDA are working to create a Civilian Conservation Corps (CCC) similar to that established by Franklin Delano Roosevelt in the United States during the 1930s Depression. Over the next two to three years the CCC will focus on establishing basic services, including roads, schools, clinics, and clean water sources in villages bordering on protected areas. This should provide the foundation for development, while also providing an infusion of capital to jump-start economic activities, including agriculture, livestock rearing, and agro-forestry. If locals realize that additional development is tied to the parks, this may provide the incentive to assist in the parks' protection.

Funding

The government of Liberia and CI propose to fund conservation and community development around the peace parks by creating the Liberia Protected Areas Trust. Liberia could invest the conservation tax on commercial forestry—approximately $1.5 a cubic meter, generating $2 million to 3 million annually—in the Trust Fund and challenge international donors to match. The donor match provides an incentive for the Liberian government to capture revenue from natural resources. The Fund will be managed by a professional financial firm and overseen by a joint Liberian/international board and an in-country secretariat.

Monitoring and Evaluation

The Trust Fund and the CCC should also finance the monitoring and evaluation necessary to establish both a baseline for park management and the accountability to ensure that funds are used effectively. This adaptive management should extend to monitoring local attitudes, beliefs, and knowledge of protected areas to help ensure the appropriate placement of parks within the network and the equitable sharing of costs and benefits with local people.

Proceeding with Caution

Creating peace parks is not without pitfalls. Fortunately these hazards can be minimized with careful planning and execution. The most obvious caveat is the possible disenfranchisement of locals if their land is expropriated. But rural people need not lose their land. Modern protected areas could allow locals to remain within enclaves or buffer zones.[24] For example, local rights can be maintained, provided they do not commercialize the extraction of timber, bush meat, or other forest products. Through the strategic placement of a network of protected areas, CI, FFI, and the FDA have been careful to propose parks in areas to minimize the opportunity cost for alternative land uses such as agriculture, forestry, and mining.[25]

If resettlement is preferred, then it should benefit from the lessons learned from other institutions, such as the World Bank (Operation Policy 4.12: Involuntary Resettlement Policy). However, if villages are moved, only those that are willing to move voluntarily should be considered, and they should be given due process and compensated fairly. The expansion of parks should not incite grievances that can lead to future conflict.

Fortunately Liberians have a history of moving to accommodate shifting patterns of swidden agriculture. In 1983 several communities were compensated for relocating in order to create Sapo Park, providing a precedent for a successful, peaceful resettlement. Local participation in decision-making is critical regardless of whether locals are accommodated within the protected area system or prefer to relocate. The involvement of the international community and key Liberian NGOs is crucial to ensure adequate consultation, instead of mere notification of resettlement.

A second concern with the establishment of peace parks is the increase in state presence in these rural areas. In the past state actors, most often military, were unwelcome and dangerous, as they often brought corruption and human rights violations, such as looting and violence.[26] Instead of soldiers, peace parks would bring park guards and conservation officers. If properly trained, the officers may be less inclined to become corrupt. For example, the United Nations recently trained 164 FDA enforcement officers, and after graduating, the officers articulated their desire to work for the benefit of Liberians and their forests.[27] However, there is no guarantee that such agents will act appropriately. The international community should continue to be involved in funding, training, and monitoring parks officers.

Conclusion

Competition for the illegal control and exploitation of natural resources has been a major driver of conflict in Liberia. The international community, through the UN and bi-lateral assistance, has made a significant investment in Liberia's reconstruction and development. Unfortunately, this assistance is yet to establish control of natural resources and reverse "business as usual." Many countries, including Liberia in the past, have returned to conflict, largely driven by competition for natural resources.[28]

If the Security Council is to achieve their objective—and if Liberians are to avoid a resumption of civil war—control of natural resources is paramount. Without peace and stability, Liberia will not develop economically. Without good governance, Liberians will remain mired in poverty, failing to realize their human rights.

Peace parks can provide strong rural authority to prevent criminal activity and reduce the opportunity for belligerents to capture revenue from natural resources. Conservation can rival other land uses in providing jobs for rural Liberians. Conservation will preserve the ecosystem services and biodiversity that many Liberians depend on for their lives and livelihoods. Peace parks will increase regional cooperation as opposed to belligerence. Peace parks will increase border security, reducing the likelihood of insurrections. Reducing regional conflict is in everyone's interest, thus the international community should be willing to finance the network of peace parks.

Natural resources have been a curse to West Africa, but that can change. The environment can be a source of cooperation and sustainable development. If the international community does not assist Liberia in creating peace parks, then we risk losing our investments in peacekeeping and the rational development of West Africa. At the joint UN–World Bank Reconstruction and Development Conference in February 2004, US Secretary of State Colin Powell warned that this was Liberia's last chance. Peace parks are a critical element in seizing that chance.

Notes

1. Globally, 34 hotspots represent areas with 75 percent of the Earth's most threatened mammals, birds, and amphibians, while covering just 2.3 percent of the planet's surface. The Upper Guinea hotspot covers most of West Africa.
2. UNSC (2003).

3. Reno (1995, 1996).

4. UNSC (2001) and Farah (2004).

5. Blundell et al. (2004).

6. Collier and Hoeffler (2004).

7. Blundell et al. (2003).

8. Blundell et al. (2005).

9. Blundell et al. (2003).

10. Richards (2005).

11. Richards (2005).

12. Oviedo (2003).

13. Blundell et al. (2004).

14. Blundell et al. (2005).

15. Chai and Manggil (2003).

16. For marine parks as a source for improved fish catches outside the protected area, see Roberts et al. (2001).

17. Bakarr (2003).

18. Sandwith (2003).

19. Mittermeier (1999).

20. Gatter (1999).

21. Christie et al., forthcoming.

22. A project operated by FFI, CI, and the Liberian government with additional support from the European Commission and Critical Ecosystem Partnership Fund.

23. Biodiversity Neutral Initiative (2005) ⟨www.biodiversitynetural.org⟩.

24. Oviedo (2003).

25. Christie and Blundell, *in prep*.

26. Blundell et al. (2003, 2004, 2005).

27. See appendix 7 of Blundell et al. (2004).

28. Collier and Hoeffler (2004).

13

Preserving Korea's Demilitarized Corridor for Conservation: A Green Approach to Conflict Resolution

Ke Chung Kim

War is often framed as a "necessary evil," by antagonists, yet sustainable peace between people and natural systems is usually achieved ultimately when parties approach conflict resolution by focusing on what is mutually valuable.[1] Armistice or merely a ceasefire does not provide a permanent peace as attested by the half-century of Korean conflict since 1953.

The Demilitarized Zone (DMZ) on the Korean peninsula was created as a military buffer by the Armistice of Korean War without a treaty for peace. The DMZ, Korea's transboundary corridor for military buffer between the two Koreas, the Democratic People's Republic of Korea (DPRK) or North Korea and the Republic of Korea (ROK) or South Korea, is protected by a thicket of sharp barbed wires edged with rows of land mines. While peace was illusory, the paucity of human activity ironically transformed the DMZ into a wild natural sanctuary for native plants and animals. In other words, the DMZ corridor, which represents a cross section of the Koreas' landscapes, offers a de facto nature reserve for conserving the last vestiges of Korea's natural heritage and the sole in-situ resource for germ plasm and genetic materials of native plants and animals.[2]

Before the advent of rapid industrialization, the Korean people lived in a relatively clean environment with beautiful landscapes on a peninsula placed between the East Sea and the Yellow Sea. This landscape is often depicted as the *Keum-Su-Gang-San* (land of embroidered river and mountains), a metaphor used even in the Korean national anthem. Throughout the cold war and to this day, military operations and economic development along with urbanization leveled the land, polluted the environment, and destroyed habitats, causing a massive loss of biodiversity.[3] Dire environmental degradation continues on both sides of the DMZ without a clear political mandate to reverse it. Environmental degradation and the loss of biodiversity not only undermine the capacity of natural ecological processes that sustain life-support systems but also harm human health and thus constrain economic development and endanger human security.[4] Korea's environmental future lies in a pan-Korean nature conservation strategy that would provide land restoration and environmental security for the peninsula.

On the day of liberation, August 15, 1945, the Japanese colonialism for almost half of the twentieth century ended, and the Korean people regained their independence. However, this independence was short-lived, and soon Korea was divided by the international agreement of four powers: China, England, Soviet Union, and the United States. The initial purpose the division was to disarm Japanese military forces on the Korean peninsula. The onset of the cold war polarized the North and the South and led to a more sinister territorial division.

The current North Korean nuclear standoff is the last straw of the historic international politics that shaped Korea's destiny. North Korea has continued to advance its nuclear technology to sustain the rogue communist regime. It is now a nuclear power poised against the democratic world, while the other participants of the Six Party Conference (comprising the United States, Japan, China, Russia, South Korea, and North Korea) struggle to find politically just means to resolve the conflict. The resolution to the contemporary Korean conflicts demands an alternate means that is linked to common critical issues pertaining to both Koreas and stakeholder nations.

The problems of environmental degradation and the loss of biodiversity have been overshadowed by relentless economic development. South and North Korea's political and economic jitters have practically eclipsed environmental priorities. This chapter describes and discusses the dynamics of Korea's Demilitarized Zone and how conservation of landscape ecosystems and biodiversity could deliver a critical alternative in the resolution of Korean conflicts surrounding international and regional politics.

Environment, Peace, and Conflict Resolution

Korea's beautiful landscapes that used to be a source of national pride for the Korean people are no longer the *Keum Su Gang San*. For the last four decades aggressive economic development and urbanization in the South and military buildup and operation along with declining economy in the North have systematically compromised the integrity of Korea's ecosystems and landscapes.

South Korea's phenomenal success in economic development came with tremendous environmental costs. Human population grew rapidly and urbanization ensued to accommodate the growing population. South Korea became one of the most populated countries in the world with 48 million people and 425 persons per square kilometer, the third highest population density in the world (average world population density 48/km^2) in 2005.

Large parts of South Korea's natural landscapes have been converted to industrial sites and urban centers, a trend that continues to this day. As a result only a few places in South Korea have escaped the massive land exploitation and severe pollution. Today's South Korea is a land of industrial and urban complexes, mingled with critical environmental degradation, polluted waters, and uncontrolled urban and commercial sprawls. Most habitats are drastically fragmented, modified, or completely destroyed, and agricultural lands are highly polluted with chemical pesticides, fertilizers, or industrial and municipal wastes, all of which contributed to the massive loss of biodiversity.[5]

Massive destruction of habitats and severe environmental deterioration have resulted in a continued loss of biodiversity, and many species of flora and fauna are now extinct or endangered outside of the DMZ and its buffer zones. The 1994 *Biodiversity Korea 2000* Report provides a conservative estimate of South Korea's biodiversity—over 20 percent of South Korea's terrestrial vertebrates, for example, are already extinct or endangered, including 14 percent of birds, 23 percent of freshwater fishes, 39 percent mammals, 48 percent reptiles, and 60 percent of amphibians. Compounding the massive loss of biodiversity is a knowledge deficit: for example, barely one-third of the insect fauna and less than 2 percent of invertebrates are currently documented.[6]

South Korea's active pursuit of the North–South reconciliation and economic exchanges has accelerated the entry of South Korean *chaebols*[7] and other business enterprises into North Korea for commercial exploitation. The historic summit of President Kim Dae-jung and North Korean leader Kim Jong-il from June 12 to 14, 2000, in Pyongyang has further enhanced the South Korean development activities in North Korea.

With a few notable exceptions, environmental management has barely been implemented in South Korean industrial practices.[8] The continued lack of environmental leadership in the Korean governments and the questionable environmental records of South Korea's chaebol and industry at large are a continuing cause for concern.[9] As a result South Korea now must confront two simultaneous challenges: (1) the cleanup of highly polluted and deteriorated environments, restoration of habitats, and rehabilitation of biodiversity in South Korea and (2) the safeguarding of the environments and building of environmental stewardship throughout the Korean peninsula, particularly, related to South Korea's industrial and development ventures in North Korea to avoid a repeat of the past environmental calamities.[10]

The last stronghold of the cold war remains on the Korean peninsula. On the northern half exists the Democratic Peoples Republic of Korea (DPRK), the last Stalinist state with a huge military force and perhaps already armed with nuclear capability. The DPRK government continues to threaten regional security and international stability in the name of its regime survival, while it barely sustains a bankrupt economy and struggles to feed its population.

Since North Korea's nuclear crisis erupted in 1994, it has been over ten years without any significant resolution. The Six-Party Talks were begun in 2003 to resolve the Korean nuclear showdown that is still in deadlock. As a nuclear-free Korea is fundamental to human security for both Koreas, avenues to amicably resolve the current standoff must be found. It appears that DPRK finally succeeded in building and stockpiling nuclear weapons.[11]

At this point all parties have agreed to avoid a military solution and are determined to find a diplomatic resolution considering the enormity of destruction and human loss if war breaks out. The situation is one of the most serious nuclear conflicts the world community has faced. In cases like this where legal recourse is unlikely, alternative dispute resolution strategies need to be explored.[12] The DMZ preservation and pan-Korean nature conservation may offer such an approach to make the Korean peninsula nuclear-free and economically sustainable for all Koreans.

The dire state of environmental degradation continues without a clear political will or public mandate. Nature conservation programs can be designed to engage the Korean people at the grassroots level on both sides of the DMZ, manifested in land restoration, wetland restoration, road repair and management, food production, and biodiversity conservation, including reforestation. A process of public involvement with these issues will improve the local economy, transportation corridors, security, and aesthetics. It may become a grassroots movement infused with industry and developmental forces that could protect biodiversity and culture and strengthen local infrastructure for development and environmental stewardship. If political and social leaderships of both Koreas are engaged through formal diplomatic negotiation for this alternative, politically neutral strategy, there may be no need for North Korea to maintain nuclear deterrence to develop its local economy.

One Korea and Two Nations: The Political Challenge

DPRK shaped its basic policy and military strategy under the one-Korea paradigm, "one-Korea and one people," ever since its inception. North Korea made a failed war based on this paradigm but has persistently applied the paradigm to successfully appease South Korean brethren, while sustaining the last roguish communist regime since the fall of Soviet Union. This strategy has finally prevailed for the North as evidenced by the last presidential election (in 2004) in South Korea—a human rights lawyer Mr. Roh Moo-hyun was elected.

The Roh government has maintained implementation of the "Sunshine Policy" (inaugurated by former President Kim Dae-jung) toward North Korea and continued actively engaging DPRK in business, culture, and sports. This strategy was originally adopted as a way to persuade North Korea to relinquish its hostility and step out of isolation.[13] Regardless of loud uncomplimentary critics, this trend may provide new opportunities for peace and stability on the Korean peninsula and in the Asian region at large if there is no further escalation of the current North Korean nuclear conflict.

North Korea, with the support of Soviet Union, has built a highly centralized, socialist state based on communism and the juché philosophy. After the birth of DPRK, generations of North Koreans were born into the cultish society of President Kim Il-sung and followed by "Dear Leader" Kim Jong-il. The North Korean people have become completely dependent on the mercy and support of the DPRK government. After the fall of the Soviet Union, however, the DPRK regime, unprepared for the loss of major economic support,[14] remained completely isolated from the outside world and also from its own people until the recent past. While maintaining one of the largest military forces in the world, it has not been able to provide the most basic needs of people without international humanitarian assistance. North Korea, now with 23 million, comprises mostly captive people living in a fear-ridden society.[15] To mitigate today's difficulties, the DPRK yearns for international recognition but needs massive humanitarian assistance from the world community, particularly from the Southern brethren.[16]

By contrast, South Korea with its open society and market economy, has become a dynamic democratic nation.[17] From a nation of poverty and disrepair, over the last four decades South Korea has transformed into a dynamic economy that is now considered one of a dozen economic powers in the world, with perhaps the tenth largest economy.[18] South Korea's phenomenal success in economic development is demonstrated by its gross national product (GNP) from the 1960s through the 1990s: per capita GNP at current prices grew from $81 in 1960 to over $10,000 by 1996, then reached $17,700 by the year 2003[19] and $20,000 by 2008.[20] This rapid economic development also brought about political stability and a mature democracy that has empowered the Korean people to regain pride and self-confidence as a nation. Economic success, however, was made possible by sacrificing future environmental security.[21] Along with it, human population grew rapidly to 48 million in 2001,[22] and urbanization ensued to accommodate the growing population. Today's South Korea is a land of industrial and urban complexes with serious environmental problems.[23]

Korea's population is expected to expand rapidly, reaching 100 million by the year 2025 compared with 48.6 million in ROK and 22.7 million in DPRK as of midyear 2003.[24] In addition economic activities continue to expand on the Korean peninsula, particularly, with South Korean commercial exploitation of the land and ecosystems in North Korea. The needs for demographic and economic expansion will require additional appropriation of land and natural resources and likely result in heightening environmental degradation and intensifying loss of biodiversity. The loss of biodiversity and the degradation of ecosystems have changed the environment in many ways and will eventually reshape Korean landscapes and affect economic and cultural development.

Although no one can predict how the current political theater will play out, the future of the Korean people must include a healthy environment, a sound economy, and a free society. A healthy environment, the *Keum Su-Gang San*, must be the foundation for sustaining a growing economy and a free society, and both North and South Korea must work together to provide the kind of environment and landscapes upon which peace and prosperity can be attained.

The Demilitarized Zone: A Unique Sanctuary

Korea's DMZ is a ready-made transboundary protected area created by the armistice as a military buffer. Although there was enormous property destruction and several millions of human lives lost, the preservation of richly biodiverse DMZ ecosystems is the single most significant and positive outcome of the Korean War. The Korea's DMZ is the 4 kilometer (2.4 mile) wide military corridor that lies along the Demarcation Line across the entire peninsula.

The DMZ was established on July 27, 1953, at the end of Korean War by the Armistice Agreement between the DPRK with the People's Republic of China (PRC) and the UN Forces, which included the ROK and the United States. The DMZ consists of the 246 kilometer (155 mile) Military Demarcation Line (MDL) and the 2 kilometer wide demilitarized zone on both sides of the MDL, embodying a mandated corridor with an area of 98,400 hectares.[25] Since then the Military Armistice Commission (MAC) has rigidly enforced the Agreement, and the buffer zones of the DMZ corridor on both sides have been highly fortified and patrolled by hostile troops, often numbering over 1.5 million.[26] To this day the military posture against the South has not been changed despite the ROK's Sunshine Policy.[27]

With the inter-Korean summit of June 12–14, 2000, the DMZ became a symbol of reconciliation and peace, while at the same time memorializing a historical blunder of global politics. For the last five decades the DMZ corridor has been guarded by military forces on both sides against human intrusion. As a result the DMZ allowed engulfed, damaged forests to regenerate and farmlands cultivated for many thousands of years to return to a natural state with flowers, grasses, and other wildlife.

Diverse habitats of different ecosystems were re-created by natural secondary succession in the areas where the forests and farmlands used to exist for many thousands of years before human inhabitation. The DMZ corridor has been rehabilitated to a luxuriant nature reserve and has also become a unique natural sanctuary for native fauna and flora. The DMZ ecosystems, including those in the narrow buffer strip of the Civilian Control Zone (CCZ, the 5 to 20 kilometer buffer zone established in 1967 occupying 1,529 square kilometers south of the DMZ), now support one of the last vestiges of Korea's natural heritage, providing important sanctuary for wildlife including Chinese egret, black bear, musk deer, mountain goat, and many endangered and practically extirpated species such as the migratory black-faced spoonbill, white-naped crane, red-crowned crane, leopard cat, and perhaps even Korean tiger. The DMZ also became the sole repository for the material basis of native species called germplasm, a necessary resource for restoring critical habitats and enriching Korea's environment.

The DMZ corridor represents cross sections of diverse ecosystems, with central and western lowlands and eastern highlands divided by Taeback Mountain and the adjacent mountain ranges running from north to south. The watershed of the Imjin River forms the central and western parts of the DMZ. The Imjin River flows into the Yellow Sea, and at its mouth it merges with the Han River. The DMZ corridor contains a diversity of landforms, and climatic, geological, and physiographic zones.[28] Most information on the biodiversity of the DMZ comes from systematic biodiversity surveys carried out by South Korean scientists in the CCZ. Findings from these surveys provide a glimpse of what to expect from the DMZ corridor (Sungchun-Munhwa-Jaedan 1996). The DMZ biota is rich in species diversity and endemism with many exciting findings, including new species, new taxonomic records, and endangered species.[29]

Preserving the DMZ Ecosystems for Conservation

The DMZ ecosystems are ready-made nature reserves, clearly delimited, demilitarized between two borders, and protected by military control.[30] Unlike other potential transboundary reserves, this land does not belong to either party but to the Korean people at large. These nature reserves are already well defined for establishing a system of protected areas under various levels of protection and are ready to be formally established as permanent protected areas, perhaps called the Korea Peace Nature Reserves System (KPNRS). This corridor would be a system of transboundary reserves made of peace parks, nature centers, peace villages, eco-farms, an international ecosystem research center, and other conservation-oriented uses on the DMZ. The reserves system could be administered by a joint commission, perhaps calling itself the Korea Peace Nature Reserves Commission (KPNRC), composed of North and South Korean representatives along with distinguished personalities and scientists from around the world.

In principle, the entire DMZ corridor and adjacent areas in the military buffer zones (i.e., CCZ for ROK) including the equivalent offshore space should be dedicated to the KPNRS. This approach should allow the Korean people to ascertain the ecological integrity of diverse ecosystems in the DMZ corridor. If individual reserves are proposed separately for specific sites as protected areas such as peace parks or other purposes at this stage, the DMZ ecosystems and their inclusive habitats would be fragmented or damaged, causing the loss of invaluable habitats and related biodiversity, the invasion of non-indigenous organisms, and the degradation of ecosystem functions. Considering the strong influence of development forces in South Korea, the government bureaucracy could easily accept any piecemeal proposals for separate reserves regardless of site-specific habitat and ecosystem characteristics before formal bilateral agreements are reached for preserving the DMZ corridor in its entirety for conservation.[31]

Should both Koreas sign the agreement to develop the KPNRS, there would be room for negotiation in defining the category and size of individual reserves. KPNRS could include different classes of broadly defined protected areas based on the characteristics of landforms and climatic, geological, physiographic features, and the status of metapopulations of those species that are endangered and threatened:[32]

• Nature reserves serve the broadly based public for long-term research and education in biodiversity science, ecology, ecosystem science, restoration ecology, and conservation.

• Protected landscapes or seascapes, including peace parks, are for natural heritage conservation and ecotourism.

• Nature villages (farming or fishing), peace villages, and eco-farms, encourage a limited number of families from North and South Korea to live together in an environmentally friendly manner.

• Special Maritime Zones can be developed in the DMZ offshore for joint controlled fishing, controlled fisheries cooperative, aquaculture, and marine environmental monitoring and assessment.[33]

Such a system would provide critical habitats and in-situ resources for native species of special concern in conservation efforts throughout the entire peninsula. It would become a unique natural laboratory for ecology and conservation science available to the global community.

Korea as a whole would gain the following from the preservation of the DMZ ecosystems: (1) continued benefits of ecological services and common natural resources, (2) cooperative protective law enforcement, (3) educational and research opportunities, (4) mutual confidence building and nationalistic spirit nurturing, and (5) enhanced land tenure. These benefits have been well documented in several successful peace parks around the world, such as the La Amistad International Biosphere Reserve between Costa Rica and Panama.[34]

Transboundary Nature Reserves (TBNRs) and Peace Parks

The importance of protected areas is well recognized in Article 8 of the UN Convention on Biological Diversity (1992), which aims to strengthen national systems of protected areas for conservation of biodiversity worldwide. Biodiversity conservation practice is usually localized and thus should depend on an ecosystem management approach that integrates major parameters involved in ecosystem structure and function. Human factors that integrate protected areas management into broader municipal and regional planning for land and water use should not be ignored. In this context effective biodiversity conservation must account for ecosystem structure and functions and ecological processes in protected area management as is usually defined by political forces, including historical, cultural, and geopolitical factors, since there are no political borders for ecosystems and species.

Biodiversity does not recognize political boundaries. Korea's biodiversity cannot be conserved in isolation. The conservation strategy must be based on the concept of common natural heritage for all people on the Korean peninsula. As a common conservation strategy, the KPNRS provides a monumental opportunity for the two Koreas to preserve the last of Korea's natural heritage, namely ecosystems and native biodiversity. It will become the crown jewel of Korea's biodiversity conservation, the pan-Korean nature conservation that calls for *Building New Keum-Su-Gang-San* as a galvanizing slogan.

The New Keum Su Gang San (NKSGS) refers to a network of protected areas that is diverse in size and category, including municipal, provincial and national parks, and green zones connected throughout the peninsula by nature corridors and greenways. This network galvanizes the conservation forces of Korea-wide grassroots groups, particularly, in the urban community of North Korea, to foster various environmental programs for flood control, soil erosion control, reforestation, and food production as well as biodiversity conservation.[35]

Korea's DMZ is a ready-made system of transboundary nature reserves (TBNRS) on the Korean peninsula—TBNR refers to "an area of land and sea that straddles one or more boundaries between states beyond the limits of national sovereignty, whose constituent parts are especially dedicated to the protection and maintenance of biological diversity, and of natural and associated cultural resources, and managed cooperatively through legal or other effective means."[36] The DMZ corridor presents a diversity of habitats and landscapes for different classes of transboundary protected areas (TBPA). The TBNRS not only preserves the property and natural heritage of common interests but also provides an important and dynamic means to coordinate conservation of ecological units and corridors. TBNRS also brings about many other benefits, including the promotion of cooperation and confidence building between the parties, and conflict resolution and cooperation across boundaries.

Peace parks are deliberately managed through legal or other effective means in promoting peace and cooperation that encompasses trust building, understanding, and reconciliation between the parties involved, and conflict prevention and resolution, all of which would contribute to cooperation between and among countries, communities, agencies, and other stakeholders.[37] Many different peace parks have been successfully established throughout the world. They support economic development, conserve biodiversity, and promote regional peace and stability.[38] Based on experiences in Southern Africa with transboundary conservation zones,[39] the former president of South Africa, Nelson Mandela, proposed in 2001 that the two Koreas should build a peace park inside the demilitarized zone to help peace take root on the Korean peninsula, the last frontier of the cold war. His idea was transmitted to the then president of the Republic of Korea, Kim Dae-jung, a fellow Nobel Peace Prizewinner who then relayed the proposal to the DPRK leadership.

In fact the first peace park is ready to be developed through the DMZ in the eastern corridor. The Sorak National Park in the South and Mount Keumgang, located close to the DMZ, in the North can readily be linked through the DMZ corridor as a peace park. Within this potential peace park, however, the part of Mount Keumgang in the Kangwon Province in the eastern corridor has already been developed into a massive resort area for the *Geumgangsan Diamond Mountains* tour by the Hyundai Asan Corporation.[40] In the western front Kaesong, an ancient city in the Gyeonggi Province, has been developed into the Kaesong Special Industrial Zone.[41] These enterprising developments represent a potentially problematic breach of the environmental stewardship that would be necessary for sustainability of these invaluable landmarks in Korea.

Paradox of Foreign Policy and Green Resolution

North Korea's nuclear conflict is not going anywhere despite DPRK's recent decision to return to the next round of stalled Six-Party Talks. Based on its long history[42] the DPRK is not going to abandon its nuclear ambition readily without getting something substantial relative to sovereignty, security, and economic development in return because it is essentially a matter of their regime survival. It is no longer a question of whether North Korea already possesses nuclear weapons but rather how and when the DPRK leadership is going to use them.

Among the stakeholder nations involved in this conflict there is a basic common understanding that the Korean peninsula should be nuclear-free and North Korea must not become a nuclear power.[43] The North Korean people have endured critical starvation and subsistence living in recent years and are faced with shortage of food and basic necessities, while living in one of the worst environments in the world. The 2005 Environmental Sustainability Index ranks North Korea last of 146 countries and South Korea twenty-second, behind Liberia and Sierra Leone.[44] With regard to attitudes of the population under the DPRK's strong communist oligarchy, the North Korean people are patriotic and loyal to their motherland and to the oligarchy of Dear Leader Kim Jong Il.[45] It is also understood by the North Koreans that their mighty conventional military forces are increasingly outdated and ineffective[46] and that the national economy has basically collapsed. Yet they are doggedly determined to overcome the threats of the free world and begin to resolve a multiplicity of national problems with the nuclear deterrence, which is considered a great achievement and pride of the nation by many sectors of North Korean society.[47]

Started under "Sunshine Policy" of the former President Kim Dae-jung and further accelerated since the 2000 Inter-Korean Summit, North Korea continues to open their gate of isolation to the world, though the opening is still subtle.[48] Despite differing policies toward North Korea, major stakeholder nations, namely China, South Korea, Japan, Russia and United States, may gradually accept and endorse the inter-Korean dialogues and commercial and industrial advancement as well as economic and humanitarian support for North Korea.[49] The North Koreans understand that the regime cannot survive without opening their country to the world community and inflow of international assistance.[50] So it is important for us to engage the North Koreans at various levels and bring them into the world community.

Koreans on both sides of the DMZ believe in the idea of eventual unification, though scenarios and timing may differ. Today's South Koreans do not believe in forced unification by the North, while they do believe in the advancement of industry and successful enterprises into North Korea. Hyundai Asan Corporation has succeeded in establishing two compounds, on Mount Keumgang and in Kaesung City, for tourism and manufacturing. Through the process of interactions between the two Koreas, North Korea is starting to experience remarkable economic change at local community levels, and this should prevent the regime collapse.[51]

Yet the official economy has basically collapsed. State-owned enterprises have mostly failed, while military expenditures have continued to rise. Remarkably under this economic crisis the private economy is expanding at a very fast rate.[52] Today North Korean citizens are being permitted to own individual market-based businesses such as government-tolerated, semi-private markets, shops, and small companies across the country. The growth of private enterprises have brought important changes at the village and community levels. There are tremendous opportunities for international NGOs and stakeholder nations to engage with private citizens in advancing the local economies and improve the environment with small institutional and private grants.[53]

North Koreans are aware of how the United States and other Western countries perceive them, but regime survival is of great concern to the DPRK people, who are willing to use anything to sustain the status quo of their socialistic government hierarchy led by their Dear Leader Kim Jong Il.[54] North Koreans also accept that "the United States is committed to a strong, independent, reunified Korea."[55] In effect all the stakeholders understand the reality of the current political stalemate. The United States and other member countries of the Six-Party Talks should instead seize the opportunity to help build the official economy, as North Korea's socioeconomic policies are rapidly changing and in the current sociopolitical climate private enterprises are expanding at the village and community levels.[56]

It is unrealistic to expect North Korea to change its military-first policy and downsize the military power very soon. Yet over time the proliferation of an official market economy along with private economic enterprises could make expenditure on huge military forces unnecessary, and the DPRK government could turn to public welfare and the local economy as new sociopolitical focal points. Such change certainly could contribute to resolving the nuclear conflict and establishing permanent peace on the Korean peninsula. Today the DMZ conservation and pan-Korean nature conservation plans offer a means to rehabilitate the environment and begin building peacefully a sustainable Korea.

TBNR and Pan-Korean Nature Conservation as a Peace Effort

Transformation of the DMZ into a system of diverse TBNRs will be a long and arduous process. Declaration by the two Koreas that it will be done can only occur after sorting out both a bilateral agreement between North and South Korea and the sanction of the United States and China for their roles in the Armistice of the Korean War. Therefore a scientific rationale for the DMZ conservation alone cannot be expected to appease the political leaders of these stakeholder nations, although environmental security and conservation are fundamental to the future of Korea's sustainability and unity. To succeed, the TBNRS must consider all relevant factors that can contribute to build a nation that is peaceful, economically dynamic, and continually sustainable.

Nature conservation involves pan-Korean issues and the DMZ conservation as common goals for the two Koreas. Conservation is particularly important for North Korea because of problems with food production, flood control, reforestation, and energy production,[57] all directly linked to biodiversity, environment, and conservation. Korea's conservation strategy must aim at the reclamation of the *Keum-Su-Gang-San*. This is an environmental goal that will require massive restorations of wildlife habitats and the rehabilitation of biodiversity throughout the peninsula, such as the DMZ ecosystems have preserved with their biodiversity intact. Nature conservation should become a national priority for both Koreas. Environmental security for sustainable development can contribute to strong economic development on the both sides of the DMZ.

Contemporary political wisdom has not considered environmental issues as a political vehicle for conflict resolution between the North and South Koreas and for resolution of Korea's nuclear issues or of inter-Korean security. The preservation of the DMZ ecosystems and their biodiversity for conservation, however, introduces the most promising route to the Koreas' conflict resolution and integration.[58]

The 1992 Rio Convention on Biological Diversity (CBD) required information exchange, collaborative research, and conservation between adjacent states. The DPRK and the ROK are signatory countries on the CBD. As biodiversity does not recognize the divided sovereignty on the Korean peninsula, it makes a good common ground on which the two Koreas could jointly meet the various terms of the Convention. Formal joint efforts for Korea's nature conservation can help foster mutual trust and confidence building between the two Koreas. The process of building the KPNRS and its subsequent management will require close, friendly relationships between the two Koreas, and this can only ultimately lead to peace, cultural revival, environmental security, and human security on the Korean peninsula.

Process and Strategy for Preserving the DMZ Ecosystems

From the political complexity and competing interests for economic development in the DMZ corridor,[59] we have to come up with a specific strategy for preserving the DMZ ecosystems as a TBNRS that ultimately leads to successfully forging an inter-Korean agreement between the two Koreas. For this reason the DMZ Forum, a nonprofit organization, was organized to promote the conservation of nature in and near Korea's DMZ in 1997. The DMZ Forum has promoted opportunistic collaboration between North and South Korea in achieving the mission to (1) galvanize civic and environmental movements for developing unity and collaboration among conservation and environmental organizations in South Korea and (2) advance the promotion of conservation values and importance of the DMZ conservation among the public in South Korea with a concern for those resided near the DMZ.

The DMZ Forum also continues to mobilize support of scientists, political leaders, governments, and international organizations to provide interdisciplinary meetings to explore preservation issues of the DMZ, and to support research in providing baseline data for developing a transboundary peace park and related protected areas in the DMZ corridor.

In 1992 North and South Korea for the first time went into a historic agreement, the Basic Agreement[60] between the two Koreas, DPRK and ROK, that affirmed the use of the DMZ for a specific peaceful purpose. Valucia (1998) stated that both Koreas expressed interest in the promotion of a biosphere reserve in the DMZ. Since then, numerous ideas have emerged for preserving the DMZ for diverse use including development and conservation. Toward the end of his administration, the former president of ROK, Kim Young Sam, proposed the preservation of the DMZ ecosystems for conservation in the Special Session on Environment and Development of the UN General Assembly on June 23, 1997, following the Seoul Environmental Declaration.

Subsequently the North Korean Nature Conservation Union (or Federation) made a positive statement concerning the preservation of the DMZ for the common interest and development of the two nations through open communication via the US Radio Free Korea (Pack 1999), though the DPRK military still regarded the DMZ strictly in terms of military buffer.

The DMZ Forum has made considerable advancement in the outreach and networking of interested conservation organizations and international agencies for the cause. The Forum has gained broad-based support of influential individuals and organizations from around the world through public lectures, individual contacts, news media, publications, and annual conferences. Since 2003 the group has directly engaged with international conservation agencies and NGOs for their active participation. With the conference as a vehicle, the DMZ Forum started working with the Gyeonggi Province and the South Korean government. The 2004 DMZ Forum conference, "International Conference on Conservation of Korea's Demilitarized Zone" (DMZ), in Seoul was sponsored by the Gyeonggi Province and focused on the designation of the DMZ for a UNESCO World Heritage Site. In 2005 the group convened the 2005 DMZ Forum conference, "International Conference on Pan-Korean Nature Conservation for Peace and Sustainability," in Seoul in mid-August.

The overall strategy for building the TBNRS on the Korean peninsula must include distinct steps to build public understanding and support for the cause[61] and offer flexibility to policy makers of the two Koreas so that they add the DMZ conservation to their respective political and socioeconomic perspectives. Both simultaneous and consecutive approaches to implementing this goal have been raised by the DMZ Forum.

Project-Based (or Bottom-up) Approach

This approach would involve a diverse number of site-specific and species-specific projects by different groups of people in South Korea aided by interested collaborators from around the world. These projects not only produce knowledge-based or conservation-affected outcomes but also help improve the community-level economy and environmental stewardship by actually engaging individual biodiversity and conservation scientists, organizations, and grassroots activists.

The project-based approach would work closely with local governments and their leaders. The DMZ Forum members would partner with other NGOs in and out of Korea, such as the KFEM (Korean Federation of Environmental Movements) and ICF (International Crane Foundation), which has promoted and supported a global coalition for conservation activities for migratory cranes. ICF collaborated with the DMZ Forum on their work in Cholwon Basin, for which an inter-Korean project was organized, and a recent submission of the GEF (Global Environmental Facility) proposal has been made to the South Korean partners.

This project-based initiative may include smaller projects such as the Cholwon Crane Festival of 2003, the inter-Korean cooperative crane census, the International Tiger Conference, the International Music Concert at Panmunjom, the International Peace Boat Event, and the education and training of North Korean specialists in conservation. Similarly an international Han-Imjin estuary reserve may be proposed for conservation of migratory birds such as the white-naped crane and black-faced spoonbill.

Whether these projects begin in South or North Korea, they must subsequently bring in the other partner to organize jointly so that the conservation or cultural objectives are met together.

Inter-Korean (or Horizontal) Approach

This approach involves the South Korean government (ROK) with support of NGOs and civic organizations. The DMZ Forum initiated this approach in South Korea in 2003 by convening a joint conference with KFEM. In 2004 the DMZ Forum initiated a formal relationship with the ROK government (Ministries of Unification and Foreign Affairs and Trade) after the Conference in Seoul. The ROK Ministry of Unification (Bureau of International Cooperation) has already succeeded in establishing the Inter-ministerial Committee on the DMZ, although only a little progress has been made by this committee to date. Future meetings at the DMZ Forum should renew the committee activities and move the DMZ conservation issues toward engagement of the upper administrative units, and ultimately the cabinet and the president. Currently the DMZ Forum is planning to submit a proposal on the DMZ conservation for eventual adoption by the ROK government and to present a national position in the negotiations of the inter-Korean meetings. The US Department of State may become involved in supporting the DMZ conservation via inter-governmental meetings with North Korea.

High-Level (Top-down) Negotiation with the DPRK Leadership

This approach involves a delegation of influential people, among them former diplomats, philanthropists, conservationists, scientists, and well-known former politicians. This group organized around high-powered international figure(s) on behalf of the DMZ Forum would meet with the DPRK leadership to promote the inter-Korean agreement for preserving the DMZ for conservation and peace and convince them of its value. In 2005 to promote this agenda Ted Turner, chair of the Turner Foundation and founder of CNN, visited Pyongyang to meet with high North Korean government officials and participated in an international conference to move forward the peace park agenda. While the meeting was inconclusive, both sides agreed that the DMZ could be instrumental in achieving cooperation between the Koreas.

Challenges and Prospective: Conclusion

The historic 2000 summit opened the possibility of positive political dialogue and interactions, and a means of reducing tension through economic cooperation that may lead to peace and eventual integration or unification. Since then the different interactions between the two Koreas have exceeded common expectations—aside from North Korea's nuclear and human right issues. If this socioeconomic relationship can continue and expand, the opportunities exist for pan-Korean environmental issues to be settled and for unification of the two Koreas. Sustainable development and environmental security are intertwined. They are the primary way to reach inter-Korean agreement, since reviving a healthy environment and a shared rich natural heritage are of paramount importance for all Korean peoples.[62]

The Korean peninsula is suffering serious environmental degradation from rapid economic development,[63] namely industrialization, urbanization, and land conversion, in the South and military buildup and operations along with a dismal economy in the North. Throughout the cold war, and to this day, military operations and economic development with their related urbanization have leveled the land, polluted the environment and water ways, and destroyed habitats, causing massive loss of biodiversity. South Korea's success in economic development has built a high standard of living and confidence. While for these reasons South Korea is often portrayed as role model for other developing countries, without a clear political mandate by the public nor a determined policy commitment on the both sides of the DMZ corridor, environmental degradation has continued. Environmental degradation and the loss of biodiversity is known to undermine the capacity of the natural ecological processes that sustain our life-support system. The degradation also harms human health, and so economic degradation will eventually constrain economic development and endanger the Korean peoples' security.[64]

South Korea's active pursuit of the North–South reconciliation and economic exchanges has accelerated the entry of South Korean industrial and business enterprises into North Korea for commercial exploitation. Since the historic 2000 Inter-Korean Summit, South Korean development and industrial ventures have accelerated despite periodic disruption in the bilateral exchanges.[65]

The lack of environmental commitments by the Korean governments and by South Korea's industries, and these industries' associated environmentally questionable practices, are cause for concern. The environmental concerns are closely linked to peace and national security. Because biodiversity and pollution do not recognize political boundaries, for the two Koreas conservation and environmental protection are transnational issues that must be addressed with common agendas. The preservation of the DMZ ecosystems promises to provide an unusual opportunity for the two Koreas to work together toward common goals and economic and environmental securities. Such strategy therefore could become an attractive vehicle for conflict resolution concerning North Korea's nuclear threats and for changing the political environment over the Korean issues toward a more flexible and optimistic future.

Although many problems of the DPRK regime require immediate attention, we must focus on the welfare of 23 million North Korean people for which the regime must prepare to improve their welfare toward building peace, security, and prosperity. The world community, including the stakeholder nations of the Six Party Talks, does not need another war to devastate the fruits of tireless economic development planted by South Korea over the last four decades, nor to sacrifice the lives of many more millions of innocent people.

While the future of the politically and diplomatically complex Asia–Pacific region may be murky and unpredictable, a peaceful resolution of North Korea's nuclear conflict that allows for an evolution of the DPRK regime and for the welfare of 23 million people may prove to be the strategy for sustained peace and sustainable development in the region. A green strategy for the North Korean conflicts through the preservation of Korea's DMZ corridor for biodiversity conservation and environmental rehabilitation can be achieved by pan-Korean nature conservation through the collective effort of all stakeholders.

As discussed in detail in this chapter, the political and economic conditions in North Korea lend themselves to the platform built by the 2000 Inter-Korean Summit: continual dialogue and economic and cultural exchanges. It appears that the political and economic conditions currently faced by North Korea offer a unique opportunity to resolve the North Korean nuclear standoff amicably by all stakeholder countries with a focus on peaceful sustainable development for the people on the Korean peninsula. Such a development process for the KPNRS could be enhanced by acceptance of the international designation of special categories for the KPBRS as Biosphere Reserve (IUCN) or World Natural Heritage site (UNESCO). Once the concept of the KPNRS is accepted by the two Koreas, the formal process of establishing the nature reserves could begin with an inaugural bilateral event that involves stakeholder nations and all potential sponsors including the United Nations and regional/global conservation organizations.

Notes

1. Solomon (2003).
2. Kim (1997).

3. Lee et al. (1994).
4. Kim and Weaver (1994).
5. Lee et al. (1994), Kim (1995, 1999), Eder (1996), and UNEP RRC.AP/UNDP (2004).
6. Lee et al. (1994).
7. *Chaebols* are large Korean business houses such as Samsung, Daewoo, and Hyundai.
8. Eder (1996).
9. Lee (2004), Kwon et al. (2002), and Kim (2006).
10. Ibid.
11. Oberdorfer (2005).
12. Burton (1995), Deutsch (1991), and Yordan (1998).
13. Levin (2002).
14. Hoare (2005).
15. Seo (2002) and *CIA—The World Fact Book* (2004).
16. Hoare (2005).
17. Oh (1999).
18. Park (2005), and Kim (2006).
19. *CIA—The World Fact Book* (2004).
20. Park (2005), and Kim (2006).
21. Eder (1996), and Kim (2006).
22. Ibid.
23. Ibid.
24. *CIA—The World Fact Book* (2004).
25. Kirkbride (1985).
26. Bedecki (1997).
27. O'Hanlon (2003) and Woo (2006).
28. Kim (1997).
29. Lee et al. (1994) and Sungchun-Munhwa-Jaedan (1996).
30. O'Neill (2004).
31. Westing (1993, 1998) and Kim (1995).
32. Kim (1995, 1997, 1999).
33. Valucia (1998).
34. Westing (1993, 1998) and IUCN WCPA Task Force Transboundary Protected Areas (2005).
35. Kim (1995).
36. Westing (1998) and IUCN WCPA Task Force Transboundary Protected Areas (2005).
37. IUCN WCPA Task Force Transboundary Protected Areas (2005).

38. For example, Westing (1993), Breymeyer and Noble (1996), Sochaczewski (1999), van der Linde et al. (2001), and Shambaugh et al. (2001).

39. Rufford Foundation (2005).

40. World Tourism Organization (2003).

41. Hyundai Asan (2004).

42. Litwak and Weathersby (2005) and Parker (2005).

43. Mead (2005).

44. Esty et al. (2005).

45. Macdonald (1996).

46. Hayes (2005).

47. Demick (2005).

48. Oh (1998).

49. O'Hanlon (2003).

50. Anderson (2005) and Shaplen and Laney (2005).

51. International Crisis Group (2005).

52. Lee and Yoon (2005).

53. Wolf (2005).

54. Kihl (2006).

55. Leach (2005).

56. Hoare (2005).

57. Hoare (2005) and Kim (2006).

58. O'Neill (2004) and Kim (2006).

59. O'Neill (2004), Kim and Wilson (2002), and Kim (2006).

60. The February 1992 Agreement on Reconciliation, Non-aggression and Exchange and Co-operation between the South and North.

61. Kihl (2005).

62. Dongguk Daehakkyo Bukhomhak (transl. Dongguk University Institute of North Room Studies) (2006).

63. Lee et al. (1994), Eder (1995), and UNEP RRC.AP/UNDP (2004).

64. Kim and Weaver (1994), Raven (1997), and Daily (1997).

65. Kim (1995, 1997, 1999, 2006) and Kim and Wilson (2002).

14

Nesting Cranes: Envisioning a Russo–Japanese Peace Park in the Kuril Islands

Jason Lambacher

I have always believed and now especially believe that the idea of nature protection will...become the basis on which friendship between peoples will be built, (and) give rise to common interests.
—Susanna Nikolaevna Fridman

Home to an extensive range of remarkably intact terrestrial and marine biodiversity, the Kuril Islands are one of the most volatile places on the Pacific's "ring of fire," both as a measure of political and physical geography (figure 14.1). Politically, the current act of a long-standing frontier drama began in the last weeks of World War II, when the Soviet Union violated a neutrality pact, declared war on Japan, and occupied the entire archipelago. Actual conflict lasted only weeks, but formal peace has been elusive for nearly sixty years, mainly due to a lingering territorial dispute over the four southernmost islands (Kunashir, Iturup, the Habomais, and Shikotan). While the Kuriles are not an area of endemic violence, a "state of war" without a peace treaty has prevented "normalized" relations. This is an outstanding fact to realist observers (in particular) who find it strange that "a few windswept islands in the Pacific" could prevent the marriage of Russia's resource endowment with Japan's capital, especially after the geopolitical changes of the cold war.[1]

Both governments have continually expressed a desire to resolve the Kuril question with recourse to law, justice, and historical facts, and build relations based on respect, trust, and mutual understanding.[2] However, justice, and historical facts in the Kuril case, are essentially contested perspectives, and there have been few opportunities to establish cooperative relations in a pragmatic context—leaving the two countries in a seemingly intractable conflict that Hasegawa calls "neither war nor peace."[3]

Kuril history demonstrates that a solution to the row will not come from the past, but rather from a spirit of political freedom infused by what Hannah Arendt calls a

Figure 14.1
Map of Kuril Islands

"pathos of novelty."[4] An International Peace Park (IPP) for the four southernmost islands represents a genuinely new strategy in a context of diplomatic stalemate, and offers an attractive, though not politically uncomplicated, way to advance peace and effective conservation. It is clear that, as both sides have acknowledged, a peace accord hinges on resolution of the Kuril conflict. This, in itself, justifies the risk of abandoning exclusive claims to Kuril sovereignty by "co-owning" the islands through joint ecological stewardship. But an IPP also has the potential to subsume nationalist animosity by triggering what Conca calls "synergies for peace," thereby helping to maintain the peace treaty through time.[5] Through the experience of pragmatic cooperation, a political process emerges to practically and strategically realize a substantive long-term peace, not just tactical short-term reconciliation. Also, because several parks, conservation areas, and wildlife refuges already exist in the islands, new land and seascapes would not need to be designated as parkland, making the political transition to an IPP more feasible to people who live nearby. Local residents of the Kuriles, long neglected by Moscow, and the economically depressed communities of northern Hokkaido would benefit from ecotourism, greater people to people contact, and an environmental regime that sustainably manages marine resources for future generations. Indeed, the democratic inclusion of local residents and NGOs explores paths to peace that have been ignored at the elite diplomatic level.

From a conservation perspective, an IPP represents an opportunity to protect a wilderness mosaic that remains relatively pristine despite the environmental degradation that afflicts much of Northeast Asia. This is especially critical because the islands are threatened by an epidemic rise in poaching by criminal networks and by development pressures like mining and industrial fishing.[6] In fact the Kuril bioregion is jeopardized *both* by continued Russian control of the islands *and* by a return to Japanese sovereignty.

Resolution of the Kuril issue without political commitment to its ecological health could lead to a more efficient exploitation of natural resources and environmental conflict in the years to come.[7] The legacy of contest between the two nations will not be easily overcome, however. Political obstacles include long historical memories, the nationalistic posture of the Russian and Japanese governments, subordination of environmental ministries in both countries, and little public knowledge about the critical need for conservation in the islands.

Establishing an IPP would also have political costs that range from abandoning fifty years of diplomatic strategy regarding sovereignty, alienating domestic groups dedicated to the settlement or return of the islands, and restricting (legal and illegal) resource extraction industries (particularly fishing). Nevertheless, the benefits of a Kuril IPP outweigh the costs, in my judgment, and the idea offers creative release from the inertia of history. The timing may be propitious as well, as the emotional nature of the Kuril issue is in many respects a generational legacy from the war that shows signs of waning.[8] Though the construction of a Kuril IPP will require leadership, risk-taking, and political outreach, we should remember that, as Morris-Suzuki importantly reminds us, frontier zones make radical political experimentation possible.[9] While not a panacea for Russo–Japanese relations, an IPP makes considerable political and ecological sense in the Kuril case and demands further awareness, debate, and political action.

Ecological Geography of the Kuril Islands

The Kuril Islands have been poetically described as "subterranean Himalayas," as the distance from ocean floor to peak is an astonishing 13,000 meters.[10] Geothermal vents, hot springs, fjords, waterfalls, giant bamboo, crater lakes, and hot black sand beaches compose the area's ruggedly beautiful physical geography. Because of frequent vulcanism, earthquakes, tsunamis, strong currents, fierce storms, ice sheets, and relentless summer fog, the islands have never been significantly transformed by human development despite settlement by the Ainu, Russians, and Japanese. The archipelago is under one of nine global flyways for birds, permitting migration between the Artic, Kamchatka, Japan, Southeast Asia, Australia, and New Zealand.[11] Abundant marine life, the product of the convergence of cold and warm sea currents and massive colonies of phytoplankton, rivals Georges Bank in the northwest Atlantic.[12] Salmon, tuna, flounder, halibut, cod, walleye pollock, perch, mackerel, saury, whale, seals, sea lions, otters, crab, scallops, shrimp, squid, sea slugs, urchin, and sea cucumber all flourish in Kuril waters, which are also home to the largest kelp forests in the world.[13] The bioregion is also noteworthy in that it has a near complete absence of foreign invasive species.[14]

Unfortunately, few Western researchers seem to know about the global significance of Kuril ecology. This is not surprising, as the islands even remained largely unknown to science until the 1990s, when the first comprehensive biological survey of the islands took place.[15] Though visible from Hokkaido, restrictions on travel and investment mean that most Japanese have never been to the islands, with the exception of a few visitors who mostly come to tend ancestor grave sites.[16] For Russians, remoteness and poor transportation links mean that the Kuriles are not a prime destination, except for transplanted locals, military personnel, and summer laborers (who comprise a peak population of fewer than 20,000 people).[17] Despite this lack of public knowledge, the environment has not been exogenous to the frontier contest. The search for fur and fish propelled expansion into the Kuril frontier for the past two centuries, and its natural abundance is a central reason why both countries seek control of the islands today.

Political Geography—A Contested Frontier

Overall, relations between Japan and Russia have been punctuated by the clash of great power rivalry, and the present iteration of the dispute continues this nationalist contest. From 1855 to 1945, the four isles were under Japanese domain; from 1945 to present, they have been under Russian control. Although the current situation does not appear in any danger of imminent eruption, political tectonics beneath the surface make the Kuril frontier unstable.

Both countries filter the conflict through the prism of language, history, and nationalist ideology. In Japan, the four disputed islands have recently been dubbed the *Hoppoo Ryoodo* (Northern Territories), and are commonly referred to as *koyuuno* (inherent territory).[18] *Koyuuno* is meant to suggest that unlike early twentieth century control of nearby Sakhalin Island, which was administered as a colonial possession, the "Northern Territories" are seen as an integral part of the Japanese home islands. Russians refer to the same islands as the "Southern Kuriles," calling the entire Kuril chain *iskonno rossiskie territorii* (timeless Russian territory).[19] *Iskonno* emphasizes the Russian argument that they settled the Kuriles first.

To complicate the matter further, a fundamental disagreement exists about which islands are even part of the Kuriles. Are they all of the islands between Kamchatka and Hokkaido, like Russia believes, or do they only include, as Japan insists, the islands north of Iturup? Are Shikotan and the Habomai islands part of the Kuril chain, or are they natural extensions of Hokkaido? Different answers draw on different political, legal, and even geological definitions, on which there is little official agreement.

Since it is not clear which country legitimately "owns" the islands, the dispute is a thicket of legal and moral argument marked by conspicuous lapses of memory about past aggressive action. Each side claims prior "discovery," but it is not possible to conclusively determine who trapped otter, extracted tribute, or settled the islands first, other than the Ainu, who still refuse to recognize both Russian and Japanese sovereignty despite virtual extermination by smallpox, forced labor, and resettlement.[20] What is clear is that both have justified claims to *some* territory in the islands. Aside from this, vague normative legitimacy succumbs to the historical legitimacy of short-lived treaties and military victory.

Seeing the Kuriles as compensation for "self-sacrifice" in war and retribution for past Japanese aggression, the Soviets invaded shortly after the bombing of Hiroshima, with the blessing of the Americans at Yalta. They depopulated Japanese from the islands, sending some back to Japan and others to join 640,000 POWs interned in forced labor camps throughout the Soviet Union.[21] Upon return, many POWs settled in eastern Hokkaido, helping to make this area a hotbed of the irredentist movement in Japan.

A 1951 peace treaty ended the American occupation of Japan, but under conditions that the Japanese have come to regret. In this treaty America forced Japan to renounce "all right, title, and claim" to the islands, though what constituted "The Kuriles" was not carefully defined. Russia argues that Japan's renunciation in the 1951 peace treaty and the Yalta Agreement confirm their legal claim to the islands. For Japan, the illegitimate agreement at Yalta, Russia's refusal to sign the 1951 peace treaty, and its particular definition of where the Kuriles begin and end constitute its core legal argument.

Kuril tremors intensified because of postwar geopolitics, and the issue quickly became the central obstacle to a peace accord. Russia offered to give Japan Shikotan and the Habomais in exchange for a peace treaty in what is now called the 1956 Declaration. The United States, eager to prevent the two countries from coming together, threatened to permanently annex the Okinawan islands should Japan agree to the deal.[22] The threat worked, and no agreement was signed. US cold war policy, as Kimura and Welch rightly argue, was "designed to prevent rapprochement between Japan and Russia" through the Kuril issue.[23]

The issue stagnated for thirty years until the Gorbachev era kindled new hope for a deal. One idea that gained traction (but ultimately failed) was for Japan to purchase the islands, the so-called Alaska Solution.[24] Reportedly, the LDP unofficially offered $28 billion in return for all four islands.[25] However, Gorbachev was successfully criticized by Yeltsin for being willing to "sell Russian territory" at a time when the Soviet Union was crumbling.[26] Ironically, Yeltsin was later criticized on the same grounds when he tried to solve the dispute.[27]

Despite these false summits, optimism surged in the 1990s. Building off a 1993 declaration that pledged a peace treaty once the Kuril issue was settled, two "no-necktie" summits at hot spring resorts in 1997 and 1998 revealed the startling news that the issue would be solved by the year 2000.[28] By 2006 this optimism has proved unfulfilled, though cooperation on terrorism, energy, and even military exchanges emerged in 2005.

But the more things change, the more they stay the same. At a meeting between Putin and Koizumi in November 2005, Putin tersely offered the return of the two smaller islands in return for a peace treaty, the same basic offer that has been repeated since 1956.[29] Koizumi's predictable response was sharply negative. Japan fears that this solution would legitimate Russia's claim to the two larger islands, and could potentially serve as precedent for other simmering island disputes with China, Taiwan, and South Korea (all of which flared up in 2005). It seems clear that without significant change in trenchant bargaining positions the status quo will assuredly persist.

The conflict appears puzzling to many outsiders. In particular, observers have long wondered how the Kuril issue could prevent seemingly boundless prospects for economic cross-fertilization.[30] This perspective underestimates and ignores the symbolic value of the islands as a site of justice, a field of great power pretension, and a crucible of national identity.[31] Although strategic and economic dimensions to the conflict are important, such as "ice-free" naval access to the Pacific for Russia, Kunashir's mineral deposits (particularly, titanium and gold), marine resources, and regional power politics, the assumptions of *realpolitik* inadequately describe the nature of the Kuril conflict.[32] Instead, the issue is better understood as one of justice with path dependent contingencies. Forsberg rightly argues that subjective conceptions of justice are especially prominent in historic border disputes such as the Kuril conflict.[33] Kimura and Welch, moreover, correctly characterize the Kuril issue as a path dependent result of historic, cultural, diplomatic, and domestic microhistories politically driven by aggrieved nationalist groups.[34]

The justice politics of territorial nationalism means that a genuine solution won't be found in the law and "facts" of the past but rather in the political freedom to create a new environment for justice and peace. As Hakamada writes, "[E]xperience between the two countries over a long time has shown that the territorial problem was not something that could be solved automatically by the two countries simply by piling up historical arguments and arguments based on international law."[35] A new approach therefore needs to unsettle nationalist thinking, defuse historical grievances, and entertain innovative solutions. With a flexible conception of sovereign territory (in this limited region) with ecological protection at its center, a Kuril IPP potentially forges a just resolution that can create a lasting peace between two important global powers.

Envisioning an IPP in the Kuriles

To represent an alternative future of peace and ecological integrity, an IPP must transform how each side understands "sovereignty" and "justice." This argument is predicated on a willingness to change perceptions of the conflict, which means thinking in political-ecological terms and advancing the prospects of environmental peacemaking. To some degree a subtle but potentially significant change of perception is already happening. For instance, the IUCN started promoting the idea of "international conservation" in the islands for endangered species protection and sustainable management of fisheries in the early 1990s.[36] More recently Japanese and Russian scientists, NGOs, and some bureaucrats have publicly broken the political taboo of questioning the sovereignty imperative and have supported a variety of transboundary conservation initiatives, including the IPP approach.[37] Importantly, this includes both countries' environmental ministries and district and regional authorities in Sakhalin.[38]

At a UNESCO-sponsored conference in 2001, Russian and Japanese scientists, in uncharacteristically sharp language, lamented that existing conservation measures in the islands were not working and that new, comprehensive efforts were needed to strengthen protection of existing parks and refuges. Both governments were directly encouraged to act "fearlessly, boldly, and with urgency."[39] In 2003 another conference organized by a consortium of grassroots NGOs, most notably the Japan-based Kuril Island Network (KIN), focused on the IPP concept specifically. This conference, which garnered considerable domestic and international press, featured speakers from KIN, the International Peace Park Foundation, the Russian Academy of Science, Birdlife Asia, the International Crane Foundation, and the Beringian Heritage Program (a Russia–Alaska joint conservation project in the Bering Sea), and is intended to be the first in a series of steps toward the formal unveiling of an IPP proposal.[40]

Although the latest developments represent innovative thinking about the issue, both governments still seem focused on political costs rather than political opportunities. The costs essentially amount to letting go of historical memories of injustice, which are always close to the social construction of national identity, and, correspondingly, to abandoning the dream of exclusive territorial sovereignty over the islands. So long as the focus is on political costs, the IPP idea will appear unrealistic and marginal. Indeed Putin's desire for greater natural resource exploitation, inability to crack down on poaching, neglect of Russia's outstanding system of nature reserves, and harassment of environmental NGOs and independent media does not, on the surface, suggest a climate favorable to the IPP concept.[41]

The Japanese political establishment, tightly focused on the sovereignty question and unable to officially countenance anything other than irredentist orthodoxy, has called the IPP concept a "legal nightmare."[42] Furthermore, even if an IPP were realized, past as prologue suggests that it would merely be the latest addition to a history of Russo–Japanese agreements that have resulted in fragile treaties, punitive victories, and unstable conditions for peace.

Skeptics might ask: How could an IPP genuinely work between *these* two countries? And further, doesn't an IPP need a stable border? Then there are the practical questions of how the bureaucracies of both countries could jointly manage differences in organization, language, culture, and budgetary politics. The nature and scope of marine conservation could also lead to serious disagreement and political backlash. Dwelling on the political obstacles and costs of a Kuril IPP gives a dose of realist caution, but ultimately leads to an eternal recurrence of frontier conflict. If the focus is instead on the political opportunities afforded by the IPP idea, then new horizons for peace and conservation open up.

Politics and Peace

With these challenges ahead, what are the political and ecological grounds for an IPP in the Kuriles? The IPP concept is a project that can lead to a politics of peace, ultimately leading to a peace treaty and better political, economic, and cultural relations, for four main reasons.

First, an IPP is a form of political compromise over the sovereignty issue. Instead of the exclusive possession of one country, the Kuriles would become shared territory in "international space." Through co-stewardship both countries could argue that they have a kind of sovereign control over the islands. This provides a considerable measure of saving face in light of the inevitable criticism from nationalists, something that is critically important in both cultures. Russia could argue that they achieved a peace treaty without giving up territory, and Japan could claim that the "Northern Territories" have "returned."

Second, the IPP idea alters the symbolic justice dimension of the dispute from conflict to cooperation. With overcoming the injustices of the past an explicit goal, the substance of justice changes from "winning" the battle of historical grievance to a future-oriented resolution that redefines what justice means in this context. This means that the activities of pragmatic cooperation become a constituent feature of justice in the Kuril issue. This cooperation could then unleash what Conca terms "synergies for peace," which include trust-building, building on a habit of cooperation, consensual knowledge, identifying mutual gains, and acting in longer time horizons.[43]

Interestingly all of these objectives have been identified in recent diplomatic rhetoric, at least in the abstract.[44] Working on the design of the park and its myriad issues—laws, enforcement, regulation of permitted activities, accommodations for ecotourism, visa controls, among others—would exercise, and test, cooperation, but it would be actual cooperation nonetheless. "Consensual knowledge" in this case means the potential for joint scientific projects and cross-cultural knowledge of different park management styles.[45] An IPP has built-in "mutual gains" in terms of ecological metrics and the reduction of inter-state hostilities.

Finally, the IPP is concerned with longer time horizons, in that preserving the wilderness character of the islands protects natural cycles and the enjoyment of the park for future generations. The process of working through the details of administration thus provides a shared framework that enhances trust and opens communication. Trust must be earned, but both sides must first commit to practical circumstances that allow trust to emerge.

Third, an IPP would involve an important role for civil society and local residents, which taps unexplored possibilities for peace. In recent years frustration with political intransigence in Moscow and Tokyo has encouraged ordinary people to vocalize their concerns about the Kuril issue, especially in Japan where hundreds of groups focus on the return of the islands.[46] Most reflect the concerns of the national government and are irredentist, but others (as mentioned previously), particularly Japan's environmental and scientific communities, are considerably more flexible in their approach to the issue. For them, and most likely for many other Japanese, the ecological appeal of preserving a "lost Japan" that escaped the modern, concrete fate of the rest of the country has powerful resonance.[47]

Russia's environmental and scientific communities, who have a truly surprising legacy of political success that stretches back to the early Soviet era, are also keen sources of support.[48] Unlike the Japanese, Russians in general aren't in any particular hurry to resolve the issue, mostly because they presently control the islands.[49] However, Kuril residents are less attached to the sovereignty question, and some have even expressed openness to Japanese sovereignty over the islands if it improves their quality of life.[50] Under the IPP concept, Kuril residents would avoid the painful options of continued neglect or transferring citizenship to Japan, as they would remain Russian citizens who live in internationally-administered space. The IPP would thus be a portal through which greater interaction between citizens could take place, helping to break down monstrous cultural stereotypes. More people to people contact would thus make more likely the creation of common interests in the region, especially through environmental protection, ecotourism, and (legitimate) trade and investment.

Finally, the international community would strongly support the idea of a nonviolent, conservation-based solution to a dangerous inter-state conflict in Asia. Global institutions, such as UNESCO-MAB, who currently restrain their advocacy of an IPP for fear of offending the Japanese and Russian governments, would likely relish the opportunity to assist such a high-profile park, assuming sufficient political support. Other transnational environmental and scientific NGOs, like the groups that sponsored the recent IPP and transboundary conservation conferences mentioned earlier, would also vigorously participate in the project. Furthermore the United States, it could be argued, has a moral obligation to contribute to a successful IPP, as its cold war policy helped to create and prolong the Kuril issue.

Ecology and Peace

The ecological argument for an IPP also rests on four main points. First, 60 percent of Kunashir, Shikotan, and the Habomai islands are already designated as *zapovedniki* (scientific wilderness reserve) or *zakaznik* (refuge).[51] Thus an "infrastructure" of conservation areas already exists. What's more, an IPP would have the advantage, from a conservation biology perspective, of having additional protected areas to the north and south of the islands, thereby creating a large swath of linked reserves simply by loosely integrating existing parks and refuges on Urup Island and in Hokkaido.

Second, an IPP would recharge a political commitment to environmental enforcement. Considered in the context of a worldwide collapse of fisheries and threats to marine biodiversity, the Kurils stand out as a notable global exception to overharvest and short-term exploitation—for the time being. However, the Russian and Japanese "fish mafia" have staggeringly increased their operations since the end of the cold war.[52] For instance, the ratio of official exports to official imports of Russian marine products to Japan is 1:9, and represents unaccounted revenue in the billions of dollars, with most of the catch coming from the Kuriles.[53] Sea urchin, squid, sea cucumber, salmon eggs, and immature crab are the most lucrative species harvested, much of which sadly and illegally ends up in Japan at inexpensive "all you can eat" sushi restaurants.[54] Further the 1990s saw the introduction of new "supertrawlers" in the region that trap untargeted species and damage the seafloor.[55]

Russia has a first-rate system of nature reserves, at least on paper, with the Kurilsky Reserve and Shikotan and Habomai refuges among the wildest and most distinctive, but they are terribly understaffed.[56] The Kurilsky Reserve once had seventy rangers and thirty researchers, but these positions have been cut by 75 percent.[57] Development projects, such as a large gold-mining project on Kunashir, threaten the northern buffer zone of the Kurilsky Reserve, though so far plans for the project have been blocked by the Sakhalin regional government.[58] An IPP would boost the enforcement of environmental laws already on the books and re-establish Russian political commitment to established protected areas. The IPP would also challenge Japan's capacity for environmental enforcement, particularly for the seas.

Marine regulations in Japan are notoriously unambitious. Kazuo Wada, chairman of Japan's Wildlife Conservation Society, notes that "Japan has no laws to protect marine mammals, nor (does it have) sanctuaries at sea."[59] It is said that the Environment Ministry "stops at the shoreline," where it is then subordinated to the powerful Ministry of Fisheries, whose main role is maximization of the marine harvest.[60] Environmentalists fear that if the islands were returned to Japan, exploitation of their rich marine resources would take precedence over conservation. An IPP would certainly include a marine sanctuary, and would force Japan's bureaucratic competence in managing marine zones for conservation purposes. It is unimaginable that there would be comprehensive restrictions to fisheries in the Kuril region, but a coherent plan for managing fisheries would be better than the lack of effective enforcement that presently exists.

The contrast between management of Russian and Japanese conservation areas could also have unexpected side benefits for conservation elsewhere in Japan. Japan's national parks are less focused on ecological preservation and more focused on revenue generation. Anyone who has climbed Mt. Fuji knows of the dissonant collision between nature and commerce of on-trail vending machines at 3,000 meters. Japanese parks could learn from Russia's sophisticated *zapovednikii* approach to nature protection (core reserves, buffer zones, transition areas), evident in the Kurilsky Reserve.[61] To be sure, balancing the restrained Russian model with Japanese mass tourism would constitute a major philosophical challenge for the park, but it could also yield surprising results in terms of a renewed focus on the goals of biodiversity conservation.

Third, an IPP in the Kuriles would encourage wider and deeper cooperation on a number of environmental issues crucial to the region's environmental health—pollution, dilapidated nuclear sites and sunken submarines, and the overharvesting of marine resources in the Sea of Japan, Sea of Okhotsk, and the North Pacific. Past oil spills, fears of nuclear contamination, and recognition of the increasing problem of poaching by criminal networks have already created a minimal degree of cooperation on these environmental issues. Success with the IPP idea has the potential to expand these nascent environmental and enforcement initiatives.

Fourth, an IPP would be an important tool for endangered species protection and strategies for ecological restoration. It is well known that we are currently living through a period of mass species extinctions, but the impact of this process is not felt evenly. In fact 75 percent of all modern era extinctions have occurred in island species.[62] Though Kuril biodiversity is relatively intact, the islands are home to a disproportionate percentage of Russia's endangered flora.[63] Also several species currently endangered in Japan, notably the iconic red-crowned crane, Stellar's sea eagle, Stellar's sea lion, tufted puffin, and Blakiston's fish owl, exist in relatively healthy numbers in the islands.[64] These populations could be used to reintroduce species back into Japan or other depleted areas of the Sea of Okhotsk bioregion.

Conclusion

The Kuril region is politically unstable, and its relatively pristine land and seascapes do not have a secure ecological future. Resolving the simmering political conflict and the impending environmental crisis in the broadest sense requires a view that looks forward to a kind of future where a commitment to the environment helps to overcome interstate hostility. Although environmental cooperation is usually treated as a dependent variable by scholars, the IPP concept has promise to develop into an independent variable in the Kuril context that can defuse tensions, increase trust, and lead to a dynamic of peace.[65] This would necessarily require a major shift away from the paradigm of aggrieved nations petulantly arguing over territory and toward an understanding of sovereignty for the islands that is international and biogeographical. To be sure, changing the "wearisome cycle of self-glorification and selective recall" will not be easy, but an IPP in the Kuril Islands can become a pathway to an alternative future of peace and conservation if there is sufficient political will, imagination, and courage to act.[66]

An IPP may not seem the most likely outcome to the Kuril issue, even if it is, in my view, the best and most comprehensive long-term resolution. There are other balance-of-power reasons that might nudge Japan and Russia to a solution of the problem—mutual recognition of the emerging power of China, for example.[67] Or, Russia might suddenly decide that the promise of copious Japanese investment is worth the return of four remote islands. However, because the problem is primarily one of justice, material compensation is an unreliable substitute for an emotional attachment to territory, especially in a context where two countries have fought five wars in the last century. The IPP concept stresses commitment to plus-sum outcomes through mutual compromise. Further, resolving the Kuril issue without an IPP-style commitment to conservation would dramatically increase resource extraction throughout the Kuriles and the region. The IPP concept thus stands out as a way of forestalling future conflicts over pollution and control of natural resources.

The red-crowned crane thrives in the disputed islands, and has long been associated with peace in East Asia. Perhaps it can become a new symbol of Russo–Japanese cooperation in the Kuril bioregion.

Notes

1. Jacob (1991: 1).
2. Rozman (2000: 6).
3. See Hasegawa (1998: vol. 2).
4. Arendt (1963: 34).
5. Conca (2002: 4).
6. Newell (2004: 376).
7. Conca (2002: 11).
8. Kimura and Welch (1998: 229).
9. Morris-Suzuki (1999: 72).
10. Stephan (1974: 21).
11. *Russian Conservation News* ⟨http://www.wild-russia.org/bioregion14/14-kurilsky/14_kurilsky.html⟩.
12. *Japan Times*, "Experts Reach Agreement to Preserve Disputed Isles," January 22, 2001.
13. Stephan (1974: 18).
14. *Japan Times*, "Naturalists Fight to Save the Kuriles," December 20, 1999.
15. See the *International Kuril Island Project*, ⟨http://artedi.fish.washington.edu/okhotskia/ikip/Results/index.html⟩.
16. *National Geographic*, "Storm Watch over the Kuriles," October 1996.
17. Lucy Craft, director, Kuril Island Network (KIN), ⟨http://www.kurilnature.org/html/news0202.html⟩.
18. Hasegawa (1998: vol. 2, 512).
19. Hasegawa (1998: vol. 2, 513).
20. See KIN ⟨http://www.kurilnature.org/html/news0303.html⟩.

21. Hasegawa (1998: vol. 1, 76).
22. Hasegawa (1998: vol. 1, 105); Hasegawa (2000: 295).
23. Kimura and Welch (1998: 230).
24. Jacob (1991: 37).
25. Shimotomai (2000: 116–18).
26. Tarlow (2000: 136).
27. Hasegawa (1998: vol. 1, 426–27); Verbitsky (2000: 266).
28. Wishnick (2002: 311).
29. *Manila Times*, "Putin Renews Kuriles Offer to Japan," November 23, 2005.
30. Jacob (1991: 1).
31. Forsberg (1996: 437) and Kimura and Welch (1998: 217).
32. Jacob (1991: 2).
33. Forsberg (1996: 445).
34. Kimura and Welch (1998: 237).
35. Hakamada (2000: 241).
36. KIN ⟨http://www.kurilnature.org/html/articles11.html⟩.
37. See Symposium transcripts for "Parks across Boundaries" ⟨http://www.kurilnature.org/html/news0401.html⟩.
38. Ibid., Valentin Ilyashenko, Russian Academy of Science.
39. UNESCO/MAB-IUCN Workshop Joint Statement 2001 ⟨http://www.kurilnature.org/html/resolution.html⟩.
40. KIN ⟨http://www.kurilnature.org/html/news0401.html⟩.
41. Newell (2004: 12).
42. KIN ⟨http://www.kurilnature.org/html/news0302.html⟩.
43. Conca (2002: 11).
44. Okada (2002: 421).
45. See Willem Van Riet, speech to IPP Symposium 2003 (Tokyo), ⟨http://www.kurilnature.org/html/news0401.html⟩.
46. Kimura and Welch (1998: 238–39).
47. KIN ⟨http://www.kurilnature.org/html/articles11.html⟩.
48. See Weiner (1999).
49. *Sakhalin Times*, "Russians Want to Delay Settlement of Kuril Problem—Poll," December 9, 2004.
50. Kimura and Welch (1998: 233).
51. *Japan Times*, "Experts Reach Agreement to Preserve Disputed Isles," January 22, 2001.
52. KIN ⟨http://www.kurilnature.org/html/news0101.html⟩.

53. Rozman (2000: 205–209).
54. KIN ⟨http://www.kurilnature.org/html/2001.7.html⟩.
55. Allison (2002: 151).
56. Newell (2004: 37).
57. *Japan Times*, "Poachers, Politics Threaten Japan's Eden," August 23, 2001.
58. *Japan Times*, "Nemuro Symposium," January 22, 2000.
59. *Japan Times*, "Rich Natural Heritage on Disputed Isles," October 18, 1999.
60. *Japan Times*, "Naturalists Fight to Save the Kuriles," December 20, 1999.
61. Guha (2000: 127).
62. Swanson (1997: 20).
63. *Russian Conservation News* ⟨http://www.wild-russia.org/bioregion14/14-kurilsky/14_kurilsky.html⟩.
64. KIN ⟨http://www.kurilnature.org/html/Hooo-ooo.html⟩.
65. See Brock (1991).
66. Stephan (1974: 184).
67. Rozman (2000: 13).

15

The Siachen Peace Park Proposal: Reconfiguring the Kashmir Conflict?

Kent Biringer and Air Marshall K. C. (Nanda) Cariappa

One of the longest military conflicts in recent history continues in the Karakoram Mountains at the western end of the Himalayas at elevations that exceed 6,000 meters. The dispute between India and Pakistan over the Siachen Glacier region of northern Kashmir has been underway since 1984, and it presents many of the salient ingredients for environmental peace-building to work. The glaciated area in question is essentially uninhabitable given the extreme climate, and yet it is of tremendous ecological value. Therefore distributive territorial factors that make the Kashmir dispute and other socioreligious conflicts more intractable are absent here. The dilemma of common aversion (pathway A in the introduction to this volume) that might otherwise catalyze cooperation has unfortunately taken a turn for the worse in this case.

Instead of using mutually detrimental destruction of the fragile glacier environment as a means of cooperation, the lack of trust and derivative linkages with other conflicts have led India and Pakistan to perpetuate hostility. This hostility is reminiscent of the way Stein (1993) has described countries to behave under conditions of extreme distrust where even dilemmas of common aversion lead to mutually deleterious noncooperative outcomes. What is perhaps even more disconcerting is that the positioning of conflict on both sides has led to escalation and the classic "conflict trap" described by social psychologists such as Brockner and Rubin (1985).

The Siachen conflict also exemplifies the peril of issue-linkage in protracted conflicts. Whereas often arbiters are tempted to conflate conflicts and package them in a comprehensive resolution strategy, such strategies can lead to deadlock and institutional paralysis. The problem largely lies in improper sequencing of trust-building measures[1] and the way in which linkages are made with other conflicts. In the case of Siachen that assumption can be easily made, since the Siachen battle derives its impetus from the Kashmir dispute, so any resolution must also resolve the larger Kashmir conflict. However, because the Siachen conflict is substantively detached, both geographically and economically, from the Kashmir conflict, linking the two conflicts together in a single resolution is not advisable.

The Siachen conflict, if anything, should instead be linked to the high-altitude border disputes between India and China. The relaxation of tensions on that front of the border, largely due to economic collaborative ties between the two Asian powers, might be considered a more useful instance of linkage politics. Furthermore a lesson on cooperation across borders for managing the region in an ecologically sound way, and additionally providing human security, can be taken from the Kashmir earthquake of October 2005.

In recent years there have been several attempts by both countries to resolve the dispute. To date these attempts have been unsuccessful, largely because of an inability for the two countries to disconnect the Siachen conflict from the larger and far more intractable Kashmir conflict. A primary goal of the peace-building effort is to withdraw the military from the region, and so eliminate the exorbitant human and financial costs of this conflict. The strong environmental interest in disengaging the military is that of minimizing the impact of troops on the highest battlefield in the world. Many individuals, besides government and the military, are addressing these issues, and they include members of academe and other nongovernmental communities. Indeed this very conflict, which has been a flashpoint in South Asian security since 1947, gave rise to the editor's desire to produce this book.[2]

Origins of the Siachen Conflict

The history of the Siachen confrontation stems, in part, from an undelineated portion of the Line of Control (LOC) that was established after the 1948 war between India and Pakistan. This line defined the separation of forces in Kashmir and has remained in effect with only minor deviations over the last fifty years. The dominant geographical feature here is the Siachen Glacier, the longest glacier in the Himalayas and indeed anywhere else in the world outside of the polar regions.

During the nearly sixty years since the partition of the Indian subcontinent, the region has often been plagued by conflict. Three major wars have been fought between India and Pakistan over that period, and one was fought between China and India. In the aftermath of nuclear testing by India and Pakistan in 1998, there have been a number of skirmishes that have led over and over to the brink of war. These include armed military maneuvers by both sides and terrorist attacks, the largest of which occurred at the Indian parliament in December 2001.

After the first India–Pakistan war in 1948, a ceasefire line (CFL) divided the princely state of Jammu and Kashmir under the terms of the Karachi agreement for most, but not all, of the disputed region. Once the CFL reached a particular point high in the Karakoram Mountains, referred to by its map coordinates as NJ9842, the agreement specified its extension as "thence North to the glaciers." Pakistan interpreted the line to proceed northeastward to the Karakoram Pass on the Chinese border, whereas the Indians construed it to go along the Saltoro Range and Siachen Glacier in a north–northwesterly direction to the Chinese border. Figure 15.1 shows a satellite mosaic of disputed Siachen Glacier region.

With only slight adjustments in the CFL after subsequent wars, Kashmir remains divided. However, in 1984, believing that Pakistan was about to occupy the region, Indian troops moved into the area of the Siachen Glacier and the Saltoro ridge, calling into question the interpretation of the phrase "thence north to the glaciers." The

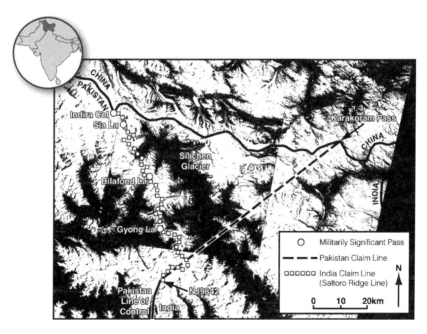

Figure 15.1
Satellite mosaic of Siachen Glacier region

result was an area of about 2,500 square kilometers (km^2) of disputed territory. The Siachen Glacier region thus became a 6,000 meter high battleground between India and Pakistan. Although many troops have been killed in the skirmishes that have occurred on this highest battleground in the world, more fatalities and casualties have been caused by the inhospitable terrain and environment.

While there are differing views on the military significance of the area, the Siachen dispute has an undeniably strong political significance. However, as India and Pakistan have worked to reach agreement on many issues over the years, Siachen has been discussed as a potential area for cooperation between the two sides through disengagement of troops from the region. In 1989 and again in 1993, a settlement on the issue was nearly reached. In 2004, Siachen was designated as one of eight topical areas for dialogue between India and Pakistan in the "composite dialogue." The costs in financial and human terms of continuing this confrontation make it an excellent candidate for bilateral cooperation while minimizing strategic and military disadvantages.

Environmental Degradation of the Himalayan Ecosystem

The future of South Asia's teeming population is at risk if the ongoing environmental degradation and ecological imbalance continues much longer. The area referred to as the Hindu Kush–Himalayan (HKH) region, and the Siachen area specifically, is being defiled by negative forms of human activity at an alarming rate.

The Himalayas are not merely a geographical feature or a range of magnificent mountains; they also embody a people's civilization. If this great range with its towering peaks were not there, the Indo-Gangetic plains of the subcontinent would not have been the "bread basket" of India. The monsoon clouds that bring life-sustaining rain would have no barriers to prevent their movement toward the Tibetan high desert (which would make the desert a misnomer). The great ranges of the Himalayas also keep the icy winds of the north from reaching the subcontinent. These mountains feed three great river systems: the Brahmaputra, the Ganges, and the Indus. It was along their valleys that great civilizations grew and flourished. Vedic hymns that are considered among the oldest spiritual texts in the world were composed and chanted in the valleys. The mountains were believed, by local inhabitants to be invested with divinity, and this belief has influenced the Indian way of thought and of life. The sylvan existence that found origin in the valleys was nurtured in the lush meadows on the mountainsides and by the sparkling, tumbling streams that gurgled from glaciers and melting snows. But as things are today, the paradisiacal vistas have disappeared in many regions.

The irrevocable damage that is being caused to the Himalayan environment, at an alarming speed, is arising from a combination of human depredations and natural causes that have become exacerbated by rapid population growth and industrialization. These forces have combined to portend an environmental crisis of colossal proportions.

The rapid melting rate of the nearly 15,000 Himalayan glaciers is sending out urgent signals that need immediately to be addressed. These glaciers comprise the largest bodies of ice and snow outside the polar caps and cover an area of nearly 32,000 square kilometers. In a report prepared for the G-8 meeting in March 2005, the World Wide Fund for Nature (WWF) estimated that the glaciers in the region are receding by 10 to 15 meters each year.[3] It is estimated that 70 percent of the water in the perennial rivers of the subcontinent is snow or glacier fed. Only about 30 percent is from the monsoons.

Recently the walls of a centuries-old monastery in Ladakh crumbled as a result of unseasonable rain, mudslides, and flooding. Climatologists have warned of huge meltwater lakes that build up behind glaciers in the high Himalayan ranges. If their icy barriers are breached due to rising global temperatures, whole communities downhill are at risk of being swept away. Environmental assessment programs have determined twenty-two and twenty-six such potentially dangerous glacial lakes in Bhutan and Nepal respectively. Catastrophic meltwater spillage could be avoided if concerted action is taken immediately.[4]

The meltwater hazards are very real and could be experienced sooner rather than later. According to Professor Syed Hasnain, the head of International Commission for Snow and Ice (ICSI), there is a possibility that the glaciers will disappear by 2035. While receding, the glaciers will leave behind unconsolidated debris that can become source material for landslides in the future, leading to further erosion of mountainsides. The likely increase in avalanches, floods, and land/mudslides will pose a very real threat of what is known as Glacier Lake Outburst Floods (GLOF), or "mountain tsunami," to downstream habitations. Should this happen, the portent is disaster for the entire northern submontane regions of the subcontinent that depend on water for irrigation, power supply, and for the very sustenance of its peoples. This could lead to further tensions in a region where water sharing is already a contentious issue. There have been several instances of GLOF damage in Bhutan, India, Nepal, Pakistan, Tibet, and China in the recent past.

Today the Himalayan region is home to 30 million indigenous peoples. It is therefore one of the richest heritage sites in the world. This fact alone could claim justification for protection under the aegis of United Nations Educational, Scientific and Cultural Organization (UNESCO) as a world heritage site. Covering an area of 3.4 million square kilometers that spreads over eight countries, it is the habitat of 116 million people. The variety of flora and fauna is considered remarkable, with one-tenth of the known high-altitude plants and animals being endemic to the region. Studies have shown that the Western Himalayan region supports cold and drought-resistant vegetation, while the areas in the East are home to subtropical trees, shrubs, and flowering plants. The mountain chain is also known for possessing the richest diversity of medicinal and aromatic plants in South Asia.

Similarly the variety of wildlife is considered among the richest in the subcontinent. Animals that are considered unique to the region are the snow leopard, Himalayan brown bear, lynx, musk deer, ibex, thar, and vulture. There is also large species diversity among vertebrates and invertebrates. This prime center of diversity in the Himalayas is now severely being threatened to extinction by habitat losses caused by de-forestation, overexploitation of resources, human settlements and encroachments, and fragmentation.

Sadly, physical, strategic, and biotic degradations are taking place all over the subcontinent. Cases in point are the mangrove swamps in the Bay of Bengal and the despoliation of the Western Ghats that are the portal to the Southwest monsoons. Deterioration of the environment will have spillover effects in all South Asian countries. This deterioration, combined with the devastation in the Himalayas and reduction of rainfall, portends the desertification of vast tracts of India, Nepal, and Pakistan. Pakistan could be hardest hit because of the limited amount of rainfall it receives and its almost complete dependence on the rivers that are nurtured and fed by the Himalayas. Continued global warming will likewise spell disaster for Bangladesh and India's island territories, which could be inundated because of rising ocean levels.

Impacts of War on Human Resources

An onerous responsibility rests on the governments of India and Pakistan, whose troops live, fight, and die in the inhospitable reaches of Siachen, the largest nonpolar glacier in the world. The presence of thousands of troops and the wherewithal to maintain them have turned the region into a vast dumping ground of the detritus of war: empty oil cans, ammunition cases, derelict vehicles, and a vast amount of human waste that is packed into drums and pushed over into the crevasses to eventually emerge in the Nubra River. There was virtually no environmental degradation in the relatively small area that encompasses the Siachen Glaciers system until the two armies confronted each other.

A region that had, prior to 1984, only known the roar of avalanches and banshee winds now reverberates to the thunder of artillery and the clacking of helicopter rotor blades as they bite their way through oxygen-starved altitudes above 4,500 meters. For over twenty years, at temperatures of −50°C, the two armies have hammered at each other with almost all they have in their respective arsenals. Over three thousand troops have lost their lives in the first ten years of the conflict; 97 percent died from the elements and related medical causes rather than from enemy action.

The cost of maintaining Indian forces is approximately one million dollars a day in India; it is a little less in Pakistan. Only humans are flown out of this bleak area. The nonbiodegradable detritus of human presence and military activity is left to pile up without much recourse. Biodegradation rates are extremely slow in this region. Army optimists have recently emerged to discredit environmental concerns by claiming that most of the waste is incinerated on the glacier and the heat generated used by the army in an ecologically efficient manner.[5] However, little empirical evidence for this is provided, and all firsthand observation accounts of the pollution on the glacier validate the worst fears about persistent pollution from biodegradable and nonbiodegradable material.[6]

The armies pollute the environment, the glacier, and the streams that form the Nubra, which then joins the Shyok as it meets the Indus deep in that mountainous wilderness. In his monumental book *Siachen: Conflict without End*, Lt. Gen. V. R. Raghavan, who had commanded the Leh Division and was responsible for the Siachen battle zone, has written: "The (Siachen) theatre of conflict, as is now widely accepted, did not offer strategic advantages.... It is clear that neither India nor Pakistan wished the Siachen conflict to assume its lasting and expensive dimensions." He quotes the late Lt. Gen. I. S. Gill, who said, "You cannot build roads on glaciers that are moving rivers of ice. We have no strategic-tactical advantage in this area, and nor has Pakistan."[7] Admiral V. Koithara, agreeing with this assessment, states: "[T]he area has...no strategic value. No military threat can be mounted from or through it." He suggests that both forces withdraw and a "wilderness reserve" be created.[8] Brigadier Gurmeet Kanwal (retired) of the Indian army and Brigadier Asad Hakeem of the Pakistani army were recently invited by Sandia National Labs to discuss ways of resolving the Siachen conflict. Both prepared a collective paper in which they agreed that a resolution of the conflict is possible provided that resolutions to other standing conflicts and issues of environmental access are sounded out as well.[9]

Concepts for Alternative Futures

As we mention above, a good case exists for framing an environmental peacebuilding solution to the Siachen dispute. The Himalayan region is of intrinsic environmental importance but also of instrumental ecological significance as a source of meltwater for rivers that provide water for millions of inhabitants downstream. The conflict clearly defies rational choice models of state behavior as both countries are losing economically in perpetuating the conflict while gaining little strategic advantage. Let us now consider how a peace park model could be applied in resolving this dispute and the agents of change who have developed this area of environmental praxis.

Siachen Peace Park

Aamir Ali, an Indian mountaineer now settled in Switzerland, proposed the idea of the Siachen Peace Park in a 1994 publication.[10] He has, with many other mountaineers and environmentalists, deplored the degradation of the magnificent Himalayan chain that stretches from the northern borders of Afghanistan to the junction of the borders between India, China, and Myanmar. His proposal for a trans-frontier peace park would not only help to prevent the further degradation of the Siachen area but would constitute a confidence-building measure for demilitarization of this volatile area. The idea found widespread support among like-minded colleagues and was endorsed at an open meeting at the India International Center, New Delhi, on June 23, 2001. The meeting addressed an appeal to the Indian Prime Minister Vajpayee on the eve of his summit with General Musharraf, the president of Pakistan, proposing that a trans-frontier peace park be established that would allow the armed forces of both countries to withdraw, under strict guarantees and surveillance, in conditions of honor and dignity. Though the summit ended in disarray, the idea of a Siachen Peace Park has found support in both India and Pakistan, and in many other parts of the world.

While there can be no doubt that establishing a transfrontier peace park covering the entire Siachen area would prevent further armed confrontation and save thousands of lives and billions of dollars, most important, it would permit the two governments to assure their respective electorates that there has been no "sell out" of interests. It could also ensure that the countries meet their constitutional obligations to protect the Siachen environment.

Further, both nations have ratified UNESCO's World Heritage Convention that encourages "identification, protection, conservation, and transmission to future generations of the cultural and natural heritage." The World Conservation Union (IUCN) believes that "protected areas along national frontiers can not only conserve biodiversity but can also be powerful symbols and agents of cooperation especially in areas of territorial conflict."[11] The loss of biodiversity and degradation of ecosystems can reshape the continental landscape, directly affecting cultural and economic development. This is precisely what has taken place in the demilitarized zone between the two Koreas as has been bought out by Ke Chung Kim in his recipe for conservation of the Korean peninsula.[12]

In concluding an article in the Mountain Research and Development journal in November 2002,[13] Aamir Ali says, "It is said on both sides of the Line of Control that to honor the blood of brave soldiers that has been spilled, not an inch of territory should be given up. One could say with even more emphasis that the sacrifice of brave men could best be honored by protecting a spectacular area consecrated with their blood." Neal Kemkar (2005) has suggested that both Pakistan and India also have legal obligations for environmental conservation that are applicable to the Siachen case and that, in international treaties, environmental protection could be a mechanism of enforcement used to ensure border security.

The Science Center Concept
Another idea has been to replace the military presence in the Himalayas by a scientific mission in the Siachen region. Behind establishing a so-called Siachen Science Center[14] is the goal to satisfy the requirement for national presence by both India and Pakistan such as would help ensure terms of a military disengagement agreement while advancing the cause of high altitude scientific study. The project could initially be conducted cooperatively by India and Pakistan. Later other regional and international participants and sponsors could be included in this effort to bring peace as well as establish transboundary environmental protection.

While the nature of an agreement on Siachen could take many forms, the science center concept assumes an agreement in which a designated area would be set aside for peaceful scientific use only. The signatories to such an agreement would seek peaceful coexistence. Other parties could become signatories in various support or participation categories. Establishing a center for scientific research in the Himalayas would provide a unique location for specialized research as well as the possibility of being integrated into other regional and international networks of scientific research stations.

International participation could take the form of providing any combination of funding, research, or operational staff, guidance, or administration.

The Siachen Science Center would consist of a scientific research facility within a designated zone in the Karakoram Range. A base camp would be established with the potential for outlying field sites where scientific instruments could be placed. Smaller scale outposts in the vicinity of the base station could also be considered. Scientists, engineers, and technicians conducting research and experiments would staff the center. Infrastructure support would have to be provided to meet administrative requirements by staff that could be a mix of bilateral, regional, or multinational personnel. The location high in the Karakoram Range in the western part of the Himalayan Mountains offers many advantages as a base for conducting a wide spectrum of scientific research in a unique geographic region. The facility has the potential to be the highest altitude manned research station in the world.

A number of scientific missions are possible. Examples include astronomy, high above much of the earth's atmosphere; geology in an area of interesting tectonics; atmospheric sciences in the complex terrain of the Himalayas; glaciology to provide insight into climatic variations throughout history; hydrologic studies to provide insight into relationships among snowfall, glacial activity, and river flows of critical water resources; life science studies of this harsh environment; physiology research to study the effects of high altitude on humans; and even psychological studies are possible, investigating the effects of a multinational group working together for prolonged periods in this hostile climatic environment. Besides scientific research, engineering knowledge could be obtained on the design, deployment, and operation of severe climate shelters, on logistical issues of supplying and maintaining a remote installation, and on characterization and operation of monitoring systems in a severe environment.

Particularly pertinent to the Siachen issue is the precedent of the Antarctic Treaty of 1959, which set aside the entire continent for peaceful scientific use only. Since entry into force in 1961, forty-three countries have become its signatories, including the seven states that originally laid claim to portions of the continent. Under terms of the treaty, all claims are held in abeyance for the term of the treaty and no new territorial claims can be submitted. India acceded to the treaty in 1983 and maintains the Maitri research station as a permanent presence there. Pakistan is not a signatory; however, it maintains the Jinnah Station Prospects for collaboration in Antarctica can certainly be expanded as discussed by Michele Zebich-Knos (chapter 9 of this volume).

The Antarctic Treaty bans any military activity in the defined area and prohibits nuclear testing. It limits national programs to those of scientific research and ensures the free exchange of information and scientists among countries. Inspection rights are granted to the facilities and operations of other countries with a presence on the continent. Provisions are made to have an open skies regime, enabling aerial observation at any time over any and all areas of Antarctica by any of the contracting parties having the right to designate observers. Regular consultative meetings of the signatory states are held and disputes are resolved by peaceful negotiation, including recourse to the International Court of Justice. While not a perfect model for South Asia, there are many features of the Antarctic Treaty that might be considered for application in Siachen.

International Karakoram Science Project (IKSP)[15]
Already underway is a multinational, interdisciplinary effort by American, Indian, and Pakistani scientists to carry forward the best possible research options and methodologies into the Karakoram Himalayas. Offices have been established in the three countries by university professors, with plans for further development if the scientific research concept succeeds. Collateral efforts include the American Association for the Advancement of Science (AAAS), Himalayan High Ice Symposium, a Global Land Ice Measurements from Space (GLIMS) Project, a University of Nebraska IKSP Expedition to K2 Mountain, and an IKSP Workshop in South Asia. The GLIMS Project is a worldwide effort supported by the US Geological Survey (USGS) and the National Aeronautics and Space Administration (NASA) to assess the global ice mass with a view to addressing the many problems that result from global warming and glacier diminution. Depletion of freshwater sources for irrigation, hazards resulting from debuttressed rock walls when glaciers melt, catastrophic meltwater floods, and many other related factors are part of the assessment process using the new ASTER satellite imagery and state-of-the-art analytical techniques.

Future plans call for joint Pakistani–Indian IKSP workshops on improving scientific knowledge in the Karakoram Mountains. The goals of these workshops are to facilitate cross-border communication and collaboration between geoscientists.

Details of this effort were presented in 2005 at the annual meeting of the American Association for the Advancement of Science. The army could also play a constructive role in the transition phase of this project by acting as rangers and engineers to coordinate the postconflict cleanup effort.[16]

Conclusion

The Siachen Glacier is ideal terrain for the instrumental use of environmental factors in peace-building. It provides all of the ingredients for a constructive application of common aversion tactics in conflict resolution. Yet the continuation of the conflict in defiance of rational choice criteria and economic incentives ostensibly supports realist critiques of environmental peace-building. However, the problem so far may be that the Siachen issue has been inextricably linked to the larger Kashmir conflict instead of sequencing it as a separate but related measure of dispute de-escalation.

A variety of concepts for a peace park or science-based initiative holds promise for reconfiguring the conflict and initiating peaceful and productive uses of the unique environment. Finding ways to end the conflict will reduce the untold human suffering, the exorbitant costs and the potentially devastating damage to the Himalayan environment, and the impact of military activity on the populations of the bordering countries. In fact millions of inhabitants of the South Asia region would benefit immeasurably from the end of armed conflict and the collective cost savings associated with this reduction. A peace park would provide a positive character to the withdrawal of both armies with dignity and honor and be a fitting monument to the soldiers of both countries to memorialize the lives lost in this conflict. In addition it would be an appropriate follow-up to the International Year for Water (2004) and to the International Year of the Mountains (2002). The ensuing cleanup of the fragile environment would help protect this endangered ecosystem against further degradation.

The Siachen Glacier and the surrounding areas form a remarkable ecosystem and are part of the world's patrimony. The proposed peace park would in effect become a Transboundary National Park that would straddle the frontier. It would be a powerful force in promoting peace, protecting the environment, and safeguarding the cultural values of indigenous peoples. This park would be unique; its size, boundaries, management plan, environmental protection, and research facilities would be negotiated by India and Pakistan for their mutual benefit. Although there can be no magic formula, the following may be possible:

- A joint declaration by the two prime ministers stressing their political commitment to the establishment of the peace park.
- A joint body to delineate the boundaries and plan the phased withdrawal of troops.
- A joint planning team with an alternating chairperson who would seek assistance and guidance from NGOs, such as IUCN and the International Mountaineering and Climbing Federation.
- A memorandum of understanding for the cooperative management of the park.
- An international treaty on the establishment of the park to be signed by the two heads of government.

The future looks brighter now that relations between India and Pakistan are improving on other matters. However, as the recent Mumbai bombings in July 2006 have shown, peace between the two nuclear rivals remains tenuous. While there continues to be disagreement about the strategic importance of Siachen as well as whether or not to delink the issue from the Kashmir conflict, the urgency of environmental action remains.[17] The cleanup of the glacier might be the first step in building trust and creating the foundation for a peace park. This could be followed by access to scientists to conduct seismological and glaciological studies, which are of common importance to ensure human security for both countries. As Prime Minister Manmohan Singh himself acknowledged in a visit to the glacier in June 2005, the potential for making the region a "peace mountain" remains active and would be a lasting legacy of positive statesmanship.

Notes

1. For a detailed exposition of sequencing see the chapter by James Sebenius, "Sequencing to Build Coalitions: With Whom Should I Talk First," in Zeckhauser et al. (1996).
2. See introduction by Saleem Ali in this volume.
3. World Wide Fund for Nature (2005).
4. "Predicting peril," Editorial, *Times of India*, February 19, 2005.
5. See, for example, General Ashok Mehta's article in *The Pioneer*, titled "No reason to leave Siachen." May 17, 2006.
6. Personal accounts of the pollution in Siachen have been documented by Harish Kapadia of the Himalayan Club in Mumbai. Mr. Kapadia has a special personal interest in this region as his son, a solider in the Indian army, was killed in battle in Kashmir. Nevertheless, rather than be radicalized by this tragedy, he is a staunch supporter of the peace park effort.

7. Raghavan (2002: 159).

8. Koithara (2004).

9. Demilitarization of the Siachen. A joint presentation by Brigadider Asad Hakeem and Brigdatier Gurmeet Kanwal at the Henry Stimson center, September 8, 2005.

10. Ali (1994).

11. Padma (2003).

12. See Ke Chung Kim, chapter in this volume.

13. Ali (2002).

14. Biringer (1998).

15. Personal communication: John F. Shroder Jr., Regents Professor of Geography and Geology, University of Nebraska at Omaha, March 2005.

16. See S. Ali (2005). Also the United Nations Environmental Program has profiled the Siachen peace park in one of their recent reports on environmental cooperation prospects. UNEP has a postconflict assessment unit that could help in coordinating the impact and cleanup effort. It has served this purpose in Afghanistan, Bosnia, and Liberia. See ⟨http://postconflict.unep.ch/index.htm⟩.

17. During June 2006, interviews with leading Indian army policy makers and retired generals were carried out by the authors of this study. Air Cmde Jasjit Singh who was once considered a "hawk" agreed with the necessity of détente with Pakistan in Siachen. He believed there were good chances of success because the ongoing cease-fire has held for over three years. Jaswant Singh voiced concerns that the Siachen issue cannot be delinked from the Kashmir conflict but did not provide any specific reasons why the decoupling was not possible.

16

Linking Afghanistan with Its Neighbors through Peace Parks: Challenges and Prospects

Stephan Fuller

Multiethinic states that border countries that have a dominant majority of one of their ethnicities have historically been zones of persistent conflict. The urge toward secessionism and the fragmentation of national identity is often acute in such cases. Afghanistan exemplifies this phenomenon most dramatically.

In many ways the Afghan identity as we know it today is a residual outpost of anticolonial forces that were able to coalesce under a common national theme. The Pashtuns have remained the dominant ethnic force in the region, but in the middle of the eighteenth century they joined forces with Tajiks and Persians to constitute the modern Afghan nation. While this common aversion to external foes united the Afghans at one level, it also led to fracturing at other levels of ideological opinion. The Afghans developed a strong sense of suspicion of any external force that must be fought for survival. This is perhaps most poignantly visible in oft-quoted stanzas of the Afghan poet Khushhal Khan Khattaq. "The very name Pashtun spells honor and glory; Lacking that honor, what is the Afghan story? In the sword alone lies our deliverance."[1]

Border cooperation in such cases of ethnic struggles for hegemony is thus particularly challenging. Yet the plight of Afghanistan following several decades of war has necessitated a need for common purposes across the country, and environmental protection along its sensitive border areas just might provide such a bonding strategy. The Afghan case presents an example of how even those countries that might be perceived by policy analysts as "failed states" are possible candidates for the implementation of peace parks. Often state failure is a result of an inability of the government to find a constructive channel for expressing identity, a phenomenon that has sometimes been referred to in the case of Afghanistan and its neighbor, Pakistan, as "nationalism without a nation."[2] The much publicized "Failed States index" developed by the Fund for Peace and the Carnegie Endowment, which ranks

Afghanistan in its highest risk category, uses twelve factors in ranking countries that are at risk of failure, and five of these have either direct environmental connections or involve activities in border zones: refugees and displaced populations, uneven development, human flight (often as a result of natural causes), demographic pressures, and external intervention.[3] Addressing these issues through the formation of peace parks might be considered a useful ingredient in a recovery strategy for the country. Indeed as political psychologist Julia Kristeva (1993) has argued, "nations without nationalism" can be a functional model that encompasses universal norms.

Postconflict Afghanistan

Humanitarian relief and reconstruction aid have been the principal priorities of international assistance underway in Afghanistan during the post conflict period from 2002 onward. It is now widely recognized that the human suffering is intertwined with twenty-five years of environmental degradation and the destruction of natural resources management institutions and procedures.

Following the completion of environmental and sustainable development reviews by both Asian Development Bank (ADB)[4] and the United Nations Environment Program (UNEP) in 2002 and 2003,[5] a variety of coordinated capacity-building and environmental programs have been initiated along with a variety of bilateral and multilateral donors. Although the challenges are many, there is an opportunity to improve human living conditions with the reconstitution of the protected areas system.

In several areas there are complementary opportunities to undertake transboundary projects with Afghanistan's neighbors. Development of some such areas may assist in finding solutions to extant security issues related to insurgency, the international movement of narcotics, and intertribal conflict—in essence as a program of peace parks for the people.

Issues and Challenges

For over two decades both civil and external forces of warfare have had a significant negative effect on the human population and economy of Afghanistan. The wars have resulted in an overwhelming and urgent need for humanitarian relief and long-term reconstruction of almost every aspect of governance and civil society.

Economic development, job creation, and poverty alleviation have become the main focus of international assistance to the newly established Transitional Government of Afghanistan (TGA). As part of the long-term reconstruction program, attention is also now appropriately being given to environmental conditions and biodiversity conservation.

The wars understandably diverted the attention of government agencies away from environmental conservation. Soil erosion and loss of rangelands, forests, and wildlife were accelerated. Systematic natural resource management as well as traditional conservation knowledge declined, and there is now the concern that postwar rehabilitation efforts may take place without consideration for the environment.

Endangered species of plants and animals are found in all of Afghanistan's representative ecosystems, ranging from the arid deserts in the southwest to the subalpine valleys in the Hindu Kush Mountains. Afghanistan's first National Park at Bande Amir and five other wildlife reserves and sanctuaries established in the 1970s after years of efforts were abandoned. Other protected area proposals intended to cover the country's needs for biodiversity conservation were completely stalled.

Local communities within and adjacent to the former protected areas are some of the least developed in Afghanistan, and they are arguably in the greatest need of assistance in the postwar development phase. Rather than simply replicating the redevelopment activities that are underway elsewhere, there is an opportunity to base some of the poverty alleviation and long-term community economic development on activities that complement and support the biodiversity conservation needs and commitments of Afghanistan.

Community development in the protected area "buffer zones" could be channeled to promote cooperation and collaborative management for the mutual benefit of both the human population and the area's biodiversity assets. There are many traditional bases for this form of collaboration that could be built upon while respecting the tenets of Islam, cultural and tribal traditions, as well as the late twentieth-century engagement of former governments of Afghanistan in conservation of globally endangered species.

The Asian Development Bank undertook an analysis of these issues in 2002,[6] which serves as the Country Environmental Assessment report as required under ADB environment policy. The environmental links identified in this report between long-term humanitarian and development assistance include attention to natural resource management subsectors as well as treatment for pollution and environmental health hazards. The ADB proposed a number of both short- and long-term interventions and a work plan for sector planning and capacity-building.

Similarly during 2002 the Geneva-based UNEP Postconflict Branch (PCOB) completed a comprehensive review of environmental conditions and problems in Afghanistan.[7] This report has been used to inform multilateral and bilateral donors about environmental conservation needs during the establishment of the transitional government of Afghanistan and subsequent reconstruction efforts. Using funding from the European Union, UNEP initiated a number of projects to re-organize government agencies and strengthen institutional work on environmental law and environmental impact assessments, as well as to improve implementation of multilateral environmental agreements (MEAs), among other priorities. Since the PCOB analysis had also confirmed the need for rehabilitation of the protected areas system and the engagement of communities in biodiversity conservation initiatives, among the MEAs adopted by the TGA was the Convention on Conservation of Biological Diversity (CBD), which was ratified by TGA in 2003. This has committed the TGA to a program of protected areas and biodiversity conservation with a strong emphasis on capacity-building and community involvement.

Included in the respective reviews and analyses of environmental conditions in Afghanistan, particularly within the prospective programs related to parks and protected areas, has been an acknowledgment of the opportunities that exist for improved security and governance if successful community-based programs are implemented. There was a call for both the re-establishment of the pre-1979 system of conservation programs and site-specific habitat protection initiatives as well as larger transboundary programs with neighboring countries. The latter category appears to present opportunities with immediate neighbors to establish a system of peace parks to protect their shared local environments.

The logic of conservation programming could be extended to other countries of central Asia. However, as the number of potential participants increases, it may become more politically problematic to achieve even modest conservation objectives, let alone any extensive opportunity for broad regional cooperation. It is clear that any practical achievements in conservation programs through the establishment of transboundary protected areas will need to be designed with very modest and adaptive criteria for success, particularly because of the fluctuating nature of the security situation in Afghanistan (in 2006, at the time of writing).

Parks, Peace, and People: Prospects for Afghanistan

As the social and political situation in Afghanistan improves, the work on a peace parks system should begin with the rehabilitation of the many pre-1979 parks that were once managed by the government. This effort holds promise for the stabilization of the Kabul government's relationship with ethnic minorities and tribal groups in outlying areas where the process of democratization, economic development, and poverty alleviation only began in 2005.

The largest protected areas included in the pre-1979 parks system are listed below (see also figure 16.1):

Pamir-i-Buzurg Although in relatively good condition because of its inaccessibility and remoteness, the "Big Pamir" reserve is now being used for grazing large numbers of domestic stock, an activity that was formerly restricted when the local people benefited from participation in the tourist hunting program and management of the reserve.

Bande Amir In recent years two of the six lakes, Bande Qumbar and Bande Pudina, have experienced temporary dry periods. Illegal hunting and fishing sometimes with explosives have persisted during intervening years, activities such as placing a flour mill and some dwellings around the lake threaten the beauty and integrity of the national park, and the area is heavily mined.

Ajar Valley Lands in the reserve are now occupied by over 300 families, and the population is increasing with repatriated families returning from Iran and Pakistan. Most of the former verdant flora in the valley has been depleted by overgrazing of domestic stock, and most woody plants have been cut for fuel and building supplies. Although the extent is not known, many hectares of reserve lands have been converted to dryland agriculture.

Ab-i-Estada Drought has affected this area, and the reserve has been dry for several years. Soil salinization is today widespread and has devastated the surrounding natural vegetation. Many households have drilled private tube wells for their agricultural lands near the lake and thus are further depriving the rehabilitation of the lake and wetlands. No flamingo breeding has been reported for several years, and Siberian Cranes have not been seen there since the late 1970s.

Dashte Nawar This is a drought-affected area today despite the small springs that dot the largely dry lake bed with pools of water. Hunting is reportedly common; nomads occupying the area now presume the lake is their property. Construction

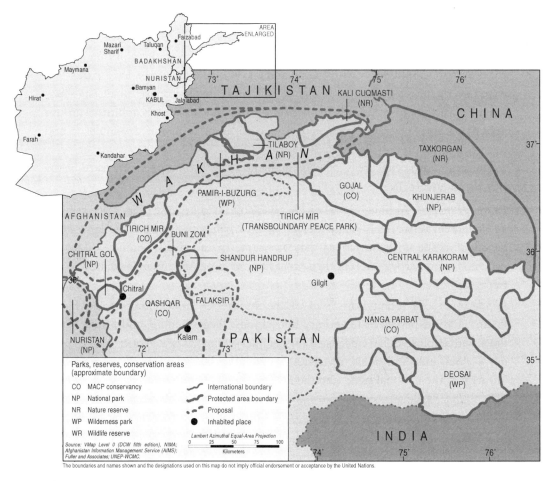

Figure 16.1
The Karakoram Constellation Proposal. Khunjerab NP, Central Karakoram NP, Deosai WP, Chitral GolNP, Taxkorgan NR, and Pamir-i-Buzurg PAs already exist; Nanga Parbat, Gojal, Tirich Mir, and Qashqar conservancy areas are under development and the others are proposals only.

of mud houses in graveyards in the immediate vicinity of the former shoreline is a serious problem.

Kole Hashmat Khan Now also a drought-affected area, the land was enlarged some years ago when the wetland was reclaimed after the cleaning of the Mastan Canal. Construction of mud houses since 1992 has considerably reduced the overall wetland area. Reeds were cut by locals and sold to private dealers during the 1990s, and much of the wetland vegetation has disappeared. Construction and other disturbances around the lake basin continue.

After a decade-long drought most of the affected areas started to experience significant recovery during 2005 as a new cycle began of high levels of rainfall. It remains to be seen if the improvement will lead to the re-establishment of a protected areas system. Already the rainfall has had the ancillary effect of raising opium poppy production, and this has dramatically complicates the complex set of "alternative livelihood" initiatives that are underway. Indirectly the re-establishment of a parks system could alleviate poverty through employment opportunities tied to the related ecotourism and adventure travel industries.

In each protected area there is opportunity for the introduction of community-wide approaches to the protected area's management. Following the preliminary studies by the TGA, UNEP, UNDP, the ADB, and other collaborators (e.g., World Conservation Monitoring Center, WCMC), the Global Environment Facility and the Asian Development Bank (to be controlled by ADB) approved funding in December 2003 for community-based biodiversity conservation programming. By definition, these funds are for the incremental costs of securing the global benefits of conservation, while the core activities such as capacity-building within the TGA and the provision of tangible community economic benefits associated with conservation are left to be funded by TGA sources or other bilateral or multilateral co-financing sources. The author is aware of several environment sector funding proposals during the period 2004 to 2006 that have been approved but not yet initiated for security reasons.[8] No doubt, there will be many other programs designed and attempted over the next few years, but it is exceptionally difficult to predict the relative degree of success that they will achieve. Certainly there is ground for optimism at the time of writing.

Prospects for Transboundary Linkages with Neighbors and the Range of Transboundary Opportunities

The types and locations of potentially protected areas around the national border of Afghanistan that might be considered for joint international management regimes widely vary in the progress of their development as wildlife reserves. This list below describes five such wildlife reserves:

Sistan Baluchistan/Hamun-i-Puzak (Iran–Afghanistan) This transboundary wetlands area has been affected by drought and has been dry for nearly a decade despite substantial rainfall and snowmelt from the Hindu Kush Mountains, which caused significant flooding along the Helmand River in early 2005. Surveys have shown that the area can recover if water is made available, but the management of water and its allocation from the Helmand River Basin have been subject to a political contest between Iran and Afghanistan for many years. The proposed GEF project for the area has not progressed beyond the conceptual stage.[9]

Northwest Afghanistan (Iran–Turkmenistan–Afghanistan) There is a longstanding proposal for a wildlife reserve in this area, encompassing an existing "protected area" in Iran. However, it is widely acknowledged that the three countries have very limited funds for managing this area actively and effectively at the present time.[10]

Tugitang Mountains (Turkmenistan–Uzbekistan–Afghanistan) An existing UNDP/GEF project involving improved wildlife management and protected areas development between Turkmenistan and Uzbekistan is underway. This could potentially be expanded southward into Afghanistan if regional security and the circumstances of governance are improved.

Kunduz Province Wildlife Reserve (Tajikistan–Afghanistan) This area was previously included in the pre-1979 Afghanistan wildlife management areas list, but it has not been actively or effectively managed during any period since. Its present status is unknown. With the active present-day interest within Tajikistan for transboundary protected areas, recognition may return to this area. All depends of course, on the resources available and the opportunities for rehabilitation of wildlife populations.

Takhar Province Wildlife Reserve (Tajikistan–Afghanistan) This area was also previously included in the pre-1979 Afghanistan wildlife management areas list, but it has not been actively or effectively managed during any period since 1979. Its present status is unknown. With the active interest shown today in Tajikistan for transboundary protected areas, the potential of this area to be managed as a wildlife reserve is also high.

It seems that the largest international peace park opportunity involving Afghanistan is the complex set of existing and proposed protected areas within and surrounding the Wakhan Corridor in the Badhakshan Province of northeastern Afghanistan (see figure 16.2).

The Wakhan area became part of Afghanistan through the power shifts of politics in the nineteenth century, and it might be regarded as an international boundary in

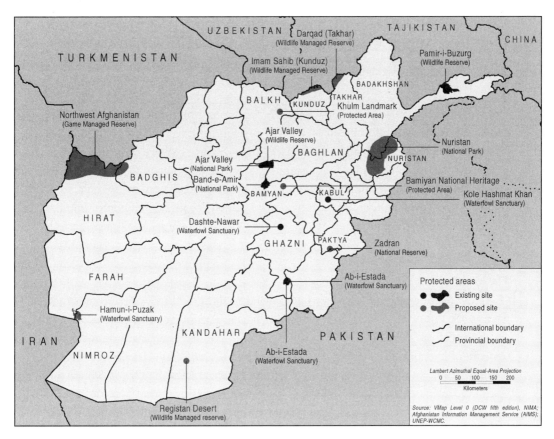

Figure 16.2
Protected Areas of Afghanistan (pre-1979, courtesy UNEP)

its own right if traditional tribal, ethnic, and linguistic divisions were actually to be honored for nation-states in central Asia. The Wakhan Corridor lies within a vast array of former trading routes that made up the historic Silk Road and its affiliations with the rising seats of political power that fluctuated over the centuries.

The idea for international cooperation on wildlife and environmental conservation in this area has several sources and has been envisioned in various forms since the early twentieth century. The Russian biologist and explorer Simon Tienshansky suggested an international program of wildlife conservancy as early as 1914.[11] More recently a variety of biological researchers from government and nongovernment organizations (from inside and outside of Afghanistan), including George Schaller of the US-based Wildlife Conservation Society, have been lobbying since as early as the 1960s for such a reserve.[12] Beginning in 1974, work in the area of northeastern Afghanistan was undertaken by FAO (funded by UNDP) on the Conservation and Utilization of Wildlife Project, but it ceased in 1979 following the Soviet invasion.

Since the regime change of 2001 to 2002, the concept of an international wildlife conservancy has focused as much on human development needs. This attention shift is due to the multilateral and bilateral donor agencies that are working toward peace and reconciliation in Afghanistan. It is now clear that any protected area in Afghanistan or in a transboundary peace program will need to involve local communities, provide for collaborative management, and be pragmatic enough to deliver whole communities from poverty by broadening economic growth opportunities—in addition to broad wildlife and ecosystem conservation initiatives.

The Wakhan Corridor Opportunity I: The Karakoram Constellation

One example of an expanded peace park concept for the Wakhan Corridor emerged from work in northern Pakistan, sponsored by the Swiss Agency for Development and Cooperation (SDC) and IUCN Pakistan. Work had begun in 1994 on the development of the Central Karakoram National Park[13] with full attention given to learning from the mistakes made by Pakistan in the establishment of Khunjerab National Park a decade earlier. In the consultative processes at that time there were extensive discussions over the prospects of cooperation with India concerning the Siachen Glacier conflict and conservation activity on the Deosai Plateau and with China concerning the Taxkorgan Wildlife Reserve. The complexity of international relations at the time, the military situation, and the incipient nature of the programming by IUCN and others in the northern Pakistan areas effectively ruled out any positive action being taken.

Although international realities frustrated much of the activity at the time, the discussions did lead to the development of a broad conservation and development program for northern Pakistan, with a major addendum on protected areas developed later. SDC provided resources for long-term conservation strategic planning (later published as the *Northern Areas Strategy for Sustainable Development*), and this allowed for more thorough investigation of a wide range of areas all across northern Pakistan, from the Khyber Pass to the Khunjerab. The areas identified as suitable for protection had potential as ecotourism sites, wilderness areas, or protected habitats. The first report was prepared and presented by the author in 1995 at the eighth

International Snow Leopard Conference[14] in Islamabad. It was expanded to include the Karakoram Constellation concept at the first IUCN World Conservation Congress in Montreal.[15] This program, along with more focused experiments on sustainable use of wildlife, provided the basis for IUCN Pakistan's later successful development of the GEF-supported Mountain Areas Conservancy Project (MACP), a $10 million initiative that is still underway today.

The four conservancy areas that are under development—Gojal, Tirich Mir, Nanga Parbat, and Qashqar—are premised on an explicit program of sustainable development priorities: economic growth through ecotourism providing local employment opportunities, and thus contributing to alleviation of poverty. (The context of this biodiversity conservation is in conjunction with Pakistan's obligations under the Convention on the Conservation of Biological Diversity.) The Gojal and Tirich Mir Conservancies are candidates for transboundary reserves with Afghanistan.

In its fully developed form the Karakoram Constellation proposal identified a larger set of transboundary protected areas opportunities, including those in Afghanistan, Tajikistan, and China. The protected areas included in the Karakoram Constellation proposal are as follows:

Nuristan National Park (Afghanistan)

Pamir-i-Buzurg Wildlife Reserve (plus Tilaboy NT and Kali Cuqmasti NR) (Afghanistan)

Pamirs National Park (Tajikistan)

Taxkorgan Nature Reserve (People's Republic of China)

Khunjerab National Park (Pakistan)

Chitral Gol National Park (Pakistan)

Gojal Conservancy Area (Pakistan MACP)

Tirich Mir Conservancy (Pakistan MACP)

Central Karakoram National Park

The noncontiguous areas included in the proposal are Qashqar Conservancy (MACP), Buni Zom NP, Shandur–Handrup NP, Nanga Parbat (MACP), and the Deosai Wilderness Park. Whatever the conservation benefits, a fundamental opportunity exists for creating a model of international cooperation, sustainable development, poverty alleviation, and regional security. The IUCN Pakistan MACP in and of itself provides an clear example of a sustainable human development program combined with protected areas and nature conservation. This may be replicated in Afghanistan when security conditions permit. Cooperation between Afghanistan and Pakistan on peace park initiatives along their extensive common border could be added to the village organizational successes of the Aga Khan Rural Support Program (AKRSP) and the results of other forms of community co-management by other countries in west and south Asia.

The Wakhan Corridor Opportunity II: The Pamirs International Conservancy

Several other proposals have been made by agencies, investigators, and consultants who have worked in the Wakhan Corridor mostly over the last two decades.[16] One of the most promising of these was the proposal for a Pamirs International Conservancy, which emerged following the establishment of the Tajik National Park (TNP) in 1993. Unfortunately, the Tajikistan civil war and extreme humanitarian relief needs, which continued into the recent past (2002), stopped most of the work on the strengthening and implementation of the Tajik system.

As originally conceived, the TNP alone included 2.6 million hectares, which is approximately 18 percent of the country contiguous with the entire Wakhan Corridor. Since the Tajik officials have promoted the idea of a Pamirs-based Biosphere Reserve and International Conservancy with several international NGOs (e.g., WCS, WWF, and IUCN). The concept was further promoted in the Khorog Declaration on the Pamirs Initiative, and modest resources have been allocated to the State Ecological Program for the period 1999 to 2008.[17]

The overall objective for the Pamirs Conservancy has been identified as "conservation of the natural and cultural environment by improving the livelihood condition of local communities at the junction region of Tajikistan, Afghanistan, Pakistan, and China." The proposed work program combines country-based operational planning and encourages support for regionally based strategic cooperation focusing on wildlife conservation. Implementation of these activities is already underway in Pakistan through MACP, China, and Tajikistan and soon even including the Lahakh in India. UN agencies have become active in discussing the concept at conferences such as the semi-regular Hindu Kush Cultural Conferences. However, this proposal has received only mixed response given the continuing focus of the international community on regional security issues in Afghanistan and Pakistan in particular.

While the government of Tajikistan has endorsed the idea of conservancy, it will most likely be indigenous nongovernment organizations that bring it to fruition. In the case of the Tajik Pamir and the Wakhan Corridor, the Aga Khan Development Network has been gradually shifting its focus from post–civil war humanitarian relief. Since 1997, the re-named Mountain Societies Development Support Program (MSDSP) based in Dushanbe, Tajikistan, has been working out a design for a new program adapted from the "lessons-learned" approach of the Pakistan-based AKRSP to the cultural and social circumstances of the Pamir Mountains. Recent successes with USAID-funded community infrastructure programs[18] have led to an explicit commitment to support a wide variety of economic development initiatives, including nature-based businesses, ecotourism, as well as the new notion of community-conserved areas enabling co-management of protected areas and of small- and medium-sized enterprises.

Clearly, the opportunity exists for cooperation between Afghanistan and Tajikistan on peace park initiatives along their extensive common border. It should also be mentioned that there has been recent interest shown for such projects in montane environments arising from the UN International Year of Mountains (2002), the World Summit on Sustainable Development (2003), and the World Parks Congress (2004) suggesting full participation of principal international organizations.[19]

Overcoming Challenges to Peace Park Implementation in Afghanistan

Any future protected areas in Afghanistan will probably not be transboundary peace parks in their own right. Basic social and economic needs of communities in proximity to buffer zones and within the protected areas must be first addressed. Pilot activities will need to be implemented to provide incentives to reduce poverty while promoting the conservation of natural resources and wildlife. The effort should begin with a strengthening of the pre-1979 parks system and a re-building of the management institutions. It can then be extended to other high-priority areas in the identified transboundary areas. Of particular interest will be the remaining forest reserves of the Nuristan Province (the locus of much military activity in 2005) and northeastward into Badhakshan and the Wakhan Corridor.

The pilot activities for the protected areas projects should all incorporate two important components. The first is to involve the local communities of the buffer zones and hinterlands in protection of the areas. The second is to create together a management plan for the protected areas.

Each conservation program must include an outline of the initial objectives, presented in a logical framework and report the outcomes or outputs of the pilot activities. The principal purpose is to prepare a fully evaluated program that can be used over the long term for the protected areas. The pilot activities should show if transboundary cooperation projects can be developed into more complete efforts when and where the political and security conditions permit.

Protected Areas Rehabilitation Activities

The technical components of biodiversity conservation activities are elaborated below based on the pre-1979 system.

Institutional Strengthening within the Ministry of Agriculture and NEPA

The group of advisors and funding for staff should come from a consortium of bilateral and multilateral institutions. The UNEP mentoring system is a successful model being developed in conjunction with the new Afghan National Environmental Protection Agency. Site-specific or area-specific projects and programs (e.g., those funded by GEF, ADB, and USAID) can proceed in parallel, but these must be centrally coordinated in order to avoid program duplication and waste of resources.

Staff Training and Capacity Development

Assessment of need for a technical training facility at a reserve is a priority activity. Needs assessment must include the training programs within the fundamental objective of strengthening the biosphere reserve model. The program might range from biodiversity conservation to community economic development to participatory planning. Some staff must specialize, but the basic curriculum should introduce a broad spectrum of social, financial, and environmental topics.

Management Information System Design

Stepwise the conservation activities must match an appropriate level of technology. Data collection must be structured to balance qualitative and quantitative information for a variety of environmental, social, and economic criteria.

Core Area Biodiversity Conservation

Technical and scientific surveys and assessments can strengthen programs in the core protected zone. Technical advisory committees can be a source of traditional ecological methods to put into practice.

Core Area Connectivity

The nature reserve is part of a network of natural areas and protected corridors for migratory species (among others). Management interventions should implement a regional approach to conservation. Education programs may be designed to create public and agency understanding of the larger ecological context of the reserve.

Buffer Zone Management

Management of land and resources involves finding an optimal mix of conservation values and sustainable resource use. Management that provides for gender-balanced participation and involves the community and stakeholders is essential for this set of activities.

In situ/Ex situ Conservation Programming

Specific programs may be needed for endangered and threatened plants and animals, particularly in circumstances where the very survival of specific genetic source is at risk. Activities may include the establishment of species-specific collection and propagation projects, collaboration with national and international research facilities, and experimentation with captive breeding programs and other technical botanical and zoological research activities.

Poverty Alleviation and Alternative Livelihoods

Poverty alleviation generally means providing alternative livelihood activities. Some such activities are listed below:

Expansion of economic/community cooperation zones Activities that serve community and economic development could include alternative means of income generation or small business assistance programs. The overall objective is improvement of the human condition and a civil society. Economic diversification is essential where there are opportunities for nonconsumptive use of resources, particularly ecotourism. Formal protected area public advisory mechanisms are essential for long-term economic development in the protected zone.

Collaborative management Activities may be designed to ensure that communities benefit from the existence of the protected area and the sustainable use of the resources within the area if this is appropriate. Local community members will be more amenable and committed to biodiversity conservation if they have a stake in the outcome and can see a tangible benefit from participation. Co-management structures and programs may be culturally and politically achievable if it is possible to provide opportunities for stakeholders to voice their opinions so that they can see what they can gain from a project's success.

Financial sustainability   Sustainable financial revenues for use in the operation and maintenance of each nature reserve should include (but are not limited to) user fees, sale of products, management fees, professional services, sustainable use revenues, ecotourism revenues, and direct and indirect subsidies.

Institutional adjustment   It is important to assess the existing capacity of the management agencies at the national and local levels, followed by a specific program of institutional strengthening. Programs to assess progress toward sustainability should be implemented and incorporated within the planning and management systems in each relevant agency.

Participatory methods can also be useful in assessing the needs of communities and developing feasible strategies and initiatives to address them. An integrated package of assistance can then be developed and implemented to include skills training to open opportunities for income generation and to help improve food security, access to health, and education. Ultimately institutional mechanisms can be built to ensure sustainability of these activities.

Ensuring Replicability as the Protected Area System Expands into the Transboundary Areas

Besides the basic program components each successful pilot protected area program should include several activities that promote and strengthen the program's replicability. Such activities are systemwide and beyond those required and relevant for a specific reserve.

Agency Programs for Replication

A specific program for the transfer of experience and information within the nature reserve system should be developed through formal courses, executive interchanges, mentoring programs, internships, and award programs in order for a transboundary system to succeed.

International Cooperation

An international context for the reserve-specific conservation programs might involve the World Commission on Protected Areas and Transboundary Protected

Areas Theme of IUCN—the World Conservation Union, which maintains working groups on both biomes and species or groups of species at risk. It is important to establish activities that foster international cooperation for management programs within a global system of protected areas. An active network of reserve management specialists should be maintained, and cooperative programming should be encouraged. Academic and agency interchanges should be understood to be essential in capacity development.

NGO Program Sponsorship
Cooperation with international nongovernmental organizations often involves sponsorship of philanthropic foundations. Organizations with a positive track record such as the World Wide Fund for Nature will often support replication of successful conservation programs.

Global Program Leveraging
Transboundary implementation activities can be used to acquire additional support from multilateral donor institutions such as the World Bank, ADB, UNDP, UNEP, FAO, and the Global Environment Facility. Philanthropic organizations and the private sector are also potential long-term financial partners.

Pragmatic First Steps: From Theory to Practice

During 2005 and 2006 there was an encouraging trend toward environmental cooperation among the multilateral and bilateral agencies.[20] Work is presently underway to jointly administer various environmental objectives that began with the formation of UNEP.[21] Conservation programs have expanded in Afghanistan to most environment and sustainable development sectors, and not just the initial group of protected areas. There is now an explicit understanding that environmental sustainability is a critical component of postwar reconstruction in Afghanistan.[22] International agencies are also initiating programs that are explicitly inclusive of the increasing civil society in the country.

In particular, in 2006, there was established in Afghanistan (with the help of the European Union, Finland, GEF, and UNEP) a comprehensive Environment Act whose framework includes biodiversity, parks, and protected areas. Through its regulations, this statute (of the Environment Act) will fundamentally reorganize and update the protected areas system and provide a modern planning and regulatory context for all existing and newly protected areas. Funding by the European Union and other donor nations[23] will enable a long-term institutional strengthening program for the Afghanistan Parks Department (part of the Ministry of Agriculture at present). An additional hierarchy of programs (initiated in 2005) funded by the Global Environment Facility, beginning with a National Capacity Self-Assessment (NCSA), will ultimately lead to complementary portfolio of countrywide and site-specific natural resource conservation programming. Included will be a National Adaptation Plan of Action (for climate change) and a National Biodiversity Strategy and Action Plan (NBSAP).

The existing ADB-financed field program for several protected areas,[24] which has been only partially implemented, will be complemented by a large USAID-financed program (to be implemented the Wildlife Conservation Society—allowing at least some of the seminal work of George Schaller to come to fruition). The field work will begin in 2006 and include preparatory work for a series of international peace parks.[25] There is further a GEF project concept related to peace parks now in the pipeline.

In yet another separate initiative, UNEP is in the very early stages of advancing a program of "environmental diplomacy" that may address the ongoing dispute over the Sistan wetlands along the Iran–Afghanistan border.[26] If successful, this model has the potential for implementation in some other historically disputed transboundary areas that have more complicated issues than those normally involved in the joint management of protected areas.

From the steps that are usually followed in coordinating an environmental program we can make some specific observations about how to proceed with a basic conservation program for Afghanistan. The list below provides a practical and effective way to establish a nature conservation program as well as transboundary cooperation.

Establish interagency steering committees and subsidiary project management systems

Prepare a stakeholder analysis and include stakeholder consultation mechanisms and processes

Use prior studies and "lessons-learned" from earlier project work in the area and in neighboring countries

Identify the capacity-building and staff-training needs

Draw up an inventory of financial and human resources needs

Make a plan for long-term capacity building and institutional strengthening in the project's implementation

Analyze all recent land-use, natural resource management, and conservation planning activities and development proposals from former cases of protected areas to determine their current status

Support community-driven economic development and alternative livelihood proposal activities, including (but not limited to):

- The design and implementation of a two-track development scheme that has relatively immediate community benefits while the planning activities begin in tandem;
- Preparation for the international management structures in the implementation of the transboundary reserves
- Development of an inventory of financial and human resources needs
- Defining preliminary boundaries of the proposed protected areas and establishing the international development requirements for effective management and sustainable use of these areas
- Preparing an international action plan and schedule for the protected areas designation, including steps for consulting and negotiating with the appropriate government agencies

Preparatory work on a pilot program can be complemented by well-developed monitoring and evaluation. An adaptive approach to strengthening the sustainability of a protected area is key to the success of a project.[27] Iterative international monitoring and evaluation programs could ensure accountability and provide feedback for the design modifications over the long term in protected areas, and institutional support can strengthen a system whether or not the protected area is internal to Afghanistan or a transboundary area. All such opportunities should be availed in due course.

Conclusions

Despite being on the brink of chaos since the overthrow of the Taliban in 2001, Afghanistan is starting to experience improved security and a stabilized economy. A constitution has been ratified by a national assembly, and democratic elections were held to select a president. A significant victory could come if the reconstruction progress continues.[28]

However, the key objectives of development and stability in Afghanistan and the region are not going to be achieved through a small program related to biodiversity conservation and protected areas, no matter how closely linked this is to community development and poverty alleviation. The most essential objectives include a regional program of partnership for economic development and trade, supported by the enlightened policies of the principal players in the security umbrella that has been erected over the country, particularly the policies of the United States and its NATO partners.

Improved trade depends on improved on international stability and cooperation, and in practical terms this means secure transportation links. Hence there is a need for partnerships with many regional countries that are presently very suspicious of one another, including Iran, Russia, China, Pakistan, and central Asia. Currently the principal requirements for physical security include road transport along the traditional routes around Afghanistan and development of other infrastructure such as gas pipeline corridors from central Asia to the south and west.

Also, because agriculture remains the single largest rural livelihood across greater central Asia, community stability requires a viable agricultural sector. Without general improvement in farm income, progress on the highest priority international objectives, such as the interdiction of the narcotics trade, is inconceivable. The international community clearly must make major progress on issues of agricultural production before turning its attention toward the rest of a long list of local and regional sustainable development initiatives. Hence the funding available through global conservation agreements will likely take secondary and tertiary priority. But no matter how difficult, environmental conservation is an essential component of sustainable development in Afghanistan.

If the diplomatic initiatives presently underway do achieve improvements in regional security, then improvements should be possible in trade, on anti-narcotics programs and crop substitution, and on agricultural productivity improvements. If this very tall order is even partially successful, then the broader agenda will likely build in projects with further security developments, improved governance, democratic institutional strengthening, additional transport and trade, culture and education, and environment and sustainable development.

In the immensely complex (and sometimes conflicting) sets of institutional arrangements and policy programs presently in place in Afghanistan, any substantive success for biodiversity conservation and protected areas programs is today predicated on the extent to which such initiatives can demonstrate that they can contribute toward overall security objectives. Unfortunately, it is true that much more enhanced security will be required before these programs can be initiated. The circular nature of this problem cannot easily be overcome.

Notes

1. Quoted by Tanner (2003: 113).
2. Jaffrelot, ed. (2002).
3. The index is published in Foreign Policy magazine and accessible at ⟨http://www.foreignpolicy.com/story/cms.php?story_id=3098⟩. Other criteria used in the index rankings are group grievance, economic decline, delegitimization of state, public services, human rights, security apparatus, and factionalized elites.
4. Azimi and McCauley (2002).
5. UNEP (2003a,b).
6. Azimi and McCauley (2002).
7. UNEP (2003b).
8. ADB (2004).
9. GEF (undated).
10. UNDP Iran, personal communication.
11. Tajik National Park, personal communication, 2005.
12. Schaller (2004).
13. Fuller and Gemin (1994).
14. Fuller (1995).
15. Fuller (1996).
16. Wildlife Conservation Society, personal communication, 2004.
17. TNP (2003).
18. Fuller (2005).
19. Hamilton and McMillan (2004).
20. ADB (2004), UNEP (2006), USAID (2005), and WCS (2005).
21. UNEP (2003a,b).
22. World Bank (2006).
23. ADB (2006).
24. ADB (2004).
25. WCS (2005).
26. A. Zaidi, UNEP, personal communication, 2006.
27. In general, see Barber et al. (2004).
28. Starr (2005).

17

Iraq and Iran in Ecological Perspective: The Mesopotamian Marshes and the Hawizeh-Azim Peace Park

Michelle L. Stevens

Perhaps the most poignant conflict of contemporary times is in the land where the earliest human agricultural civilization evolved—Mesopotamia. This is the heartland of the Middle East, which has been at the crossroads of numerous civilizational expansions and hence identities have shifted over time.[1] Much of this ancient land is now encompassed by the states of Iraq and Iran, which are divided on linguistic and cultural grounds as Arab and Persian, respectively. This linguistic divide has also been exacerbated by sectarian differences that have existed between Shia and Sunni Muslims in this region.

So far peace park development has occurred largely between countries that do not have any active conflict. The conflict between Iran and Iraq over the Shat al Arab boundary extends back to the Peace Treaty of 1639 between Persia and the Ottoman Empire. Control of the Shat al Arab waterway has been a source of contention ever since. The Mesopotamian marshes have been a no man's land between countries that have been in conflict for 400 years. The area is even more in need of a buffer today, with the Gulf being a principle location for oil exports for both Iran and Iraq. Conservation in the form of establishing some form of transboundary peace park may provide an opportunity for environmental peace-building around common environmental and cultural resources. The marshes provide a no man's land between countries, helping prevent conflict in these areas. The wreckage of war throughout the Shat al Arab region makes this and the extensive al Ahwar marsh area a priority for peacekeeping as well as biodiversity conservation.

Investment in conserving and rehabilitating the Mesopotamian marshes may help reduce the cost of peacekeeping and humanitarian relief by attacking the roots of conflict and violence. Enhancing the intrinsic relationship between resource restoration and environmental management, linked to improvement of human health and the provision of sustainable livelihoods, may help stabilize the region in the long term and reduce the likelihood of future conflicts. A transboundary protected area (TBPA) or transfrontier conservation area (TFCA) may provide an instrument for conflict mitigation. Despite ongoing conflict Iraqis have been involved in reflooding the marshes and monitoring ecological and hydrologic marsh rehabilitation since termination of the Baathist regime.

There is little doubt that environmental restoration of this region has been a priority for all Iraqis and indeed even their bellicose neighbors in Iran. This was exemplified at presentations at a unique conference hosted by the Iranian government in May 2005, titled "Environment, Peace and the Dialogue among Civilizations and Cultures." Prospects for environmental cooperation with Iraq, particularly in the marshlands region was a salient topic of discussion, and the conference was attended by the highest echelons of leadership including the president of Iran.[2] However, moving beyond such forums in both Iran and Iraq requires grassroots support as well as viable governmental infrastructure to implement conservation planning and implementation. This chapter attempts to provide some context for a peace park effort in perhaps the most troubled part of the world. Violence against the environment in this region has impacted an ancient indigenous culture, which has in turn become even more resolute to restore the landscape. Restoration of the landscape can be an important unifying theme in peace park formation and can also play a more figurative role in restoring ties between neighboring communities in the process.

The Marshlands of Mesopotamia: A Unified Ecosystem

The al Ahwar[3] marshes of southern Iraq and Iran encompass the largest wetland ecosystem in the Middle East and western Eurasia, historically covering 15,000 to 20,000 square kilometers of interconnected lakes, mudflats, and wetlands. Also called the Mesopotamian marshes, the area is considered by Muslims, Christians, and Jews as the site of the legendary Garden of Eden. The marshes are a biodiversity center of global importance, have supported the traditional lifestyles of approximately 500,000 indigenous people—the Marsh Arabs or Ma'dan—and support important agricultural production of rice, wheat, millet, and dates.

The marshes and their inhabitants have suffered a variety of negative consequences over millennia as a result of regional conflicts, with perhaps the darkest chapter of their history occurring in just the last two and half decades. In what the United Nations Environmental Program has declared "one of the world's greatest environmental disasters," more than 90 percent of the marshlands were desiccated as a result of upstream damming of the Tigris and Euphrates rivers, military operations during the Iran–Iraq war in the 1980s, downstream drainage projects, and ecocide and genocide undertaken by the Iraqi Baathist government in the 1990s.[4] The opportunity now exists for reversing the recent degradation of the marsh ecosystem through restoration and conservation projects. To the extent that these projects can involve both Iraq and Iran in a cooperative relationship, the al Ahwar marshes may be able to play a part in resolving conflict, instead of bearing its brunt.

Regional Geographic Context

The Tigris–Euphrates catchment area is a highly variable musaic of 950,876 square kilometers, with the headwaters in the mountains of Turkey and Syria in the north and Iran in the east. The Tigris and Euphrates rivers flow for about 1,300 and 1,000 kilometers respectively, through the great Mesopotamian[5] alluvial plain, and join to form the Shatt al Arab River, which flows for 190 kilometers into the Persian Gulf. The Karkheh River, originating in the Zagros Mountains of Iran, joins the Tigris River above the Tigris–Euphrates confluence.

The rivers of the Mesopotamian plain support four marshland areas: the Al Hammar, Central and Al Hawizeh marshes in Iraq, and the Al Azim marsh in Iran (figure 17.1). The Hawizeh–Azim marshes are the highest quality marshes remaining in the larger al Ahwar ecosystem complex; they are fed by the Tigris River (the two major distributaries are the Al Mausharah and Al Kahla), and by the Karkheh River. The Hawizeh–Azim marshes are the focus of this chapter because they are transboundary marshes straddling the Iraq–Iran border with globally significant cultural and heritage values.

The verdant green marshes are surrounded by vast tracts of desert to the south and mountains to the north and east. The region is hot and dry, with only 10 centimeters of annual rainfall. Evapotranspiration rates are 25 times that of precipitation, sucking water from marshes, lakes, and irrigated agricultural areas.[6] The marsh ecosystem is therefore entirely dependent on moisture from outside Iraq: the headwaters of the Tigris–Euphrates Rivers in Turkey and Syria and the Karkheh River in Iran. Prior to upstream dam construction, the inter- and intra-annual hydrologic variability resulted in a fivefold fluctuation in the size of the marshes.[7] The flood pulse in spring released large quantities of water in a short time period; this replenished alluvial soils, provided nutrients, and flushed accumulated salts from

Figure 17.1
Hawr al Ahwar Marshes: Hawizeh, Central, and Hammar marshes

the system. Spring floods are the boundary condition or driving force that shape marsh ecosystem dynamics. Upstream damming and water diversions have had a negative impact on water supply, water quality, and replenishment of the marsh surface.

The Al Ahwar Ecosystem and Its Biodiversity Heritage

The Al Ahwar ecoregion is a complex of shallow freshwater lakes, marshes, and seasonally inundated plains dominated by reeds (*Phragmites australis*), fragments of riparian forest, and vast orchards of date palms along the Shatt al Arab. The wetlands are made up of a mosaic of permanent and seasonal marshes, shallow and deepwater lakes, and regularly inundated mudflats.

A major haven of regional and global biodiversity, the marshlands support significant populations and species of wildlife.[8] Two-thirds of west Asia's wintering wildfowl, estimated at from one to ten million birds, are believed to winter in the marshes. Some of the largest world concentrations of several waterfowl species have been reported from the area.[9] The Mesopotamian marshes are a globally important center of endemism for birds, one of only eleven such wetland centers identified worldwide.[10] The marshes are the only breeding area of the rare and endemic Iraq babbler (*Turdoides altirostris*) and Basra reed-warbler (*Acrocephalus griseldis*).[11]

The Shatt al Arab is rich in nutrients and is the main source of nutrients entering the northwest portion of the Persian Gulf.[12] These nutrients are responsible for much of the area's marine productivity. The marsh ecosystem sustains an economically important local and regional fishery, providing spawning habitat for migratory fin fish and penaid shrimp species that use the marshes for spawning migrations to and from the Gulf. The total fish catch in Kuwait nets $19.41 million per annum, and pomfret alone nets $9.477 million or 49 percent of the catch.[13] Preliminary economic assessments of activities associated with the marsh dwellers are estimated at more than $300 million per annum.[14]

Al Ahwar's Cultural Heritage

The al Ahwar marshes are the homeland of a distinct cultural group—the Marsh Arabs.[15] Primarily Shi'ite Muslim, they consider their ancestral territory and cultural identity to straddle the present Iraq–Iran border, and there are strong kinship ties between marsh dwellers in the two countries.

Traditionally Marsh Arabs lived in a watery landscape, sleeping in reed homes, traveling in their boats or *mashoofs*, and welcoming travelers in their *mudhifs*, large structures woven of reeds in a style that dates back to the Sumerian culture. Water buffalos played a role in the culture very similar to that of the camel in Bedouin Arab culture.[16] Life in the marshes traditionally centered around gathering reeds, caring for water buffalo, fishing, hunting for birds, and seasonal agricultural work in date palm plantations, rice fields, and other local crops.

During ethnographic interviews I conducted with ex-patriot Iraqis in San Diego, California, I found that the marshes are a considered a cultural icon, similar to the Statue of Liberty.[17] All interviewees expressed the desire that the marshes continue to exist and to thrive. One Iraqi said, "We grow like a bird in the marsh. Everything is in front of us. We canoe inside the marshes for reeds for the animals and for fish." They expressed a great desire to have the marshes restored, saying "the marshes are like our body, our blood. You cannot miss one part. It all should stay as marsh." Interviewees estimated that 90 percent of the people from the marshes would want to go back, if they had autonomy and their own way of life. They also made it clear that people returning to the marshes would want clean water, health care, education, and other modern conveniences.

The Impact of Regional Conflict

Rivalries between various kingdoms of Mesopotamia (modern Iraq) and Persia (Iran) have occurred for more than 7,000 years. Although events before the 1970s have some relevance to the current discussion, the specifics of this earlier history are beyond the scope of this chapter. The most salient facts are the division of Muslims in the region into two sects, Shi'a and Sunni (a split that dates to AD 661), and a more recent history of control by imperialist powers—the Ottomans and then the British. In the context of the region's oil resources and the potential wealth they represent, these historical facts set the stage for the more recent conflicts affecting the marsh ecosystem and its inhabitants.

The principal cause of the 1980 war between Iran and Iraq was control of the Shatt al Arab waterway, which has been a source of contention since the 1639 peace treaty between Persia and the Ottoman Empire. After World War I, border delineations on the part of the British first placed the dispute in an Iran–Iraq context. The Algiers Agreement of March 1975 was established along the *thalweg* principle (midriver) and was later rejected by Iraq.

In 1971 Iraq claimed sovereignty rights over islands in the Gulf near the mouth of the Shatt al Arab, an important channel for oil exports. Iraq invaded Iran in 1981, and the transboundary marshes were transformed into a frontline combat zone. Iraq used Tabun, a nerve agent, causing some 5,000 deaths and the flight of over a hundred thousand Iraqi civilians to Iran and Turkey. War was disastrous for both countries, stalling economic development, disrupting oil exports, and creating approximately 2.5 million refugees. Most fighting ended in 1988, when Iran accepted the UN-sponsored peace treaty.

Mounting war debts incurred by the war with Iran, the falling world price of oil, and the arguable provocation of a buildup of American troops in Saudi Arabia led Iraq to invade and annex Kuwait on August 2, 1990. The Gulf War lasted six weeks, with an official cease-fire accepted and signed April 6, 1991. The conflict created between two and three million refugees and resulted in severe environmental damage as the retreating Iraqi army blew up and set fire to Kuwait's oil wells; oil spills affected large areas of the region.

An *intifada*, or uprising, against Saddam Hussein followed the cease-fire after the Gulf War, during which the Iraqi government lost control of 14 out of Iraq's 18 provinces.[18] Retribution by the Baathist regime was both brutal and swift, and fell particularly heavy on the marsh region. Saddam saw the Shi'ite Marsh Arabs as disloyal. The marshes had always been a place for people to escape government authority and disappear; strategically they were a difficult-to-control environment. The al Ahwar wetlands were drained, leaving a scarred landscape encrusted with salt. Every living thing was destroyed—villages were firebombed and leveled, people killed or displaced, the fish and precious water buffalo poisoned, and the rice fields fallowed. The Marsh people who survived were forcibly evicted and became environmental refugees.[19] By April 2003, only 83,000 of an estimated 500,000 people remained in the marshes; the rest had been forcibly displaced, "dumped on a drained, barren, sweltering piece of land."[20] The AMAR international charitable foundation (AMAR) provided primary health care, education, and sanitation for these refugees.

Causes of Marsh Desiccation

The destruction of the Mesopotamian marshland ecosystem began well before the 1990s. Upstream dam and diversion activities in Turkey, Syria, and Iran prior to the 1980s reduced flows in the Tigris and Euphrates, ended the annual flood pulses, and trapped the rivers' life-giving sediment behind the dams.[21] More dams have continued to be built. The cumulative impacts of the construction of more than thirty large dams, particularly those recently built in the headwater region of Turkey under the Southeast Anatolia Project (GAP), have been enormous.[22]

During the Iran–Iraq war in the 1980s, the marshes were partially drained, mined, and damaged, as upstream diversions continued. Embankments were constructed that dried out the northwestern shores, which had traditionally been an important rice cultivation area.

In the early 1990s, the Iraqi government initiated a massive drainage program that included building a huge drainage canal[23] running between the Tigris and Euphrates rivers, draining water under the Euphrates River and out the Shatt al

Arab.[24] This was accompanied by construction of barrages, dams, dikes, and barriers to divert water away from the marshes. In early 2000, the Hawr al Hawizeh had been reduced to one-third of its 1973 to 1976 area, and the Hawr al Azim had been reduced to less than 50 percent, as Iran started to impound water behind its large dam on the Karkheh River.[25] Dam construction to control water flows greatly reduced and in some years eliminated the flood pulse. This resulted in increased turbidity, and organic loads—evidenced by a sediment plume extending out into the Gulf—have had adverse impacts on fin fish and penaid shrimp productivity and on the marine food chain in general,[26] while decreased freshwater flows have increased the salinity of both groundwater and surface water in the marsh region. Desiccation of the marshes has resulted in warming of the microclimate by an estimated 5° Celsius and increased both the frequency and severity of dust storms.

The Potential for Restoration and Conservation of the Marshes

Recent empirical research shows that less than 10 percent of the original marshes of Iraq remain as fully functioning wetlands.[27] Successful marsh rehabilitation is most likely to occur in the Hawizeh marsh, due to rapid colonization of native flora and fauna in the newly reflooded areas. High water quality, low salinity, and the presence of permanent lakes and dense vegetation give hope that this area can function as a refugia for sensitive species.

There are other encouraging signs. According to United Nations estimates, 30 to 40 percent of the al Ahwar marshes have been re-inundated since 2002.[28] Some areas are rejuvenating beautifully, with lush growth of reeds, and rebounding of some fish populations. Most important, people are returning to the marsh ecotonal area, living on the edges of the marshes. They are gathering reeds and rebuilding their homes.

If there is hope for restoring the marsh ecosystem, it may lie with the indigenous Marsh Arabs. They are powerfully motivated to re-inhabit the area and manage it sustainably. The intimate connection between a functioning marsh ecosystem and the cultural identity of Marsh Arabs is expressed well by the Iraqi poet Dr. Rasheed Bander al-Khayoun[29]:

The people of al Ahwar need water in the marshes.... Their spiritual need surpasses the material need, since draining the marshes means putting the boats out of service and an end to regional poetry specific to al-Ahwar, and to singing, which can only be performed in that theatre of water and reeds and rushes. Indeed, draining the marshes means the death of a way of life that people have practiced for tens of centuries. There is no doubt that the people desperately want their environment to return to its natural state.... All the people dream of is the marshes full with fishes, birds, cows and buffalos with modernized passageways and islands, because it is this vision that is in harmony with their spiritual heritages as found in their songs, poems and tales.

Eco-cultural Restoration

The marshes are a culturalized landscape, formed over thousands of years by agricultural and traditional management practices such as the selective harvesting of reeds, the use of fire, and hunting and fishing. These intermediate-scale disturbances have long been key to ecosystem structure and function. Reeds were used for water buffalo fodder, weaving of mats for sale and for use in the home, and to replenish construction materials. Traditional management of the marshes included selective harvesting and burning of reeds on a seasonal basis, multiple species management (reeds, fish, waterfowl, bird eggs, rice), burning senescent vegetation to stimulate new growth, spatial and temporal restriction of fish harvest during spawning, and landscape patchiness management.

Because the marsh ecosystem is adapted to human management, any effort to restore the ecosystem must also be an effort to re-establish Marsh Arab culture and make use of the marsh dwellers' traditional management practices. Thus maintaining the integrity, identity, and culture of the Marsh Arab society must be pre-eminent in restoration planning, and this must include supporting the return of Marsh Arabs to the area. "The future of the 5,000-year-old Marsh Arab culture and the economic stability of a large portion of southern Iraq are dependent on the success of this restoration effort," write Richardson and coauthors,[30] but the converse is equally true: the success of the restoration effort depends on Marsh Arab culture and the economic stability of a large portion of southern Iraq.

Creating a Transboundary Conservation Area

Given the region's history of conflict, the need for funding for conservation and sustainable development, the marshes' straddling of an international border, and the pressures for water diversion and resource-extraction activities, the Hawizeh–Azim marshes need some kind of internationally recognized protected area status.

Any consideration of a transboundary conservation area would need to include a core area of wildland or traditionally managed marshes, with buffer zones of agricultural production. Local stakeholders and site conditions would inform development of zone characteristics, and conservation plans would need to integrate cultural uses. The Ecosystem Approach, adopted by the contracting parties on the Convention on Biological Diversity, offers the possibility of balancing biodiversity conservation, sustainable use, and equitable sharing of the benefits of genetic resources, and it puts a human perspective at the center of the process.[31] The Ecosystem Approach stresses management at appropriate spatial and temporal scales, and points to transboundary cooperation as key to maintaining the ecological integrity of the area and an adequately diverse and sufficiently large gene pool.

Parallel conservation measures, ecosystem surveys, ethnographic interviews, and adaptive resource management will make sustainable biodiversity conservation far more likely. Harmonization of the relevant legislation and regulations across each component of the transboundary park designation, including development of specific infrastructure for marsh management, will assist in sustainable park establishment and management.

Key challenges to establishment of any protected area status for the Hawizeh–Azim include engagement of all appropriate stakeholders, including traditional marsh dwellers and women.[32] Currently the New Eden project with Nature Iraq is conducting a socioeconomic survey in 250 villages, selected to represent the populations most affected by the condition of the marshes. The Amar Appeal has also been instrumental in marsh surveys, primarily with a human health focus.[33]

Given the violence and conflict in the area, empowerment, education, and employment opportunities are essential to build local support for any type of transboundary reserve or park. Marsh security will depend on inventorying and eliminating land mines, unexploded ordinance, and military toxins from the environment. Legitimate and well-informed decision making structures need to be instituted, including resource management strategies for sustainable fisheries. International support and transboundary cooperation will depend on a combination of integrated approaches to conservation, restoration, resource management, and sustainable development. This is definitely significant for regional security, especially given the vast oil resources and global multinational interest in those resources. There is always the potential of wars erupting over access to, and control over, vital natural resources such as oil, natural gas, and water.

Another key challenge to restoration, conservation, and sustainable development of al Ahwar is water management in Iraq. According to Dr. Azzam Alwash, director of the Nature Iraq project, "From the Iraqi perspective, the new Iraqi constitution gives the power over water management to the federal government, but there is ambiguity as to how the power of the federal government is to interact with the powers of the local governments and the various regional entities that will be created in the future."

Iran has immediate influence on Iraqi politics because of history and geography as well as economic, ethnic, religious, and paramilitary ties. According to Geoffrey Kemp, US Institute of Peace Special Report, 2005, "Iran has two fears as the nascent government of Iraq begins to emerge: (1) chaos and civil war among the Shiite factions, Sunnis, former Baathists, Kurds, and Turks; and (2) creation of a stable, pro-Western secular democracy that enjoys good relations with the United States, Saudi Arabia, Jordan, and Israel. A stable Iraq is a competitive threat for primacy in oil exports." Oil revenue, as well as development of natural gas reserves, are a critical component in Iran's capacity to ride out its disputes with the United States. A transboundary conservation area provides a moat or buffer between the two countries in area that has been a traditional source of conflict. It also helps maintain the unopposed usage of the Shatt waterway.[34]

Transboundary Park Alternatives

There are several possible international instruments to support conservation and cooperative adaptive management of the Hawizeh-Azim marshes. As a first step, the Iraq Council endorsed participation in the Ramsar Convention on Wetlands with the designation of the Hawizeh marshes as the first Ramsar wetland of international importance in 2006. The Iraq Cabinet needs to ratify this for there to be final approval, but it is considered to be a nearly automatic step. These approaches can be combined to maximize benefits to local communities, resource conservation, and regional security. They include the following: creation of a Shared Peace Park or Transboundary Biosphere Reserve,[35] designation as Ramsar Sites (Wetlands of International Significance),[36] and designation as a World Heritage Site.[37] Other possible designations at the country level include National Park, National Monument, Habitat/Species Management Area, and Managed Resource Protected Area. Iran and Iraq are most likely to be successful in obtaining funds and resources if they jointly nominate the marsh area under one of the international instruments mentioned above. Given the political situation in the two countries at this time, it is more likely that separate designation of conservation status for the Al Azim and Al Haweizeh marshes is a more timely first step to the environmental peacemaking process.

Shared Peace Park
Parks for peace are transboundary protected areas that are formally dedicated to the protection and maintenance of biological diversity, natural and associated cultural resources, and the promotion of peace and cooperation. The advantage of peace park status is that there is an explicit purpose to promote peace and cooperation between governments. The first step is to collaboratively, with strong grassroots input, develop a common vision. Local genesis of a vision for the Hawizeh–Azim Peace Park will help formulate the linkages between the environment and regional security. Peace parks in other parts of the world have a proven strategic value in bringing parties together and accessing new funding sources for sustainable development.

Cooperation between Iran and Iraq could potentially result in coordination and co-management of this globally significant area, and future establishment of a demilitarized zone between the countries. The two countries could host joint events, field days, and festivals, building on recent events such as the "Environmental Peace building in Iran" conference held in 2005, two international meetings of the Regional Organization for Protection of the Marine Environment (ROPME) held in the Gulf region,[38] and two conferences on the marshes held in Iraq.[39] In the long run the romantic legacy of Wilfred Thesiger,[40] Gavin Maxwell, and Gavin Young may help create a marketable draw for ecotourism, once the region is safe and stable. Working jointly on peace park status will enable the countries to address land and natural resource rights—including the vast oil reserves under the marshes—to the mutual advantage of both nations. The Iran–Iraq war of the 1980s illustrates the cost in lives and livelihood of war to each nation.

Ramsar Wetland of International Importance
The Hawizeh–Azim wetlands are habitat for endemic, vulnerable, endangered, and critically endangered species.[41] These wetlands are critical to the regional conservation of biological diversity, particularly considering the extensive drainage and destruction of the Mesopotamian Marsh ecosystem and local culture; as such they are good candidates for inclusion on the list of "Wetlands of International Importance" under the Ramsar convention.[42]

Under the Ramsar Convention, designation of the Azim and/or Hawizeh marshes requires appropriate legal and institutional frameworks that are essential to avoid misunderstandings or disputes between countries. While Iran is a signatory to the convention, Iraq is not. It is anticipated that Iraq will join the Convention this year (2005), and designate the Hawizeh Marsh as its first site.[43] Iran and Iraq have been engaged in dialogue to give transboundary conservation status to the marshes, facilitated through the UNEP Iran–Iraq Dialogue. If both countries become signatories to the wetland treaty, its provisions will help ensure the sustainable wise use and conservation of the wetland ecosystem. Designation as a Wetland of International Significance makes it possible to obtain funds from the Global Environmental Facility (GEF). GEF disperses funds for research-oriented monitoring and management that come from the World Bank, the United Nations Development Program (UNDP), and United Nations Environmental Program (UNEP). Designation under the Ramsar Convention should be paired with regional protected area status of some kind in each country.

World Heritage Site
The World Heritage Convention, promoting Transboundary Conservation Cooperation through UNESCO World Heritage and Biosphere Reserves, links ecological conservation and the preservation of cultural properties. The Convention recognizes the way in which people interact with nature, and the fundamental need to preserve the balance between the two. The Convention has been signed by all countries in the Tigris–Euphrates watershed (Iraq, the Islamic Republic of Iran, Kuwait, Syrian Arab Republic, and Turkey). Signatories to the Convention pledge to protect not only the World Heritage Sites on its own territory but also to protect its national heritage. Nomination of the Hawizeh–Azim marshes as a World Heritage Site would be a logical choice due to their natural and cultural heritage value. The benefits to establishment of a World Heritage Site to the participant countries include technical assistance, professional training for staff, support for public education, and encouragement of participation of the local population in preservation of their cultural and natural heritage.

Hawizeh is currently under review as a World Heritage Site (WHS). Benefits of WHS designation include international status as well as associated United Nations funds and technical support. It often takes a very long time to get WHS designation, and it is now years away for the Hawizeh and Azim marshes.

Coordination of International Efforts

Restoration and sustainable use of the Mesopotamian marshlands needs to be anchored in a regional framework, as the water supply for the marshes largely originates from outside of Iraq, and the impacts of marshland desiccation extend to the marine environment of the Persian Gulf. Several regional planning efforts are underway, coordinated by the Center for Restoration of the Iraq Marshlands, UNEP, the Regional Organization for the Protection of the Marine Environment (ROPME), UNDP, UNESCO, the Kuwait Institute of Scientific Research (KISR), the Nature Iraq/New Eden Project, the Canada Iraq Marshlands Initiative (CIMI), Italian Ministry of Environment and Territory (IMET), Environmental NGP Roundtable and Regional Environmental Forum, Iraqi Universities, Iraqi stakeholders, and funding and technical assistance from donor countries. Comprehensive planning efforts currently underway include a Master Plan for Integrated Water Resources Management, a Sustainable Restoration Plan for the Mesopotamian Marshes, and a Feasibility Study for Potable Water.[44]

To date, only about $30 million has been designated for restoration, while billions have been designated for other Iraq reconstruction activities. The US Agency for International Development (USAID) has contributed $4 million, the Canadian International Development Agency (CIDA) $3 million, and the Italian government approximately $10 million. The Japanese government, under the banner of "Support for Environmental Management of the Iraqi Marshlands," has contributed $11.4 million.

UNEP promotes the sustainable restoration of the Mesopotamian marshlands by pursuing two high-priority areas for cooperation: (1) promoting bilateral dialogue and cooperation between Iran and Iraq on the shared Hawizeh–Azim marshlands, which may include developing an integrated management plan for the shared wetlands, and (2) re-engaging Iraq with other countries in the region by including it within the activities of ROPME, a possible starting point for broader discussions involving all Tigris–Euphrates countries in cooperation at the river basin scale.

UNEP hosted an Iran–Iraq Technical Meeting on the shared al-Hawizeh/al-Azim marshlands in May 2004. The Centre for the Restoration of the Iraqi Marshlands (CRIM) under the Ministry of Water Resources emphasized a holistic restoration vision based on ecological, sociocultural, and economic criteria. Both delegations reaffirmed the need to increase international awareness and support of the marshlands, and in particular to enhance the role of international financial institutions including the World Bank and the Global Environmental Facility. They agreed that there is a need to collaborate internationally and regionally through the Ramsar Convention on Wetlands, ROPME, UNESCO, and other multilateral Environmental Agreements.

A March 2005 ROPME/UNEP conference on the restoration of the Mesopotamian Marshes emphasized the functional and cultural linkages between the Mesopotamian marshes and the marine environment, underlining the importance of restoration and cooperation to the people of the ROPME sea area and beyond in terms of economic benefits and environmental security. The meeting endorsed the potential benefits of membership of the Ramsar Convention on Wetlands and recognized the advantages of Wetland-of-International-Importance designation to both Iran and Iraq.

Iraq's Marsh Stakeholders: Sustainable Development and Restoration

The Eden Again/New Eden Project sponsored a National Conference on Development and Needs Assessment of the Marshes was held in Iraq in December 2004, with over 400 people attending.[45] The level of national pride and interest in this effort was indicated by the participants call for December 2 to be designated the annual national day of the Iraqi marshes, and proposed that an international conference on the restoration of the marshes be held in Iraq. A "Rehabilitation of Southern Iraqi Marshes" Conference was held at Basrah University on April 11–12, 2005. Iraq Prime Minister Ibrahaim Al-Jaafari sent a telegram to be read at the conference, congratulating the participants in their efforts to restore the marshes, and affirming the marshes are humanity's heritage. Prime Minister Al-Jaafari pledged the support of the Iraqi government to restore the marshes. Seven Iraqi ministries were represented; the event was widely reported in the Arabic press (and not mentioned in the English press). More than eighty-six papers were presented by Iraqi scientists; there was broad attendance by the indigenous Marsh Arabs at the meetings, with several exhibits related to their traditional lifestyle. The beauty of the marshes and the Marsh Arab way of life were reflected in an inspirational art gallery. In 2006 the second Canadian–Iraq Consultation Workshop on Iraqi marshlands was held, with sixty representatives of the national ministries, local councils and Iraqi NGOs participating. The objectives of the meeting were to discuss long-term development planning, stakeholder outreach, and governance for the Mesopotamian marshes of Iraq.

Future Threats to Sustainable Water Supply

An adequate water supply is the most crucial factor in restoring and sustaining the marsh ecosystem and the downstream marine environment of the northwestern Gulf, but such a supply is by no means assured.

The Karkheh Dam, Iran's largest reservoir, is under construction and will divert water away from the marshes. Iran is also building a 34 kilometer-wide dike along the border of the two countries that will bisect the marshes. Additional threats are emerging from planned hydroelectric dam construction on the tributaries of the Tigris River in Iraq. The largest of these is the Beckme Dam in north Iraq, with a capacity of 14.4 square kilometers on the Great Zab River. Water flow currently sustaining the marsh ecosystem would be diverted for agriculture. Further blockage of sediment flow by the dams could result in bank and channel erosion and marsh subsidence. In a recent meeting in Rome on October 21, Italy joined Iraq in a $300 million hydroelectric mega-dam building program intended to make Iraq an electricity exporter. It will be difficult to discourage Turkey and Syria from building dams in the headwaters of the Tigris and Euphrates if Iraq and Iran are draining their wetlands for hydroelectric production and irrigation.

Conclusion

Perhaps the most compelling conflict of contemporary times can also be an exemplar of the power of environmental peacemaking. Idyllic as this may seem, the fact that restoration projects in the marshlands have continued despite the Iraqi insurgency suggests that there is indeed a visceral respect for conservation among Sunni and Shia communities of this region. Despite the hawkish stance of Iran on most matters of international relations, the willingness of the administration to engage in dialogue on the environment at the highest level (as exemplified by the Tehran conference of 2005) challenges many realist assumptions. Similar to the cases in Afghanistan discussed in chapter 16, the establishment of such a park would not end all territorial conflicts, but it could provide a means of building trust and a cooperative nexus between disparate communities.

Designation of the Hawizeh–Azim marshes as an internationally recognized conservation area or transboundary peace park is important for the restoration and sustainable management of this cultural and ecological treasure. It will allow Iran, Iraq, and the international community to protect regional biodiversity, maintain traditional cultural integrity, support regional economic growth, and promote peace and cooperation among the countries in the watershed. Peace park or conservation status would have the added benefit of providing a buffer between two countries that have been in conflict over the Shat al Arab waterway for centuries. Establishing a conservation park would ensure vital shipping access for Iraq, which is seen as vital to economic and security interests. Cooperation and mutually beneficial goals are crucial for regional security in an area with limited sweet water and the world's largest oil reserves.

The majority of the people on both sides of the Iran–Iraq border want the marshes restored and whole. In the words of an Iranian ecologist, "We share the joy of winter-hosting millions of migratory birds in our south-western wetlands, which cannot be assumed separate from the Mesopotamian wetland system, and know the value and role of the mother lakes and reed-beds in southern Iraq." Iraq poet Dr. Rasheed Bander al-Khayoun adds, "The al-Ahwar has its own heritage and is distinguished from the region, from Iraq and Iran, and the rest of the world. Because of the brevity of the period of destruction, the heritage has remained, and the spiritual heritage is still there. The poetry, songs, poems and arts will return with the water."

Notes

1. Lewis (2001: 10).
2. International Institute for Sustainable Development, *Dialgoue among Civilizations Bulletin*, vol 108, no. 1, May 2005.
3. The term *al ahwar* is derived from Aramaic and means "whiteness" or "the illumination of sun on water."
4. UNEP (2001).
5. Mesopotamia means, literally, *the land between two rivers*.
6. Mean annual precipitation ranges from 10 to 40 centimeters in the wetlands. Annual evaporation is about 275 centimeters.
7. Prior to dam construction, considerable variation in stream flow occurred as a result of annual precipitation. Between 1923 and 1946 the flow of the Euphrates River averaged 14 billion cubic meters per year (bcm/yr), but ranged from 5 to 20 bcm/yr. Similarly the flow of the Tigris River averaged 37 bcm/yr, but ranged from 16 to 59 bcm/yr over the same time period.
8. UNEP (2001).
9. Scott (1995).
10. Evans (2001), Scott (1995), and Stattersfield et al. (1998).

11. Clay Rubec, CIMI, 2006. Ongoing surveys are being conducted by Nature-Iraq, the New Eden Project, and the Canada–Iraq Marshlands Initiative.

12. Personal communication, Faiza al Yamani, September 2005.

13. Personal communication, Faiza al Yamani, Kuwait Institute of Scientific Research, September 2005.

14. The main categories of economic activities of the marsh dwellers are fishery, hunting, manufacturing of handicraft articles of cane, buffalo breeding, maintenance of small domestic animals, and growing wheat, rice, and other crops in small and microscopic plots for domestic consumption. Alexander Tkachenko (in Clark and Magee, 2001).

15. The area's inhabitants are commonly known as Ma'dan, Marsh Arabs or marsh dwellers. Ethnically the population's composition is intermingled between the Persians to the east and Arab Bedouins to the west. The Marsh Arab way of life is largely based on the traditions of the Arab Bedouin (Thesiger 1964).

16. Thesiger (1964).

17. M. Stevens, unpublished field notes, 2002–2005.

18. The uprising was carried out primarily by Kurds in the north of Iraq and Shi'ites in the south, and was publicly supported by President Bush.

19. People of the Altiraba villages alone were displaced 14 to 17 times at gunpoint. Nicholson and Clark (2002).

20. Baroness Emma Nicholson, personal communication, 2005. About 200,000 Iraqis became refugees in Iran, about 95,000 of whom went to camps in Khuzistan in southwestern Iran. Other refugees were relocated to the United States, United Kingdom, Germany, and Australia.

21. UNEP (2001).

22. More than twenty additional dams are planned for the Twin Rivers watersheds, or are currently under construction.

23. The drainage canal has been called the Third River, Leader River, Saddam River, or Main Outfall Drain (MOD).

24. Mitchell (2002).

25. UNEP (2001).

26. Al Yamani et al. (2004).

27. Richardson et al. (2005).

28. UNEP (2005) and Eden Again (2005).

29. Carnegie Council on Ethnics and International Affairs (2004).

30. Richardson et al. (2005).

31. Smith and Maltby (2003).

32. UNEP (2001); Eden Again (2005).

33. *Nature Iraq Newsletter*, vol. 2, no. 1, p. 2, Winter 2006.

34. Kemp (2005).

35. ROPME/UNEP High-Level Meeting on the Restoration of the Mesopotamian Marshlands (al Ahwar) in Bahrain, 2005, recommended that a transboundary analysis of coastal and marine environmental issues in the Arabian Peninsula and Iran include evaluation of a shared peace park or transboundary reserve.

36. Ramsar (2000).

37. Partow and Maltby (2004).

38. ROPME/UNEP High-Level Meeting on the Restoration of the Mesopotamian Marshlands (al Ahwar) in Bahrain (2005).

39. The Canada Iraq Marshlands Initiative and New Eden/Eden Again (2004 and 2006); several field training courses in support of national assessment of key biodiversity areas; a national conference on Development and Needs Assessment of the Marshes (2004); a special session on the Mesopotamian Marshes with Iraqi scientists at the Ecological Society of America meetings in Montreal, Canada (2005); and National Wetland Management Training Courses (2005–2007).

40. Thesiger (1964), Maxwell (1957), and Young (1977).

41. Species in the marshes are on the IUCN Red List, the appendixes of the Convention on International Trade in Endangered Species of Wild Flora and Fauna (CITES), and the Convention on Migratory Species (CMS or Bonn Convention).

42. Ramsar defines wise use of wetlands as their "sustainable utilization for the benefit of mankind in a way compatible with the maintenance of the natural properties of the ecosystem." Community involvement and participation in co-management of wetland sites have been recognized as essential throughout the history of the Ramsar Convention.

43. The Canada Iraq Marshlands Initiative has offered to facilitate and advise Iraq on this process, and has offered to assist in sponsoring an Iraq Delegation to the Ramsar ninth Conference of the Parties Meeting in November 2004.

44. Eden Again/New Eden (2004).

45. Azzam Alwash, Eden Again, personal communication, 2004.

18

Conclusion: Implementing the Vision of Peace Parks

Saleem H. Ali

The material covered in this book has come from numerous disciplines, spanning a range of geographic and cultural contexts. However, interconnections among ecosystems that collectively constitute our biosphere have allowed for a common theme to emerge that communities in conflict can cooperate on resource conservation. There are, of course, necessary prerequisites for such cooperation to occur to prevent scarcity from being co-opted by short-term competitive interest, particularly over territorial assets. Peace park formation provides an opportunity to transform territorial conflicts into a means of collective action toward a long-term vision of ecological, economic, and social sustainability.

Yet we cannot be too sanguine about the win-win prospects of this concept in resolving conflicts or sustaining peace. A salient theme in this book has also been that forming a peace park is an inherently political process and that micro-conflicts over resource use and control may indeed arise. Such conflicts, however, primarily occur when parties are unable to appreciate the gains of long-term cooperation over resource management.

Figure 18.1 shows at a conceptual level how national identities might be transformed by peace park formation. Based on the findings of this book we conclude that the most basic definition of security in terms of food and shelter continues to prevail and must be the focus of strategic development of peace parks. Conducting a peace and conflict assessment, as delineated in the first chapter, provides a level playing field for stakeholder engagement that must be an essential prerequisite to any conservation project that is likely to have conflict linkages. However, differentiating between passive consultation and active negotiation, wherein the community has decision-making capacity, is critically important for success of such efforts.

As exemplified by the chapters on Africa and Central America, peace park development has met with considerable resistance because of a perceived elitism in many

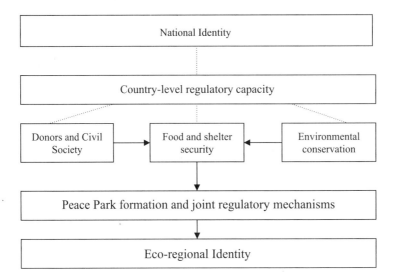

Figure 18.1
Factors for success of conservation parks in developing regions

conservation programs. The role of donors and civil society must be to use the community's basic needs as the fundamental means of building trust. At the same time the role of environmental factors in ensuring security must be equally stressed in the planning of such ventures—only then will we be able to overcome the usual stereotyping of environmental concerns as luxury issues.

Thus the first step of any implementation effort at establishing peace parks must be coupled with concomitant investments in environmental education and civic engagement. The role of science in peace-building is exemplified by the chapter on Antarctica and how such an epistemic focus could translate into larger political cooperation between nations.

The formation of peace parks is a complex process and appropriate planning is essential for us to move forward without spurring micro-level conflicts and attrition over process issues. Therefore an active role by environmental planners is important to galvanize action and to help in the realization of environmental issues in peace-building.

Balancing competitive and collaborative approaches or trade-offs between efficiency and efficacy has been a major preoccupation of planners. Historically one of the primary aims of centralized planning was to maintain physical security for cities, regions, and nation states. The inclusion of environmental criteria as a means of achieving this aim was particularly focused on the availability of essential resources such as food and water. Just as the environmental security literature exemplified by figure I.1 (pathway B) in the Introduction to this book was often predicated on assumptions of scarcity, much of the earlier approaches to planning also revolved around scarcity. However, the planner would subsume the causality mechanisms of scarcity into a prescriptive pathway for policy change. This could be accomplished either by finding alternative resources or by reconfiguring the problem whereby the interconnections and dependencies might be leveraged as a means of avoiding scarcity.

Since environmental planning is inherently concerned with future outcomes of present decisions, it tends to be less likely to be mired in historical grievances between communities that often tinge security discourse. Even within academe, planners are often seen balancing their social science pedigree with pragmatic and expedient solution development—hence the term "praxis" is often used to define their goals.[1] While there are many connotations of the term, we define praxis as the confluence of theory and practice—a constructive enterprise that attempts to inform day-to-day decisions about environmental planning with structural insights. In order to accomplish this most effectively, academicians must inevitably partner with practitioners and learn from each other's knowledge and experience.

Planners have always struggled with competing demands of developing idealized notions of how the world should work while designing and implementing schemes that fit realistic constraints. Similarly security practitioners in government agencies and private institutions are often confronted with similar day-to-day challenges that require an immediacy and flexibility in their operations. Planners and security professionals thus have much to learn from each other and environmental conduits provide an excellent means of accomplishing this task.

Table 18.1 shows ways in which environmental planners can approach this task. The first step, which is highlighted in all the chapters, is that environmental concerns must be framed in terms of a "dilemma of common aversions." This term implies that harm to the environment is a mutually destructive outcome that rational actors in a conflict would wish to avoid. This recognition can be facilitated by sharing of information between parties at various levels, usually facilitated by a neutral party. The chapter on game theory showed us how this is analytically possible, and the case examples in various contexts such as Liberia also exemplified the role of external entities such as the United Nations in facilitating this outcome.

The second level of linkage whereby environmental concerns are connected to other issues is very appealing in peace park formation as economic development

Table 18.1
Environmental planning for cooperative behavior

Concept	Approach	Action	Initiative	Function
Framing conflict as a dilemma of common aversion	Provide information on joint harms of noncooperation	Institute long-term engagement between parties to monitor environmental harms	Joint audits of environmental criteria and data collection for ecosystem-based planning efforts	Establishes neutral cognitive base for discussion of derivative issues
Linking environmental concerns to other issues	Provide a bargaining opportunity for sides where none was perceived to exist	Negotiate comprehensive agreements rather than individual contracts on specific issues	Interdisciplinary commissions for problem-solving that are facilitated by a mutually agreeable mediator	Enlarges "the pie" for positive solutions and adds flexibility for integrative bargaining
Using environmental concerns as a trust-building tool	Provide forums for joint participation in conservation initiatives	Develop conservation plans that would be inclusive of adversaries	Good neighbor compacts on riparian conservation and sister-city lesson drawing arrangements	Provides a mutually satisfying experience for parties to exemplify rewards of cooperation

ventures such as ecotourism are presented. We observed in the case analyses in part II of the book that in all the cases where wildlife tourism was involved peace park formation was facilitated. However, there is also a danger that these economic interests will marginalize poor communities living at the periphery. Thus the linkage must be made at an economic development level that encompasses local projects rather than aggregate indicators of wealth generation.

Finally, the instrumental use of these processes to facilitate trust at a larger level between parties must be carefully calibrated to the context of the case. The Korean case illustrated this both in theory (from chapter 2) as well as in practice in the case description of chapter 14. Indeed, in other complex cases such as the Mesopotamian marshlands peace park proposal, the willingness of Iran to engage at the highest level in conversations about the environment with Iraq is a positive sign.

Nevertheless, major challenges remain in convincing governments of the value of peace parks as a means of conflict resolution.

Overcoming the Label of Low Politics

Skeptics will still argue that cooperation on environmental issues between adversaries can be relegated to low politics and will not translate into a larger resolution of the conflict. In this view, environmental conservation would at best be a means of diplomatic maneuvering between mid-level bureaucrats and, at worst, be a tool of cooptation by the influential members of a polity. Such realist critics of functionalism give examples of cooperation on water resources between adversarial states like India and Pakistan or Jordan and Israel without translating into broader reconciliation.[2] Thus it could be argued that water and environmental issues are not as important as to play an instrumental role. However, a more positive framing of the case might reveal that water resources in this context are so important that even adversaries must show some semblance of cooperation over them.

Furthermore the instrumental impact of environmental issues in building peace must be considered over longer time horizons. The process by which environmental issues can play a positive role in peace-building is premised on a series of steps:

1. A unified information base on a mutual environmental threat
2. Recognition of the importance of cooperation to alleviate that threat
3. A cognitive connection and trust development due to environmental cooperation
4. Continued interactions due to environmental necessity

5. Clarification of misunderstandings as a result of continued interactions
6. De-escalation of conflict and resultant peace-building

Given the necessity for certain environmental resources and a growing realization that environmental issues require integrated solutions across borders, the likelihood for their instrumental use in peace-building has gone up in recent years. There is a growing commitment to "bioregionalism" or the realization that ecological management must be defined by natural delineations such as watersheds and biomes rather than through arbitrary national borders. Numerous joint environmental commissions between countries and jurisdictions have taken root all over the world in this regard. We have seen this played out in various ways at international forums where bioregionalism and common environmental sensitivities have transcended traditional notions of state sovereignty.

Regional environmental action plans such as those in the Mediterranean, the Caribbean, and the Red Sea are examples in this regard. Aquatic management arrangements provide instructive insights regarding environmental peace-building. One of the earliest contributions to the development of the study of environmental peace-building was Peter Haas's work on the context of the Mediterranean Action Plan.[3] However, since they involve nonhabitation, their strategic role in peace-building is quite different from terrestrial peace parks. The distributive and conflicting aspects of aquatic systems are only apparent when fisheries or navigation rights are at play. Otherwise, the ecosystems services that are provided by the aquatic system are best managed through cooperative means.[4] The Red Sea marine peace park shared by Israel, Jordan, and Egypt presents an interesting example as well where the term "peace park" has been used to describe a joint conservation management regime for ecosystem services. However, here the challenge is that the establishment of the peace park has not really led to a reduction of tensions among the players, and the subsequent implementation has been limited.

While we are a long way from having global governance of environmental issues, the momentum is clearly in the direction of giving environmental protection that directly impacts human lives and livelihoods the same moral ascendancy as "human rights."

Policy Lessons beyond Boundaries

The literature on environmental security has largely been moderated by social scientists with little contribution from ecologists or conservation managers. The result is often an elegant but relatively ineffective model with little application. This book has attempted to bridge the divide between academics and practitioners by providing contributions from both and an opportunity to contribute to each other's work.

When dealing with matters as emotive as environmental protection and conflict mitigation, one can't help but feel a sense of urgency and advocacy for a phenomenon that holds promise in harmonizing these two worthy goals. However, policy makers must constantly balance their allegiance to various constituencies, some of whom may consider conservation as an inherently low priority compared to peacebuilding.

The specter of peace parks and other environmental peacemaking efforts offers a potential win-win solution to such policy quandaries. The major concern in undertaking such efforts is to avoid micro-conflicts in the conservation efforts. In particular, the historical dispossession of land for conservation and the co-optation of environmental measures for creating enclaves that might disconnect communities need to be cautiously considered. These are, of course, questions of proper implementation rather than a critique of the idea of peace parks themselves. For proper implementation the peace park effort must first go through a phase of local review and be transparent in all its proceedings. Clarity of process is particularly important in conflict settings to avoid the spread of conspiracy theories that can lead to suspicion and rumor-mongering, which often spoil even the most sincere efforts.

In addition the role of the military in such zones should be considered as a facilitator of the process rather than a hindrance. Hence demilitarization might not be the first step but rather a transformation of the military's role to one of a ranger force—so as to assuage security and employment concerns while accomplishing conservation tasks. If there has been environmental damage as a result of the conflict, the military can certainly play in important role in the cleanup effort. The United Nations Environment Programme also has a postconflict assessment unit that is gaining traction with environmental cleanup efforts in high-profile conflicts such as those in the Balkans, the Middle East, and East Africa.[5]

Some of the challenges and potential solutions for implementing peace parks are summarized in the sidebar, with the proviso that much of these issues will inevitably require further refinement for individual cases.

Clearly, the recommendations in the sidebar require considerable care for implementation and must be coupled with traditional means of conflict resolution through diplomatic exchanges and political bargaining. However, the opportunity exists to use the environment as a means of bringing adversaries together in a way that may reconfigure the conversation in constructive ways.

> **Summary of Challenges and Solutions to Peace Park Formation**
>
> **Endogeniety, or the perception that conservation is a consequence rather than a constituent of peace-building**
>
> *Solution* Dialectical policy—consider conservation as a trust-building activity in a feedback loop
>
> **Preexisting local conflicts undermine peace-building by labeling it as co-optation and dispossession**
>
> *Solution* Resolve micro-conflicts beforehand, acknowledge past grievances and make process transparent to local residents
>
> **Conservation agencies are external to security decision apparatus**
>
> *Solution* Make conservation a strategic asset in foreign policy matters with participation of scientists and environmental agency staff in deliberations
>
> **International NGOs that are hesitant to interfere in border issues**
>
> They may fear denied access or political retribution—or that a confrontational approach will lead to their marginalization.
>
> *Solution* NGOs should play an epistemic role—exchanging knowledge between parties and mediating for community members on all sides

Finding Ends beyond Entrapment

Conflict situations can lead to a sense of cynicism and entrapment. Parties in protracted conflicts such as the Indo–Pak case or those in the Middle East tend to feel that too much time and too heavy a price has been paid, so any sign of compromise is unacceptable and will in any event be tantamount to losing face. This is a classic psychological trap, which I always ask my students in environmental conflict resolution to recognize. Often, in my classes, I use the example provided by the late Jeffrey Rubin, an eminent conflict psychologist and mountaineer. Dr. Rubin used the example of a wolf trap that was operated by Canadian trappers in the winter to explain the process of entrapment. Trappers would use small bait attached to a knife's edge and buried in the snow. A wolf that tried to eat the bait would cut its tongue and would taste some of its own blood and go on licking the knife's blade, eventually bleeding to death. Such is the peril of psychological entrapment—for it seems so compelling and is yet so cruel and condemned to failure.

The positive economic impact of peace park formulation is often quantifiable, based on increased tourism potential as well as the willingness of donors to invest in such a program. Integrated planning for peace parks must thus include a clear assessment of livelihoods and how those would be sustainable through the development of a peace park. The incorporation of conservation provisions and access to peace park areas through visa waivers or on-site processing of visas for the conservation zones can also be proposed.

Once these multiple factors are collectively considered, there is greater likelihood for policy success. As with many complex interactions of human behavior and the environment, we must resist an expectation of instant solutions. Such expectations can consequently lead to instant dismissals of otherwise worthwhile policies that have not been given time to mature. Peace parks constitute a new vision of how global conflicts can be addressed and hence will proceed through growing pains before reaching cognitive acceptance and practical results. However, there is substantive theoretical backing for their efficacy as well as emerging applied examples of their success that we should consider with a sense of optimism.

Notes

1. Minca, ed. (2001).
2. Lowi (1995).
3. Haas (1992).
4. The primacy of valuing ecosystem services collaboratively is being developed by ecological economists. See Daly and Farley (2004). While many neoclassical economists often disparage this approach, it is gaining increasing attention in policy circles, especially in Europe.
5. See the Web site for the postconflict assessment unit for reports produced by these projects ⟨http://postconflict.unep.ch/⟩.

References

Adams, W. M., and D. Hulme. 2001. If community conservation is the answer in Africa, what is the question? *Oryx* 35(3): 193–200.

Addison, T., G. Mavrotas, and M. McGillivray. 2005. Aid, debt relief and new sources of finance for meeting the millenium development goals. *Journal of International Affairs* 58(2): 113–27.

Adler, E., and P. Haas. 1992. Epistemic communities, world order, and the creation of a reflective research program. *International Organization* 46(1): 101–46.

Ali, A., ed. 1994. *Environmental Protection of the Himalaya: A Mountaineers View*. New Delhi: Indus.

Ali, A. 2002. A Siachen peace park: The solution to a half-century of international conflict? *Mountain Research and Development Journal* 22(4): 316–19.

Ali, S. H. 2003. Environmental planning and cooperative behavior. *Journal of Planning Education and Research* 23: 165–76.

Ali, S. H. 2005. Siachen: Ecological peace between India and Pakistan. *Sanctuary Asia*, February: 76–77.

Allen, P. 1999. Vast national park envisioned in SW Arizona. *Tucson Citizen*, March 10, p. B1.

Allison, T. 2002. Crisis of the region's fishing industry: Sources, prospects and the role of foreign influence. In *Russia's Far East: A Region at Risk*, ed. by J. Thornton and C. Ziegler. Seattle: National Bureau of Asian Research/University of Washington Press.

Alwash, A. 2005. Eden again/New eden project. Paper presented at ROPME meetings, March, in Bahrain.

Al-Yamani, F. Y., M. Al-Husaini, and J. Bishop. 2004. The effects of the Mesopotamian marsh drainage and river diversion on the oceanography and fisheries of the northwestern Arabian Gulf. Paper presented at SWS 25th Annual Meeting, Seattle, WA. Kuwait Institute of Scientific Research.

Anastasiou, H. 2002. Communication across conflict lines: The case of ethnically divided cyprus. *Journal of Peace Research* 39(5): 581–96.

Anderies, J., M. Janssen, and E. Ostrom. 2004. A framework to analyze the robustness of social-ecological systems from an institutional perspective. *Ecology and Society* 9(1): 18.

Anderson, D. 2005. *The North Korean crisis*. [Napsnet] *Policy Forum Online*, March 17, pp. 1–9.

Anderson, J., and L. O'Dowd. 1999. Borders, border regions and territoriality: Contradictory meanings, changing significance. *Regional Studies* 33(7): 593–604.

Anderson, M. 1999. *Do No Harm: How Aid Can Support Peace—Or War*. Boulder, CO: Lynne Rienner.

Andreas, P. 2003. Redrawing the line: borders and security in the twenty-first century. *International Security* 28(2): 78–111.

Arendt, H. 1963 (1990). *On Revolution*. London: Penguin Books.

Argentine National Antarctic Institute (Direccion Nacional del Antartico Instituto Antartico Argentino). 2005. ⟨http://www.dna.gov.ar/INGLES/DIVULGAC/INDEX.HTM⟩.

Arizona State Senate. 2003a. Bill status overview: Sonoran desert peace park. SM 1001. Available at ⟨http://www.azleg.state.az.us/legtext/46leg/1r/bills/sm1001o.asp⟩.

Arizona State Senate. 2003b. Fact sheet for S.M. 1001: Sonoran desert peace park. Available at ⟨http://www.azleg.state.az.us/legtext/46leg/1r/summary/s.sm1001nrt.doc.htm⟩.

Arizona State Senate. 2003c. Minutes of committee on natural resources and transportation. February 18. Available at ⟨http://www.azleg.state.az.us/legtext/46leg/1r/comm_min/senate/021803%20nrt.dot.htm⟩.

Arizona State Senate. 2003d. Sonoran desert peace park. SM 1001. Available at ⟨http://www.azleg.state.az.us/legtext/46leg/1r/bills/sm1001p.htm⟩.

Asian Development Bank. 2006. *Project Summary: Regional Transboundary Peace Park*. Kabul: ADB, 2 pp.

Asian Development Bank. 2005. Afghan National Ecotourism Strategy (Draft). (TA-4541) Manila, 67 pp.

Asian Development Bank. 2004. Technical Assistance to the Islamic Republic of Afghanistan for Natural Resources Management and Poverty Reduction (TAR: AFG 38039). Manila, 16 pp.

Austin, J. E., and C. E. Bruch. 2003. Legal mechanisms for addressing wartime damage to tropical forests. In *War and Tropical Forests: Conservation in Areas of Armed Conflict*, ed. by S. V. Price. Binghamton, NY: Haworth Press, pp. 167–99.

Axelrod, R. 1984. *The Evolution of Cooperation*. New York: Basic Books.

Azimi, A., and D. McCauley. 2002. *Afghanistan's Environment in Transition*. Manila: Asian Development Bank.

Bagre, A. S., H. Bary, A. Ouotara, M. Ouedraogo, and D. Thieba. 2003. Challenges for a viable decentralization process in rural Burkina Faso. *Royal Tropical Institute Bulletin* 356: 63. Amsterdam: Royal Tropical Institute.

Bakarr, M. I. 2003. Conservation on the frontier. *Tropical Forest Update* 13(2): 3–5.

Baldus, R. D., et al. 2001. Experiences with community based wildlife conservation in Tanzania. Tanzania Wildlife Discussion Paper 29. GTZ, Dar es Salaam.

Baldus, R. D., et al. 2003a. Seeking conservation partnerships in the Selous Game Reserve, Tanzania. *Parks* 13(1): 50–61.

Baldus, R. D., et al. 2003b. The Selous–Niassa wildlife corridor. Tanzania Wildlife Discussion Paper 34. GTZ, Dar es Salaam.

Baldus, R. D. 2004a. Lion conservation in Tanzania leads to serious human–lion conflicts. Tanzania Wildlife Discussion Paper 41. GTZ, Dar es Salaam. Available at ⟨www.wildlife-programme.gtz.de/wildlife⟩.

Baldus, R. D. 2004b. *Promoting Cross-border Cooperation in the Management of Natural Resources in Shared Ecosystems*. Dar es Salaam, Tanzania: GTZ Wildlife Programme; GTZ Water Sector Reform Programme; InWEnt.

Bannon, I., and P. Collier, eds. 2003. *Natural Resources and Violent Conflict: Options and Actions*. Washington, DC: World Bank.

Barber, C. V., K. R. Miller, and M. Boness, eds. 2004. *Securing Protected Areas in the Face of Global Change: Issues and Strategies*. Gland, Switzerland: IUCN.

Baregu, M. 1999. The DRC war and the second scramble for Africa. In *The Crisis in the Democratic Republic of Congo*, ed. by M. Baregu. Harare: SAPES Books, pp. 36–41.

Barrett, C., K. Brandon, C. Gibson, and H. Gjertsen. 2001. Conserving tropical biodiversity amid weak institutions. *BioScience* 51(6): 497–502.

Barrett, C. B., and P. Arcese. 1995. Are integrated conservation development projects (IDCPs) sustainable? On conservation of large mammals in sub-Saharan Africa. *World Development* 23(7): 1073–84.

Barrow, E., et al. 2001. The evolution of community conservation policy and practice in east Africa. In *African Wildlife and Livelihoods: The Promise and Performance of Community Conservation*, ed. by D. Hulme and M. Murphree. Oxford: James Currey Ltd., pp. 59–73.

Basnet, K. 2003. Transboundary biodiversity conservation initiative: An example from Nepal. *Journal of Sustainable Forestry* 17(1–2): 205–27.

Beauregard, R. 1995. Theorizing the global–local connection. In *World Cities in a World System*, ed. by P. L. Knox and P. J. Taylor. Cambridge: Cambridge University Press.

Bedeski, R. E. 1997. Arms control inspections, the armistice agreement, and new challenges to peace on the Korean peninsula. In *Special Report, Policy Forum Online (#7), Northeast Asia Peace and Security Network*, July 23.

Bedi, J. S. 1998. World's highest, biggest junkyard. *The Tribune*, August 29, p. 3.

Beilin, Y. 2004. *The path to Geneva: The quest for a permanent agreement, 1996–2004*. New York: RDV Books.

Beinart, W., and P. Coates. 1995. *Environment and History: The Taming of Nature in the USA and South Africa*. London: Routledge.

Beltran, J. Undated. Indigenous and traditional peoples and protected areas. In *Principles, Guidelines and Case Studies*. Gland, Switzerland: IUCN–World Conservation Union, p. 144.

Berthold-Bond, D. 2000. The ethics of "place": Reflections on bioregionalism. *Environmental Ethics* 22(1): 5–20.

Besançon, C., and C. Savy. 2005. Global list of internationally adjoining protected areas and other transboundary conservation initiatives. In *Conservation beyond Borders: A Transboundary Vision for Protected Areas*, ed. by R. Mittermeier, C. Mittermeier, C. Kormos, T. Sandwith, and C. Besançon. Washington, DC: CEMEX/Conservation International.

Bhumpakphan, N. 2003. Management of the Pha Taem protected forest complex to promote cooperation for trans-boundary biodiversity conservation between Thailand, Cambodia, and Laos (phase I): Wildlife ecology technical final report. Bangkok: Faculty of Forestry, Kasertsart University.

Biodiversity Support Program. 1999. *Study on the Development of Transboundary Natural Resource Management Areas in Southern Africa: Highlights and Findings*. Washington, DC: Biodiversity Support Program.

Biodiversity Support Program. 2001a. *Beyond Boundaries: Transboundary Natural Resource Management in West Africa*. Washington, DC: Biodiversity Support Program.

Biodiversity Support Program. 2001b. *Beyond Boundaries: Transboundary Natural Resource Management in Central Africa*. Washington, DC: Biodiversity Support Program.

Biodiversity Support Program. 2001c. *Beyond Boundaries: Transboundary Natural Resource Management in Eastern Africa*. Washington, DC: Biodiversity Support Program.

Biringer, K. L. 1998. Siachen science center: A concept for cooperation at the top of the world. *CMC Occasional Paper, SAND98-0505/2*. Albuquerque, NM: Sandia National Laboratories.

Blake, G. H., W. J. Hildesley, M. A. Pratt, R. J. Ridley, and C. H. Schofield, eds. 1995. *The Peaceful Management of Transboundary Resources*. London: Graham and Trotman.

Blum, J. 1994. The deep freeze: Torts, choice of law, and the Antarctic treaty regime. *Emory International Law Review* 8: 667–99.

Blundell, A. G., A. Bodian, D. Callamand, R. Carisch, C. Fithen, and H. Kelley. 2003. UN report on impact of timber sanctions on Liberia. UN Security Council S/2003/779. Available at ⟨http://daccessdds.un.org/doc/UNDOC/GEN/N03/447/41/PDF/N0344741.pdf?OpenElement⟩.

Blundell, A. G., A. Bodian, D. Callamand, C. Fithen, and T. Garnett. 2004. UN report on compliance with UN sanctions on Liberia. UN Security Council S/2004/955. Available at ⟨http://daccessdds.un.org/doc/UNDOC/GEN/N04/626/19/IMG/N0462619.pdf?OpenElement⟩.

Blundell, A. G., D. Callamand, C. Fithen, T. Garnett, and R. Sinha. 2005. UN report on compliance with UN sanctions on Liberia. UN Security Council S/2005/360. Available at ⟨http://daccessdds.un.org/doc/UNDOC/GEN/N05/363/39/IMG/N0536339.pdf?OpenElement⟩.

Bob, C. 2005. *The Marketing of Rebellion: Insurgents, Media and International Activism*. Cambridge: Cambridge University Press.

Bogdan, R. C., and B. Sariknoff. 1998. *Qualitative Research in Education: An Introduction to Theory and Method*, 3rd ed. Boston: Allyn and Bacon.

Bohr, A. 2004. Regionalism in central Asia: New geopolitics, old regional order. *International Affairs* 80(3): 485–502.

Boo, E. 1990. *Ecotourism: The Potentials and the Pitfalls*. Washington, DC: WWF.

Border Briefs. 1997. Arizona, Sonora step up binational desert protection. *BorderLines*, April 5(4). Available at ⟨http://americaspolicy.org/borderlines/1997/bl34/bl34bb.html⟩.

Brandon, K., K. H. Redford, and S. E. Sanderson, eds. 1998. *Parks in Peril: People, Politics, and Protected Areas*. Covelo, CA: Island Press.

Brazos Films. 2003. *Land of Contrasts: Big Bend National Park* (video).

BRC. 1999. Huge support for Sonoran desert park proposal. Phoenix: Behavior Research Center. News Release, November 3.

Brenner, N. 2001. The limits to scale? Methodological reflections on scalar structuration. *Progress in Human Geography* 25(4): 591–614.

Breymeyer, A., and R. Noble. 1996. *Biodiversity Conservation in Transboundary Protected Areas*. Office for Central Europe and Eurasia, Office of International Affairs, National Research Council. Washington, DC: National Academy Press.

Bridges, W. 1991. *Managing Transitions: Making the Most of Change*. Cambridge, MA: Perseus Books.

Broch-Due, V., and R. A. Schroeder, eds. 2000. *Producing Nature and Poverty in Africa*. Uppsala: Nordiska Afrikainstitutet.

Brock, L. 1991. Peace through parks: The environment on the peace research agenda. *Journal of Peace Research* 28(4): 407–23.

Brockington, D. 2002. *Fortress Conservation: The Preservation of the Mkomazi Game Reserve, Tanzania*. Oxford: James Currey.

Broome, B. 2004. Reaching across the dividing line: Building a collective vision for peace in Cyprus. *Journal of Peace Research* 41(2): 191–209.

Broome, B. J. 1998. Overview of Conflict Resolution Activities in Cyprus: Their Contribution to the Peace Process. *Cyprus Review* 10(1): 47–66.

Brosius, J. P., and D. Russell. 2003. Conservation from above: An anthropological perspective on transboundary protected areas and ecoregional planning. *Journal of Sustainable Forestry* 17(1–2): 39–66.

Broyles, B. 2004. The desert sisters. *Earth Island Journal*, Autumn 19(3): 35–38.

Brunner, R. 1999. Parks for life: Transboundary protected areas in Europe. Final report prepared for IUCN/WCDA Parks for Life Coordination Office.

Budowski, G. 2003. Transboundary protected areas as a vehicle for peaceful co-operation. Paper prepared for the workshop on transboundary protected areas in the governance stream of the 5th World Parks Congress, Durban, South Africa.

Bureau for International Narcotics and Law Enforcement Affairs. 1998. *International Narcotics Control Strategy Report, 1997: Canada, Mexico, and Central America*. Washington, DC: US Department of State.

Burton, J. W. 1995. Conflict prevention as political system. In *Beyond Confrontation: Learning Conflict Resolution in the Post Cold War Era*, ed. by J. A. Vasquez, et al. Ann Arbor: University of Michigan Press.

Burton, J. W. 1990. *Conflict: Human Needs Theory*. New York: St. Martin's Press.

Bush, K. 1998. A measure of peace: Peace and conflict impact assessment (PCIA) of development projects in conflict zones. Working Paper 1. The Peacebuilding and Reconstruction Programme Initiative and The Evaluation Unit, IDRC.

Bush, K. 2003. *Hands-on PCIA: A Handbook for Peace and Conflict Impact Assessment (PCIA)*. Available at ⟨http://action.web.ca/home/cpcc/en_resources.shtml?x=46859⟩.

Buzan, B., and R. Little. 1999. Beyond Westphalia? Capitalism after the "fall." *Review of International Studies* 25(5): 89–104.

Calcagno, A. 2002. Instilling comprehensive management and technological approaches for the sustainable management of La Plata river basin water resources. *Ecohydrology and Hydrobiology* 2(1–4): 93–103.

Carabias, J., and B. Babbitt. 1997. Letter of intent between the Department of Interior (DOI) of the United States and Secretariat of Environment, Natural Resources and Fisheries (SEMARNAP) of the United Mexican States. Available at ⟨http://www.cerc.usgs.gov/fcc/protected_agreement.htm⟩. (January 30, 2001.)

Carnegie Council on Ethics and International Affairs. 2004. The Marsh Arabs of Iraq: The legacy of Saddam Hussein and an agenda for restoration and justice. Edited transcript of panel discussion held at New York University, October 26.

Carret, J.-C., and D. Loyer. 2004. Madagascar protected area network sustainable financing: Economic analysis perspective. World Bank.

Castro-Chamberlain, J. J. 1997. Peace parks in Central America: Successes and failures in implementing management cooperation. In *Parks for Peace: Conference Proceedings, Parks for peace, International Conference on Transboundary Protected Areas as a Vehicle for International Co-operation*, Cape Town, South Africa: IUCN and UNEP, pp. 49–60.

Center for Strategic and International Studies. December 2004. Progress or peril? Measuring Iraq's reconstruction. *Post-conflict Reconstruction, The PCR Brief*. Washington DC: CSIS.

Cerovsky, J. 1993. Transboundary protected areas. In *Parks for life: Report of the 4th World Congress on National Parks and Protected Areas*, ed. by J. A. McNeely. Nairobi: United Nations Environment Programme, pp. 154–56.

Cerovsky, J., ed. 1996. *Biodiversity Conservation in Transboundary Protected Areas in Europe*. Prague: Ecopoint Foundation.

CESVI. 2002. The Sengwe corridor concept paper. Draft April 3, 2002. Report to DNPWM and Beitbridge and Chiredzi RDCs, Sustainable development and natural resource management in Southern Zimbabwe, Southern Lowveld Poject CESVI/MAE/AID 5063.

Chai, P. P. K., and P. Manggil. 2003. Thinking outside the box. *Tropical Forest Update* 13(2): 15–17.

Chandler, D. 2002. *From Kosovo to Kabul: Human Rights and International Intervention*. London: Pluto Press.

Chapin, M. 2004. A challenge to conservationists. *Worldwatch* (November–December), pp. 17–31.

Chaumba, J., I. Scoones, and W. Wolmer. 2003. New politics, new livelihoods: Changes in the Zimbabwean lowveld since the farm occupations of 2000. Sustainable Livelihoods in Southern Africa Research Paper 3. Institute of Development Studies, Brighton.

Chester, C. C. 2003. Responding to the idea of transboundary conservation: An overview of publics reaction to the Yellowstone to Yukon (Y2Y) conservation initiative. *Journal of Sustainable Forestry* 17(1–2): 103–26.

Chester, C. C. 2006. *Conservation across Borders: Biodiversity in an Interdependent World.* Washington, DC: Island Press.

Cheng, D. 2004. Monitoring large mammals of the northern plain landscape in Cambodia. Paper presented at 25th Annual Wildlife Seminar, December 24–26, 2004, Bangkok, Thailand. Faculty of Forestry, Kasetsart University.

Chikanga, K. 2004. Zimbabwe denies scuppering development of transfrontier park. Business Report, April 18, 2004, Available at ⟨www.peaceparks.org⟩.

Cho, I.-K., and D. Kreps. 1987. Signaling games and stable equilibria. *Quarterly Journal of Economics* 102(2): 179–222.

Christie, I. T. 2004. Overview of tourism in Mozambique. MIGA Swiss Program. World Bank AFTPS. August 25.

Christie, T., and A. Blundell. 2006. *The Design of a Protected Area Network for Liberia: An Economic, Social, and Ecological Analysis.* Washington, DC: CABS-Conservation International.

CIA-The World Factbook. 2004. *The World Factbook—Korea, North.* Available at ⟨http://www.cia.gov/cia/publications/factbook/geos/kn.html⟩.

Cisneros, J. A., and V. J. Naylor. 1999. Uniting la frontera: The ongoing efforts to establish a transboundary park. *Environment* 41(3): 12–20+.

Clapham, C. 1996. *Africa and the International System: The Politics of State Survival.* Cambridge: Cambridge University Press.

Codex dds. 2001–2004. Development of the Great Limpopo transfrontier park: Institutional arrangements. Available at ⟨www.greatlimpopopark.com⟩.

Collier, P., and A. Hoeffler. 2004. The challenge of reducing the global incidence of civil war. Copenhagen Consensus Challenge Paper. Available at ⟨http://www.copenhagenconsensus.com/Files/Filer/CC/Papers/Conflicts_230404.pdf⟩.

Collinge, C. 2005. The difference between society and space: Nested scales and the returns of spatial fetishism. *Environment and Planning D: Society and Space* 23(2): 189–206.

Commission for Environmental Cooperation. 1997. *Ecological Regions of North America: Toward a Common Perspective.* Montreal.

Commission on Global Governance. 1995. *Our Global Neighborhood: The Teport of the Commission on Global Governance.* Oxford: Oxford University Press.

CONANP. 2003. *Qué son las ANP?* Comisión Nacional de Áreas Naturales Protegidas. 1 July. Available at ⟨http://conanp.gob.mx/anp/⟩.

Conca, K. 2002. The case for environmental peacemaking. In *Environmental peacemaking*, ed. by K. Conca and G. D. Dabelko. Baltimore: Johns Hopkins University Press, pp. 1–22.

Conca, K., and G. Dabelko, eds. 2003. *Environmental Peacemaking.* Baltimore: Johns Hopkins University Press.

Conley, A., and M. A. Moote. 2003. Evaluating collaborative natural resource management. *Society and Natural Resources* 16(5): 371–86.

Cooke, B., and U. Kothari, eds. 2001. *Participation: The New Tyranny.* London: Zed Books.

Cooperrider, A., and D. S. Wilcove. 1995. *Defending the Desert: Conserving Biodiversity on BLM Lands in the Southwest*. New York: Environmental Defense Fund.

Cornelius, S. 2000. Transborder conservation areas: An option for the Sonoran desert? *Borderlines* 8(6): 1.

Cumming, D. H. M. 1999. *Study on the Development of Transboundary Natural Resource Management Areas in Southern Africa—Environmental Context*. Washington, DC: Biodiversity Support Program.

Dabelko, G. D., and A. Carius. 2004. Institutionalizing responses to environment, conflict, and cooperation. In *Understanding Environment, Conflict, and Cooperation*. Nairobi: United Nations Environment Programme, pp. 21–33. Available at ⟨http://www.wilsoncenter.org/topics/pubs/unep.pdf⟩.

Daily, G. C., ed. 1997. *Nature's Services: Societal Dependence on Natural Ecosystems*. Washington, DC: Island Press.

Daly, H., and J. Farley. 2004. *Ecological Economics*. Washington, DC: Island Press.

Danby, R. K. 1997. International transborder protected areas: Experience, benefits, and opportunities. *Environments* 25(1): 1–14.

Danish, K. W. 1995. International environmental law and the "bottom-up" approach: A review of the Desertification Convention. *Indiana Journal of Global Legal Studies* 3(1): 133–76.

de la Harpe, D. 2001. The Malilangwe Trust: Development through conservation. Annual Report 2000.

De Villiers, B. 1999. *Peace Parks, The Way Ahead*. Pretoria: HSRC Publishers.

Deere, C. L., and D. C. Esty, eds. 2002. *Greening the Americas*. Cambridge: MIT Press.

Demick, B. 2005. North Koreans romance development of nuclear weapons. *The China Post*, July 12. Available at ⟨http://www.chinapost.com.tw/p_detail.asp?id=65217&GRP=C&on News⟩.

Department of Environmental Affairs and Tourism (DEAT). Undated. *DEAT's poverty relief projects and how they work*. Available at ⟨www.environment.gov.za⟩. Accessed January 18, 2005.

Dergham, R., and J. Singer. 2000. *Special Policy Forum Report: A Syria-Israel summit—Prospects for Peace*. Washington, DC: Washington Institute for Near East Policy. Available at ⟨http://www.ciaonet.org/pbei/sites/winep_peace2001.html⟩.

Deudney, D. H., and R. A. Matthew, 1999. *Contested Grounds: Security and Conflict in the New Environmental Politics*. Albany: State University of New York Press.

Deutsch, M. 1991. Subjective features of conflict resolution: Psychological, social and cultural influences. In *New Directions in Conflict Theory*, ed. by R. Vayrynen. London: Sage.

DFID. 2003. Concept paper: Alleviation of poverty in buffer zones of protected areas in Afghanistan. Kabul, Afghanistan.

Diehl, P., ed. 2001. *The Politics of Global Governance: International Organizations in a Changing World*. Boulder, CO: Lynne Rienner.

Dilsaver, L. M., ed. 1994. *America's National Park System: The Critical Documents*. New York: Rowman and Littlefield.

Dimmitt, M. A. 2000. Biomes and communities of the Sonoran Desert Region. In *A Natural History of the Sonoran Desert*, ed. by S. J. Phillips and P. Wentworth Comus. Tucson: Arizona-Sonora Desert Museum Press, pp. 3–18.

Dione, O., et al. 2005. *Niger River Basin: A Vision for Sustainable Management*. Washington, DC: World Bank.

Dongguk Daehakyo BukhanHak Yeongguseo (English trans.: Dongguk University Institute of North Korean Studies). 2006. *DMZ Saengtae wa Hanbando Pyunghwa* (English trans.: The DMZ Ecology and Peace of the Korean Peninsula). Seoul, Korea: Acanet.

Douglas, M., and A. Wildavsky. 1983. *Risk and Culture*. Berkeley: University of California Press.

Duffield, M. 2001. *Global Governance and the New Wars: The Merger of Development and Security*. London: Zed Books.

Duffy, R. 1997. The environmental challenge to the nation-state: Superparks and national parks policy in Zimbabwe. *Journal of Southern African Studies* 23(3): 441–51.

Duffy, R. 2000. Shadow players: Ecotourism development, corruption and state politics in Belize. *Third World Quarterly* 21(3): 549–65.

Duffy, R. 2002. *A Trip Too Far: Ecotourism, Politics and Exploitation*. London: Earthscan.

Duffy, R. 2002. Peace parks: The paradox of globalisation. *Geopolitics* 6(2): 1–26.

Duffy, R. 2006. Global governance and environmental management: The politics of transboundary conservation areas in Southern Africa. *Political Geography* 25(1): 87–112.

Dzingirai, V. 2004. *Disenfranchisement at Large: Transfrontier Zones, Conservation and Local Livelihoods*. Harare, Zimbabwe: IUCN ROSA.

Eden Again Project. 2003. Building a scientific basis for restoration of the Mesopotamian Marshlands. In *Findings of the International Technical Advisory Panel Restoration Planning Workshop*. The Iraq Foundation. Available at ⟨http://www.edenagain.org/reports.html⟩.

Eden again/New eden. 2004. The new eden project: Final report. Presented at the United Nations CSD-12, April 14–30, in New York. The Iraq Foundation and the New Eden Working Team. Available at ⟨http://www.edenagain.org/reports.html⟩.

Eder, N. 1996. *Poisoned Prosperity: Development, Modernization, and Environment in South Korea*. Armonk, NY: M.E. Sharpe.

Elliffe, S. 1999. Guidelines for the release/development of dormant state or community assets for ecotourism development in the context of community involvement, land issues and environmental requirements. Paper presented at Community Public Private Partnerships Conference, November 16–18, Johannesburg.

ESRI. 1995. *Understanding GIS: Arc/Info method*. New York: Environmental Systems Research Institute.

Esty, D. C., M. Levy, T. Srebotnjak, and A. de Shertinin. 2005. *2005 Environmental Sustainability Index: Benchmarking National Environmental Stewardship*. New Haven: Yale Center for Environmental Law and Policy.

Fairhead, J. 2001. International dimensions of conflict over natural and environmental resources. In *Violent Environment*, ed. by Nancy Lee Peluso and Michael Watts. Ithaca: Cornell University Press, pp. 213–36.

Fakir, S. 2003. From sweet talk to delivery: Community participation in transfrontier conservation areas (TFCAs). In *Transboundary Protected Areas: Guidelines for Good Practice and Implementation*, ed. by L. E. O. Braack and T. Petermann. Schortau, Germany: Internationale Weirbildung und Entwicklung (InWEnt), pp. 15–24.

Fall, J. J. 2003. Planning protected areas across boundaries: New paradigms and old ghosts. In *Transboundary Protected Areas: The Viability of Regional Conservation Strategies*, ed. by U. Manage Goodale. New York: Food Products Press, pp. 75–96.

Fall, J. J. 1999. Transboundary biosphere reserves: A new framework for cooperation. *Environmental Conservation* 26: 252–55.

Fall, J. 2003. Planning protected areas across boundaries: New paradigms and old ghosts. *Journal of Sustainable Forestry* 17(1–2): 81–103.

Farah, D. 2004. *Blood from Stones*. New York: Broadway Publishers.

Felger, R. S., B. Broyles, M. Wilson, G. P. Nabhan, and D. S. Turner. 2006. Six grand reserves, one Sonoran Desert. In *Dry Borders*, ed. by R. S. Felger and B. Broyles. Salt Lake City: University of Utah Press.

Ferreira, S. 2004. Problems associated with tourism development in Southern Africa: The case of transfrontier conservation areas. *GeoJournal* 60(3): 301–10.

Fladeland, M., et al., eds. 2003. *Transboundary Protected Areas: The Variability of Regional Conservationist Strategies*. New York: Haworth Press.

Fonds Européen de Développement. 2003. Expertise technique pour l'appui à la formulation de la stratégie et des actions d'intégration du pastoralisme et de la conservation de la biodiversité aux périphéries du Par W du fleuve Niger. Mission de recherche, rapport trimestriel.

Ford Foundation. 2002. The displaced Makulekes recover community land and wildlife assets. *Sustainable Solutions*: 46–51.

Forsberg, T. 1996. Explaining territorial disputes: From power politics to normative reasons. *Journal of Peace Research* 33(4): 433–49.

Fortna, V. P. 2003. Scraps of Paper? Agreements and the Durability of Peace. *International Organization* 57(2): 337–72.

Friedman, J. W. 1971. A noncooperative equilibrium for supergames. *Review of Economic Studies* 38(1): 1–12.

French Polar Institute. 2005. Available at ⟨http://www.ifremer.fr/ifrtp/pages/institut_polaire.html⟩.

French Southern and Antarctic Lands (*Terres Australes et Antarctiques Françaises—TAAF*). 2005. ⟨http://www.taaf.fr⟩.

Frome, M. 1985. *Promised Land: Adventures and Encounters in Wild America*. New York: Morrow.

Fukuyama, F. 1992. *The End of History and the Last Man*. London: Hamilton.

Fuller, S., and M. Gemin. 2004. *Proceedings of the Karakoram Workshop*. Karachi, Pakistan: IUCN.

Fuller, S. 1995. Opportunities for transfrontier protected areas in northern Pakistan. *Proceedings of the 8th International Snow Leopard Conference*. November 12–16. Islamabad, Pakistan, pp. 203–15.

Fuller, S. 1996a. The Karakoram constellation. Paper presented at the 1st World Conservation Congress, Montreal.

Fuller, S. 1996b. Opportunities for snow leopard habitat conservation in new protected areas in northern Pakistan. In *Proceedings of the 8th International Snow Leopard Conference, Islamabad, Pakistan*.

Fuller, S. 2004. Technical assistance proposal: Protected areas and poverty alleviation project: Afghanistan. Asian Development Bank, Manila.

Fuller, S. 2005. Final Evaluation Report. Community Action Investment Program. Dushanbe: AKF/MSDSP, 77 pp.

Gaan, N. 2002. Water and conflict in west bank and Golan Heights: Are environmental issues and national security linked? *India Quarterly* 58(3–4): 129–44.

Gaigals, C., and M. Leonhardt. 2001. *Conflict-Sensitive Approaches to Development: A Review of Practice*. London: International Alert, Saferworld and IDRC.

Gallagher, K. P. 2004. *Free Trade and the Environment: Mexico, NAFTA and Beyond*. Stanford: Stanford University Press.

Galt, A., T. Sigaty, and M. Vinton, eds. 2000. The world commission on protected areas. *2nd Southeast Asia Regional Forum, Pakse, vol. 2: Papers Presented*. December 6–11, 1999, in Lao PDR. Vientiane: IUCN-The World Conservation Union.

Gasana, J. K., P. P. K. Chai, and Y. Trisurat. 2005. The management of protected areas in borderlands: Understanding the process of transboundary. Available at ⟨http://www.tropicalforests.ch/files/thematic_issues/transboundary_conservation.pdf⟩.

Gelderblom, C., B. W. van Wilgen, and N. Rossouw. 1998. Proposed transfrontier conservation areas: Preliminary data sheets and financial requirements. CISR Division of Water, Environment and Forestry Technology, Stellenbosch, South Africa.

George, A. 2004. Lifeless lake or exotic ecosystem. *New Scientist* 183: 6–8.

GFA Terrasystems. 2002. Development of the communities of Limpopo National Park, August 2002.

Gibson, L. J., and B. Evans. 2002. *National Parks as Economic Development: An Exploratory Note*. Tucson: University of Arizona.

Gill, S. 1998. New constitutionalism, democratisation and global political economy. *Pacifica Review* 10(1): 23–38.

Gilligan, C. 1982. *In a Different Voice: Psychological Theory and Women's Development*. Cambridge: Harvard University Press.

Glassner, M., and C. Fahrer. 2004. *Political Geography, 3rd ed*. New York: Wiley.

Gleditsch, N. P., and H. Urdal. 2004. Don't blame environmental decay for the next war. *International Herald Tribune*, November 22, p. 3.

Global IDP Project. 2003–2004. Calls for Resumption of Peace Talks Fall Flat. Available at ⟨http://www.db.idpproject.org⟩.

Global Witness. 2004. Dangerous liaisons: The continued relationship between Liberia's natural resource industries, arms trafficking and regional insecurity. A briefing documents submitted by Global Witness to the UN Security Council, December 8, 2004. Available at ⟨http://www.globalpolicy.org/security/issues/liberia/2004/1208dliaisons.pdf⟩.

Godwin, P. 2001. Wildlife without borders: Uniting Africa's wildlife reserves. *National Geographic* 200(3): 2–29.

Gomez, A., et al. 2006. The Last of the transboundary wild areas: Opportunities for conservaton and international cooperation. Paper presented at the annual meeting of the Society for Conservation Biology, San Jose, CA.

Goodale, U. M., M. J. Stern, C. Margoluis, A. G. Lanfer, and M. Fladeland, eds. 2003. *Transboundary Protected Areas*. New York: Food Products Press.

Grant, A., and F. Søderbaum, eds. 2003. *The New Regionalism in Africa*. Aldershot: Ashgate.

Granovetter, M. 1985. Economic action and social structure: The problem of embeddedness. *American Journal of Sociology* 91(3): 481–510.

Griffin, J., D. Cumming, S. Metcalfe, M. t'Sas-Rolfes, J. Singh, E. Chonguiça, M. Rowen, and J. Oglethorpe. 1999. *Study on the Development of Transboundary Natural Resource Management Areas in Southern Africa*. Washington, DC: Biodiversity Support Program.

Griffiths, J., D. Cumming, J. Singh, and S. Metcalfe, et al. 1999. *Study on the Development and Management of Transboundary Conservation Areas in Southern Africa*. Washington, DC: BSP/USAID.

Guha, Ramachandra. 2000. *Environmentalism: A Global History*. New York: Longman.

Guha, R., and J. Martinez-Alier, eds. 1997. *Varieties of Environmentalism*. London: Earthscan.

Haas, P. 1992. *Saving the Mediterranean: The Politics of International Environmental Cooperation*. New York: Columbia University Press.

Hadjipavlou-Trigeorgis, M. 1998. Different relationships to the land: Personal narratives, political implications and future possibilities in Cyprus. In *Cyprus and Its People: Nation, Identity, and Experience in Unimaginable Community 1955–1997*, ed. by V. Calotychos. Boulder, CO: Westview Press, pp. 35–52.

Haggard, S., and B. A. Simmons. 1987. Theories of international regimes. *International Organization* 41(3): 491–517.

Hahn, R., et al. 2005. Environmental baseline study for the ruvuma interface. In *Prefeasibility Study into Sustainable Development and Conservation*. Dar es Salaam: GTZ.

Hakamada, S. 2000. Japanese–Russian relations in 1997–1999: The struggle against illusions. In *Japan and Russia: The Tortuous Path to Normalization, 1949–1999*, ed. by G. Rozman. New York: St. Martin's Press.

Hamilton, L., and L. McMillan, eds. 2004. *Guidelines for Planning and Managing Mountain Protected Areas*. Gland, Switzerland: IUCN.

Hammill, A., and C. Besançon. 2003. Promoting conflict sensitivity in transboundary protected areas: A role for peace and conflict impact assessments. Paper prepared for the workshop on transboundary protected areas in the governance stream of the 5th World Parks Congress, September 12–13, in Durban, South Africa.

Hampton, M. P., and M. Levi. 1999. Fast spinning into oblivion? Recent developments in money laundering policies and offshore finance centres. *Third World Quarterly* 20(3): 645–56.

Hanks, J. 1998. Transfrontier conservation areas for Zimbabwe: An opportunity for job-creation, regional cooperation and biodiversity conservation. Bob Rutherford Trust Annual Memorial Lecture, London.

Hanks, J. 2003. Transfrontier conservation areas (TFCAs) in Southern Africa: Their role in conserving biodiversity, socioeconomic development and promoting a culture of peace. *Journal of Sustainable Forestry* 17(1–2): 127–48.

Hardin, G. 1968. Tragedy of the commons. *Science* 162: 1243–48.

Harmon, D., eds. 2001. Crossing boundaries in park management. In *Proceedings of the 11th Conference on Research and Resource Management in Parks and on Public Lands*. Denver: George Wright Society.

Harrison, S. 2005. DPRK Trip report. [Napsnet] *Policy Forum Online*, May 10, pp. 1–4.

Hasegawa, T. 1998. *The Northern Territories Dispute and Russo–Japanese Relations. Volume 1: Between War and Peace, 1697–1985*. Berkeley: University of California Press.

Hasegawa, T. 1998. *The Northern Territories Dispute and Russo–Japanese relations. Volume 2: Neither War nor Peace, 1985–1998*. Berkeley: University of California Press.

Hasegawa, T. 2000. Japanese perceptions of the Soviet Union and Russia in the postwar period. In *Japan and Russia: The Tortuous Path to Normalization, 1949–1999*, ed. by G. Rozman. New York: St. Martin's Press.

Hasenclaver, A., P. Mayer, and V. Rittberger. 1996. Interests, power, knowledge: The study of international regimes. *Mershon International Studies Review* 40(2): 177–228.

Hays, P. 2005. Defense intelligence agency says North Korea has nuclear armed missiles. Northeast Asia Peace and Security Network. [Napsnet] *Special Report*, May 3, pp. 1–5.

Helvetas, M. 2002. Establishment of a community-based tourism enterprise in Canhane, Massingir District, Gaza Province, Great Limpopo TFCA.

Herstein, I., and J. Milnor. 1953. An axiomatic approach to measurable utility. *Econometrica* 21(2): 291–97.

Hewson, M., and T. J. Sinclair, eds. 1999. *Approaches to Global Governance Theory*. Albany: State University of New York Press.

Hillstrom, K., and L. C. Hillstrom, eds. 2004. *The World's Environments—Latin America and the Caribbean: A Continental Overview of Environmental Issues*. Santa Barbara: ABC-CLIO.

Hoare, J. E., and S. Pares. 2005. *North Korea in the 21st Century: An Interpretative Guide*. Kent, England: Global Oriented Ltd.

Hoekstra, H., and W. F. Fagan. 1998. Body size, dispersal ability and compositional disharmony: The carnivore-dominated fauna of the Kuril Islands. *Diversity and Distributions* (4): 135–49.

Hofer, H., et al. 2004. Distribution and movements of elephants and other wildlife in the Selous–Niassa wildlife corridor, Tanzania. Eschborn. *Deutsche Gesellschaft für Technische Zusammenarbeit/TOEB F-iV/12e*.

Homer-Dixon, T. 1998. *The Environment, Scarcity and Violence*. Princeton: Princeton University Press.

Hufbauer, G. C., D. C. Esty, D. Orejas, L. Rubio, and J. J. Schott. 2000. *NAFTA and the Environment: Seven Years Later*. Washington, DC: Institute for International Economics.

Huggins, G., E. Barendse, A. Fischer, and J. Sitoi. 2003. *Limpopo National Park: Resettlement Policy Framework*. Pretoria: South African Government.

Hughes, D. M. 2002. Going transboundary: Scale-making and exclusion in Southern-African conservation. Available at ⟨http://hdgc.epp.cmu.edu/misc/Going%20Transboundary%20-%20Hughes%5B1%5D.pdf⟩.

Hulme, D., and M. Murphree, eds. 2001. *African Wildlife and Livelihoods*. London: Heinemann.

Hunter, C. 1997. Sustainable tourism as an adaptive paradigm. *Annals of Tourism Research* 24(4): 850–67.

Husserl, E. 1900. *Logical Investigations* (Logische Untersuchungen), vol. 1, English trans. 1970 by J. N. Findlay. New York: Routledge.

Huth, P. K. 1996. *Standing Your Ground: Territorial Disputes and International Conflict*. Ann Arbor: University of Michigan Press.

Hutton, J., W. A. Adams, and J. C. Murombedzi. 2005. Back to the barriers? Changing narratives in biodiversity conservation. *Forum for Development Studies* (2): 341–70.

Hyundai, A. 2004. *Hyundai Asan Inter-Korean Cooperative Business*. Seoul, Korea: Hyundai Asan Corporation.

International Association of Antarctic Tour Operators. 2004. IAATO Overview of Antarctic tourism. *2003–2004 Antarctic Season* ATCM XXVII-IP 63.

International Crisis Group. 2005. North Korea: Can the iron fist accept the invisible hand? [Napsnet] *Policy Forum Online*, May 19, pp. 1–3.

International Gorilla Conservation Programme. 2004. Mountain gorillas: Some social and biological data. Available at ⟨http://www.mountaingorillas.org/pdf/gorilla_profile.pdf⟩.

International Rescue Committee. 2003. Mortality in the Democratic Republic of Congo: Results from a nationwide survey. Available at ⟨http://intranet.theirc.org/docs/drc_mortality_iii_full.pdf⟩.

InWEnt. 2003. Welcome and workshop opening. Paper presented at the TBPA Approaches and Processes in East Africa, Mwoka, Tanzania.

ITTO/RFD. 2000. Project document (PD 15/00 Rev.2 F): Management of the Pha Taem protected forests complex to promote cooperation for trans-boundary biodiversity conservation between Thailand, Cambodia, and Laos (Phase I). Bangkok: International Tropical Timber Organization and Royal Forest Department.

ITTO/RFD/FA. 2004. Project phase II proposal: Management of the emerald triangle protected forests complex to promote cooperation for trans-boundary biodiversity conservation between Thailand, Cambodia, and Laos (Phase II). Bangkok: International Tropical Timber Organization, Thailand's Royal Forest Department, and Cambodia's Forest Administration.

IUCN (The World Conservation Union). 2001. *IUCN Regional Office for Central Africa: Progress and assessment report.* Available at ⟨http://www.iucn.org/pareport/docs/brac_report.pdf⟩.

IUCN. 2004. Etude sur l'état des lieux descriptif et analytique du complexe de parc W (Bénin, Burkina Faso, Niger), d'Arly (Burkina Faso), de la Pendjari (Bénin). Rapport de consultation. 70 pp.

IUCN WCPA (The World Conservation Union, World Commission on Protected Areas). 2005. WCPA task force transboundary protected areas. IUCN/WCPA/TPA. Available at ⟨http://www.iucn.org/themes/wcpa/theme/parts/parks.html⟩.

IUCN-ROSA. 2002. *Rethinking the Great Limpopo Transfrontier Conservation Area and TBNRM Developments in Southern Africa.* Harare: IUCN-ROSA.

IUCN-The World Conservation Union. 1984. *Categories, Objectives and Criteria for Protected Areas.* Gland, Switzerland: IUCN.

Jacob, J. D. C., ed. 1991. *Beyond the Hoppo Ryodo: Japanese–Soviet–American Relations in the 1990s.* Washington, DC: AEI Press.

Jaeger, E. C. 1957. *The North American Deserts.* Stanford: Stanford University Press.

Jaffrelot, C., ed. 2002. *Pakistan: Nationalism without a Nation.* London: Zed Books.

Jameson, J. 1996. *The Story of Big Bend.* Austin: University of Texas Press.

Janssen, M., J. Anderies, and E. Ostrom, 2004. Robustness of social-ecological systems to spatial and temporal variability. Working paper. Workshop in Political Theory and Policy Analysis, Indiana University, Bloomington.

Joint Management Board (JMB). 2002. Great Limpopo Transfrontier Park, Joint management plan.

Joquin, Sr., A. 1988. The Tohono O'odham Nation: A position statement. In *Simposio de Investigación sobre la Zona Ecológica de El Pinacate*, ed. by Environment Committee of the Arizona-Mexico Commission, Sonora, Mexico: Hermosillo, pp. 13–14.

Jordan, L. J. 2005. U.S. to add 500 agents along vulnerable Arizona border. *San Francisco Chronicle*, March 30, p. A5.

Joyner, C. C. 1998. Governing the frozen commons: The Antarctic regime and environmental protection. Columbia, SC: University of South Carolina.

Joyner, C. C., and E. R. Theis. 1997. *Eagle over the Ice: The U.S. in the Antarctic.* Hanover, NH: University Press of New England.

Kang, S. M., M. S. Kim, and M. Lee. 2002. The trends of composite environmental indices in Korea. *Journal of Environmental Management* 64(2): 199–206.

Katarere, Y., R. Hill, and S. Moyo. 2001. A critique of transboundary natural resource management in Southern Africa. Paper 1. *IUCN-ROSA series on transboundary natural resource management*. Available at ⟨http://www.iucnrosa.org.zw/tbnrm/publications/book1.pdf⟩.

Kayizzi-Mugerwa, S. 2005. Report of the commission for Africa: What is new? *Journal of Development Studies* 41(6): 1126–1132.

Kemkar, N. 2005. Environmental peacemaking: Ending the conflict between India and Pakistan on the Siachen Glacier. *Stanford Environmental Law Review* 25(1): 1–55.

Keohane, R. O. 1984. After hegemony: Cooperation and discord in the world political economy. Princeton: Princeton University Press.

KfW. 2004. 2003. Annual report on cooperation with developing countries.

Kihl, Y. W. 2005. *Transforming Korean Politics: Democracy, Reform, and Culture*. Armonk, NY: M.E. Sharpe.

Kihl, Y. W., and H. N. Kim, eds. 2006. *North Korea: The Politics of Regime Survival*. Armonk, NY: M.E. Sharpe.

Kim, K. C. 2006. DMZ, Saengmyong kwa Pyunghwa eui Gongkan (Original English draft title: DMZ, A Space for Life and Peace in Korea). In *DMZ Saengtae wa Hanbando Pyunghwa* (English trans.: The DMZ Ecology and Peace of the Korean Peninsula), Ed. by Dongguk Adebayo BukhanHak Yeongguseo (Dongguk University Institute of North Korean Studies), Seoul: Acanet, pp. 59–92.

Kim, K. C., and E. O. Wilson. 2002. The land that war Protected. *New York Times OP-Ed*, December 10, p. A31.

Kim, K. C. 1999. The environment: A cause for concern. Special issue, *Digital Chosun*, July 12, p. 3.

Kim, K. C. 1997. Preserving biodiversity in Korea's demilitarized zone. *Science* 278: 242–43.

Kim, K. C. 1995. *Nature conservation for the Korean posterity: A network of protected areas and greenways. Collected Papers of the International Symposium on National Parks and Protected areas in Korea*, May 11–12, Seoul. Unpublished document.

Kim, K. C., and R. D. Weaver, eds. 1994. *Biodiversity and Landscapes: A Paradox of Humanity*. New York: Cambridge University Press.

Kimura, M., and D. Welch. 1998. Specifying interests: Japans claim to the northern territories and its implications for international relations theory. *International Studies Quarterly* 42(2): 213–43.

Kirkbride, W. A. 1985. *Panmunjum: Facts about the Korean DMZ*. Seoul: Hollym.

Koch, E. 2001. Tales of white elephants. *The Courier*, July–August.

Kocs, S. 1995. Territorial disputes and interstate water, 1945–1987. *Journal of Politics* 57(1): 159–75.

Koithara, V. 2004. *Crafting Peace in Kashmir: Through a Realist Lens*. New Delhi: Sago Publications.

Korea Development Institute (KDI). 1975. *Korea's Economy: Past and Present*. Seoul: Korea Development Institute.

Korea Economic Institute of America (KEI). 1996. *Korea's Economy 1996*, vol. 12. Washington, DC: Korea Economic Institute of America.

KPMG. 2002. Integrated tourism development plan. Great Limpopo Transfrontier Park Technical Committee, February.

Krasner, S. D., ed. 1983. *International Regimes*. Ithaca: Cornell University Press.

Krasner, S. 2004. Sharing sovereignty: New institutions for collapsed and failing states. *International Security* 29(2): 85–120.

Kraul, C. 2005. Rice mends fences with Mexico in run-up to summit. *San Francisco Chronicle*, March 11, p. A-2.

Krauze, E. 1997. *Mexico: Biography of Power*. New York: Harper-Collins.

Kristeva, J. 1993. *Nations without Nationalism*. New York: Columbia University Press.

Kwon, D.-M., M.-S. Seo, and Y.-C. Seo. 2002. A study of compliance with environmental regulations of ISO 14001 certified companies in Korea. *Journal of Environmental Management* 65(4): 347–53.

Laird, W., J. Murrieta-Saldivar, and J. Shepard. 1997. Cooperation across borders: A brief history of biosphere reserves in the Sonoran Desert. *Journal of the Southwest* 39(3–4): 307–13.

Lake Vostok DEEP ICE Project. 2005. Dr. Sergey A. Bulat, Principal Investigator, Petersburg Nuclear Physics Institute, Russian Academy of Sciences. Available at ⟨http://salegos-scar.montana.edu/⟩.

Landau, S. 1999. National park urged for desert. *Arizona Republic*, March 10, p. B1.

Langholz, J. 2006. Crouching tiger hidden terrorist: Security threats posed by international peace parks. Monterey CA: Monterey Institute for International Studies.

Lanjouw, A. 2003. Building partnerships in the face of political and armed crisis. In *War and Tropical Forests: Conservation in Areas of Armed Conflict*, ed. by S. V. Price. Binghamton, NY: Haworth Press, pp. 93–114.

Lawler, E., and J. Yoon. 1996. Commitment in exchange relations: Test of a theory of relational cohesion. *American Sociological Review* 61(1): 89–108.

Leach, J. A. 2005. Prospects for U.S. Policy toward the Korean peninsula in the second Bush admiistration. Speech in the CSIS and Chosun Ilbo Conference, May 17. Available at ⟨http://www.house.gov/leach/JLKoreaCSISspeech.doc.rtf⟩.

Lee, H. 2004. Environmental awareness and environmental practice in Korea. *Korea Journal* 44(3): 165–84.

Lee, Y.-S., and D. R. Yoon. 2005. The structure of North Korea's political economy: Changes and effects. [Napsnet] *Special Report*, March 15, pp. 1–3.

Lejano, R. 2006. Frameworks for policy analysis: Merging text and context. New York: Routledge.

Leonhardt, M. 2001. Towards a unified methodology: Reframing PCIA (response paper). In *Berghof Handbook for Conflict Transformation*. Berlin: Berhof Research Center for Constructive Conflict Management. Available at ⟨http://www.berghof-handbook.net/leonhardt/⟩.

Levin, N. D., and Y.-S. Han. 2002. Sunshine in Korea: The South Korean debate over policies toward North Korea. Pittsburgh: RAND.

Lewis, B. 2001. *The Multiple Identities of the Middle East*. New York: Schocken Books.

Limpopo Tourism Consortium. 2004. Tourism development plan for Parque Nacional do Limpopo as part of the development of the Greater Limpopo Transfrontier Park. Prepared for Parque Nacional do Limpopo, February.

Linde, H. van der, J. Oglethorpe, T. Sandwith, D. Snelson, and Y. Tessema. 2001. *Beyond boundaries: Transboundary natural resource management in sub-Saharan Africa*. Washington, DC: US Biodiversity Support Program.

Lindley, D. 2004. UNDOF: Operational analysis and lessons learned. *Defense and Security Analysis* 20(2): 153–64.

Litfin, K. 1998. The Greening of Sovereignty in Global Civic Poltics. Cambridge: MIT Press.

Litwak, R. S., and K. Weathersby. 2005. The Kims' obsession: Archives show their quest to preserve the regime. [Napsnet] *Special Report*, June 30, pp. 1–5.

Lock, N. 1997. Transboundary protected areas between Mexico and Belize. *Coastal Management* 25(4): 445–54.

Lowi, M. 1995. *Water and Power: The Politics of a Scarce Resource in the Jordan Basin*. Cambridge: Cambridge University Press.

MacDonald, D. S. 1996. *The Koreans: Contemporary Politics and Society*, 3rd ed. Boulder, CO: Westview Press.

MacMahon, J. A. 2000. Warm deserts. In *North American terrestrial vegetation*, ed. by Michael G. Barbour and W. D. Billings. New York: Cambridge University Press, pp. 285–322.

Maltby, E., ed. 1994. An environmental and ecological study of the Marshlands of Mesopotamia. (Draft consultative bulletin.) Wetland Ecosystems Research Group, University of Exeter. London: AMAR Appeal Trust.

Maluleke, M. L. 2003. Presentation by Mashangu Livingston Maluleke on behalf of the Makuleke Community to the World Parks Congress 2003.

Mamdani, M. 2001. When victims become killers: Colonialism, nativism, and the genocide in Rwanda. Princeton: Princeton University Press.

Marod, D. 2003. Management of the Pha Taem protected forest complex to promote cooperation for trans-boundary biodiversity conservation between Thailand, Cambodia, and Laos (Phase I): Forest ecology technical final report. Faculty of Forestry, Kasertsart University, Bangkok.

Marquand, R. 1999. Glaciers in the high Himalayas melting at a rapid rate. *Christian Science Monitor*, November 5.

Massey, D. 1991. The political place of locality studies. *Environment and Planning A* 23(2): 267–81.

Matthew, R., M. Halle, and J. Switzer, eds. 2002. *Conserving the Peace: Resources, Livelihoods and Security*. Geneva: IUCN.

Matthew, R., M. Halle, and J. Switzer. 2002. A growing threat? In *Conserving the Peace: Resources, Livelihoods and Security*, ed. by R. Matthew, M. Halle, and J. Switzer. Geneva: IUCN, pp. 4–21.

Maxwell, G. 1957. *A Reed Shaken by the Wind: A Journey through the Unexplored Marshlands of Iraq*. Penguin: Harmondsworth.

Mayoral-Phillips, A. J. 2002. *Transboundary Development in Southern Africa: Rhetoric and Reality*. New York: Ford Foundation.

McGrew, A. G., and P. G. Lewis, et al., eds. 1992. *Global Politics*. Cambridge: Polity Press.

McNeely, J. A. 2002. Biodiversity, conflict and tropical forests. In *Conserving the Peace: Resources, Livelihoods and Security*, ed. by R. Matthew, M. Halle, and J. Switzer. Winnipeg, Canada: International Institute for Sustainable Development.

Mead, W. R. 2005. Should nukes bloom in Asia? [Napsnet] *Policy Forum Online*, June 23, pp. 1–4.

Mendez, E., and R. Garduno. 2005. Piden Diputatodos a Fox Promover con Bush Plan de Empleo Temporal. In *La Jornada*, March 18. Available at ⟨http://www.jornada.unam.mx/2005/mar05⟩.

Meyers, D. W. 2003. Does "smarter" lead to safer? An assessment of the US border accords with Canada and Mexico. *International Migration* 41(4): 5–44.

Miller, K., and L. Hamilton, eds. 1999. Bioregional approach to protected areas. *Parks* 9: 1–6.

Minca, C. 2001. *Postmodern Geography: Theory and Praxis*. Oxford: Blackwell.

Mittermeier, R., et al. 2005. *Transboundary Conservation: A New Vision for Protected Areas*. Washington, DC: Conservation International.

Mitrany, D. 1930. Pan-Europa: A hope or a danger? *Political Quarterly* (September–December). 457–78.

Mitrany, D. 1975. *The Functional Theory of Politics*. London: Martin Robertson.

Morris-Suzuki, T. 1999. Lines in the snow: Imagining the Russo Japanese frontier. *Pacific Affairs* 72(1): 57–77.

Mpanduji D. G. 2004. Population structure, movements and health status of elephants and other wildlife in the Selous–Niassa Wildlife Corridor, southern Tanzania. PhD dissertation. Free University of Berlin.

Mpanduji, D. G., et al. 2002. Movement of elephants in the Selous–Niassa Wildlife Corridor, southern Tanzania. *Pachyderm* 34: 18–31.

Mpanduji, D. G., et al. 2003. Immobilization and evaluation of clinical parameters from free-ranging elephants in southern Tanzania. *Pachyderm* 35: 140–45.

Mtisi, S. M. S., and J. Chaumba. 2001. Zimbabwe country study: Case studies of Sangwe and Mahenye communal areas in Chiredzi and Chipinge districts. Mapping phase report.

Mulroy, K. 1993. *Freedom on the Border*. Lubbock: Texas Tech University Press.

Murombedzi, J. 1992. Decentralization or recentralization? Implementing CAMPFIRE in the Omay communal lands of the Nyaminyami District. *CASS Occasional Paper*. Harare: University of Zimbabwe. Cited in V. Dzingirai, 2004, op. cit.

Murphy, C. 2000. Global governance: Poorly done and poorly understood. *International Affairs* 76(4): 789–804.

Nabhan, G. P. 1989. *Enduring Seeds*. Berkeley: North Point Press.

Nabhan, G. P. 1996. Completion of the Sonoran desert biosphere reserve network along the U.S./Mexico border: A proposal for the U.S. MAB Directorate. International Sonoran Desert Alliance. Seville, Spain. Files of the International Sonoran Desert Alliance, Ajo, Arizona.

NACEC. Undated. *Our Programs and Projects: Conservation of Biodiversity*. Montreal: North American Commission on Environmental Cooperation.

Najam, A., ed. 2003. Environment, Development and Human Security. Washington, DC: University Press of America.

National Parks Conservation Association. 2003. Big bend national park: A state of the parks report. Available at ⟨http://www.npca.org/media_center/reports/default.asp⟩.

Ndunguru, I. F. 1989. Big animals and big problems: The background to the Ruvuma village wildlife project. SCP Discussion Paper 6. GTZ, Dar es Salaam.

Neumann, R. P. 1998. *Imposing Wilderness: Struggles over Livelihood and Nature Preservation in Africa*. Berkeley: University of California Press.

Newell, J. 2004. *The Russian Far East: A Reference Guide for Conservation and Development*. McKinleyville, CA: Daniel and Daniel Publishers.

Newstead, C. 2005. Scaling Caribbean (in)dependence. *Geoforum* 36(1): 45–58.

Nicholson, E., and P. Clark, eds. 2002. The Iraqi marshlands: A human and environmental study. Amar International Charitable Foundation, London.

Nietschmann, B. 1997. Protecting indigenous coral reefs and sea territories, Miskito Coast, RAAN, Nicaragua. In *Conservation through Cultural Survival: Indigenous Peoples and Protected Areas*, ed. by S. Stevens. Washington, DC: Island Press, pp. 193–224.

Nijman, J. 1999. Cultural globalisation and the identity of place: The reconstruction of Amsterdam. *Ecumene* 6(2): 146–64.

Nordstrom, C. 2004. *Shadows of War: Violence, Power, and International Profiteering in the Twenty-first Century*. Berkeley: University of California Press.

Nudler, O. 1990. On conflicts and metaphors: Toward an extended rationality. In *Conflict: Human Needs Theory*, ed. by J. Burton. New York: St. Martin's Press.

Nugent, P., and A. Asiwaju, eds. 1996. *African Boundaries: Barriers, Conduits and Opportunities*. London: Pinter.

O'Brien, R., A.-M. Goetz, J. A. Scholte, and M. Williams. 2000. *Contesting Global Governance: Multilateral Economic Institutions and Global Social Movements*. Cambridge: Cambridge University Press.

O'Hanlon, M., and M. Mochizuki. 2003. *Crisis on the Korean Peninsula: How to Deal with a Nuclear North Korea*. New York: McGraw-Hill.

O'Neill, T. 2004. DMZ: Korea's dangerous divide. In *The Two Koreas*, vol. 76, ed. by J. Peloso. New York: H.W. Wilson, pp. 13–23.

Oberdorfer, D. 2005. Dealing with the North Korean nuclear threat. Northeast Asia Peace and Security Network. [Napsnet] *Policy Forum Online*, June 14, p. 9.

Office of International Affairs. 2005. Mexico. U.S. National Park Service. Available at ⟨http://www.nps.gov/oia/around/mexico.htm⟩.

Oh, J. K.-c. 1999. *Korean Politics: The Quest for Democratization and Economic Development*. Ithica: Cornell University Press.

Oh, K. 1998. North Korea's foreign relations in transition. In *Two Koreas in Transition: Implications for U.S. Policy*, ed. by I. J. Kim. Rockville, MD: In Depth Books, pp. 131–47.

Okada, K. 2002. The Japanese economic presence in the Russian far east. In *Russia's Far East: A Region at Risk*, ed. by J. Thornton and C. Ziegler. Seattle: National Bureau of Asian Research/University of Washington Press.

Ostrom, E. 1990. *Governing the Commons*. New York: Cambridge University Press.

Oviedo, G. T. 2003. Re-uniting communities with their landscapes. *Tropical Forest Update* 13(2): 8–9.

Paasi, A. 2004. Place and region: Looking through the prism of scale. *Progress in Human Geography* 28(4): 536–46.

Pack, J. S. 1999. The Statement of the secretary general of the Korean Nature conservation union (public/personal communication). August 19 (Juche 88). Pyongyang.

Padma, T. V. 2003. Achieving peace in the Himalayan peaks: A transboundary park. *Terra Green* online magazine. Available at ⟨http://static.teriin.org⟩, issue 45.

Paolini, A. J., A. P. Jarvis, and C. Reus-Smit. 1998. *Between Sovereignty and Global Governance: The United Nations, State and Civil Society*. Basingstoke: Macmillan.

Park, Y.-c. 2005. Korea becomes world's 10th largest economy. *English Chosun Ilbo: Biz/Tech*, 1. Available at ⟨http://English.chosun.com/w2idata/html/nes/200505/200501280023.html⟩.

Parker, J. 2005. Restore US nukes to South Korea. [Napsnet] *Policy Forum Online*, March 8, pp. 1–8.

Parks for Peace, International Conference on Transboundary Protected Areas as a Vehicle for International Co-Operation. 1997. *Parks for Peace: Conference Proceedings*. September 16–18. Somerset West (near Cape Town), South Africa.

Partow, H., and E. Maltby. October 2004. Trans-boundary context of marshland restoration. Presentation at Mesopotamian Marshes and Modern Development Conference at Cambridge, Harvard University. UNEP.

Patton, D. R. 1992. *Wildlife-Habitat Relationships in Forested Ecosystems*. Portland OR: Timber Press.

Pearson, G. 1998. Organ Pipe Cactus National Monument tri-national management challenges. US Department of the Interior and Instituto Nacional de Ecología (SEMARNAP).

Pearson, F. 2001. Dimensions of conflict resolution in ethnopolitical disputes. *Journal of Peace Research* 38(3): 275–87.

Pedynowski, D. 2003. Prospects for ecosystem management in the crown of the continent ecosystem, Canada–United States: Survey and recommendations. *Conservation Biology* 17(5): 1261–69.

Peet, R., and M. Watts, eds. 1996. *Liberation Ecologies: Environment, Development and Social Movements*. London: Routledge.

Peet, R., and M. Watts. 1996. Liberation ecology: Development, sustainability, and environment in an age of market triumphalism. In *Liberation Ecologies: Environment, Development and Social Movements*, ed. by R. Peet and M. Watts. London: Routledge, pp. 1–45.

Peluso, N. L., and M. Watts. 2001. Violent environments. In *Violent Environments*, ed. by N. L. Peluso and M. Watts. Ithaca: Cornell University Press, pp. 3–38.

Peluso, N. L. 1993. Coercing conservation? The politics of state resource control. In *The State and Social Power in Global Environmental Politics*, ed. by R. D. Lipschutz and K. Conca. New York: Columbia University Press, pp. 343–52.

Peri, Y. 2004. *Peace with Syria*. Washington, DC: Middle East Institute. Available at ⟨http://www.mideasti.org/articles/doc289.html⟩.

Phillips, A. 1998. Biosphere reserves and protected areas: What is the difference? In *Proceedings of a Workshop on Biosphere Reserve-Myth or Reality?* Gland and Cambridge: IUCN–The World Conservation Union, pp. 7–10.

Plumptre, A., A. Kayitare, H. Rainer, M. Gray, I. Munanura, N. Barakabuye, S. Asuma, M. Sivha, and A. Namara. 2004. The socio-economic status of people living near protected areas in the central Albertine Rift. *Albertine Rift Technical Reports* 4: 127.

Poku, N. 2001. *Regionalization and Security in Southern Africa*. London: Palgrave.

Prescott, S. T., ed. 2003. *Federal Land Management: Current Issues and Background*. New York: Nova Science.

Princen, T., and M. Finger. 1994. *Environmental NGOs in World Politics: Linking the Global and the Local*. London: Routledge.

Projet Parc ECOPAS. 2003. Elaboration d'une préstratégie nationale de gestion de la transhumance dans la zone périphérique du Parc W. Rapport de consultation. 49 pp.

Rabinovich, I. 1998. *The Brink of Peace: The Israeli-Syrian Negotiations*. Princeton: Princeton University Press.

Rabinovich, I. 2004. *Waging Peace: Israel and the Arabs, 1948–2003*. Princeton: Princeton University Press.

Raghavan, V. R. 2002. *Siachen: Conflict without End*. New Delhi: Viking.

Ramsar Convention Bureau. 2000. *Ramsar Handbooks for the Wise Use of Wetlands*. Ramsar, Iran.

Ramsbotham, A., M. S. Bah, and F. Calder. 2005. Enhancing African peace and security capacity: A useful role for the UK and the G8? *International Affairs* 81(2): 325–39.

Ramutsindela, M. 2004. Glocalisation and conservation strategies in 21st Century Southern Africa. *Tijdschrift voor Economische en Sociale Geografie* 95(1): 61–72(12). Available at ⟨http://www.ingentaconnect.com/content/bpl/tesg⟩.

Ramutsindela, M. 2004a. *Parks and People in Postcolonial Societies: Experiences in Southern Africa*. Dordrecht: Kluwer.

Ramutsindela, M. 2004b. Glocalisation and nature conservation in 21st century Southern Africa. *Tijdschrift voor Economische en Sociale Geografie* 95(1): 61–72.

Raven, P. H., ed. 1997. *Nature and Human Society: The Quest for a Sustainable World*. Washington, DC: National Academy Press.

Redekop, V. 2002. *Deep-rooted Conflict and a Culture of Peace*. Ottawa, Canada: Novalis.

Redford, K., and S. E. Sanderson. 2000. Extracting humans from nature. *Conservation Biology* 14(5): 1362–64.

Reid, J., ed. 2004. *Rio Grande*. Austin: University of Texas Press.

Reinicke, W. 1989. *Global Public Policy: Governing without Government?* Washington, DC: Brookings Institution.

Renner, M. 2004. Environmental threats. *International Herald Tribune*, December 3, p. 3.

Reno, W. 1995. Reinvention of an African patrimonial state: Charles Taylor's Liberia. *Third World Quarterly* 16(1): 109–20.

Reno, W. 1996. The business of war in Liberia. *Current History* 95(601): 211–15.

Reychler, L. 1999. *The Conflict Impact Assessment System (CIAS): A Method for Designing and Evaluating Development Policies and Projects*. Ebenhausen: CPN.

Richards, P. 1996. *Fighting for the Rain Forest: War, Youth and Resources in Sierra Leone*. Oxford: James Currey.

Richards, P. 2005. The Mano River conflicts as forest wars. *European Tropical Forest Research Network* 43–44: 29–32.

Ripley, J. D., T. H. Lillie, S. E. Cornelius, and R. M. Marshall. 2000. The U.S. Department of Defense embraces biodiversity conservation through ecoregional partnerships in the Sonoran Desert. *Diversity* 15(4): 3–5.

Roberts, C. M., Bohnsack, J. A., Gell, F., Hawkins, J. P., and Goodridge, R. 2001. Effects of marine reserves on adjacent fisheries. *Science* 294(5548): 1920–23.

Robertson, P., and S.-Y. Tang. 1995. The role of commitment in collective action: Comparing the organizational behavior and rational choice perspectives, *Public Administration Review* 55(1): 67–80.

Rosenau, J. N. 1990. *Turbulence in World Politics: A Theory of Change and Continuity*. New York: Harvester Wheatsheaf.

Rosenau, J. 1995. Governance in the 21st century. *Global Governance* 1(1): 13–43.

Rosenau, J. N. 1999. Toward an ontology for global governance. In *Approaches to Global Governance Theory*, ed. by M. Hewson and T. Sinclair. Albany: State University of New York Press, pp. 287–301.

Rosenau, J., and E.-O. Czempiel, eds. 1992. *Governance without Government: Order and Change in World Politics*. Cambridge: Cambridge University Press.

Ross, M. H. 1995. Interests and identities in natural resources: Conflicts involving indigenous peoples. *Cultural Survival Quarterly* (Fall): 74–76.

Roth, D. L. 1992. Mexican and American policy alternatives in the big bend region—An updated study of the proposed Mexican national park in the Sierra Del Carmen. Masters thesis. University of Texas at Austin.

Rothman, J., and M. Olson. 2001. From interests to identities: Towards a new emphasis in interactive conflict resolution. *Journal of Peace Research* 38(3): 289–305.

Rothman, J. 1997. *Resolving Identity-Based Conflict in Nations, Organizations and Communities*. San Francisco: Jossey-Bass.

Rothwell, D. R. 2000. Polar environmental protection and international law: The 1991 Antarctic protocol. *Environmental Journal of International Law* 11(3): 591–614.

Royal Forest Department (RFD). 2004. *The Management Plan of the Pha Taem Protected Forests Complex*. Bangkok: Royal Forest Department.

Rozman, G., ed. 2000. *Japan and Russia: The Tortuous Path to Normalization, 1949–1999*. New York: St. Martin's Press.

Rubinstein, A. 1985. A bargaining information with incomplete information about time preferences. *Econometrica* 53(5): 1151–72.

Rufford Foundation. 2005. Peace parks foundation. Available at ⟨http://www.ruffortd.org/Projects/PeaceParksFoundation.html⟩.

Runte, A. 1987. *National Parks: The American Experience*. Lincoln: University of Nebraska Press.

Samuels, L. 2005. Water accord ends 12-year dispute between United States, Mexico. *Dallas Morning News*, March 10, p. 3.

Sandwith, T. 2003. Is it worth the effort and expense? *Tropical Forest Update* 13(2): 6–7.

Sandwith, T., C. Shine, L. Hamilton, and D. Sheppard. 2001. *Transboundary Protected Area for Peace and Cooperation*. Gland, Switzerland: IUCN.

Sandwith, T., C. Shine, L. Hamilton, and D. Sheppard, eds. 2001. *Transboundary Protected Areas for Peace and Cooperation*. World Commission on Protected Areas (WCPA) Best Practice Protected Area Guidelines Series 7. Gland, Switzerland: IUCN–The World Conservation Union.

Sandwith, T. 2003. Overcoming barriers: Conservation and development in the Maloti–Drakensberg mountains of Southern Africa. *Journal of Sustainable Forestry* 17(1–2): 149–70.

SANParks. 2000. Bidding memorandum for the tender of concession sites. 2nd Draft, September 25, 2000. South African National Parks.

SANParks. 2001. Prequalification memorandum for the second phase of the concession programme. South African National Parks.

Satloff, R., and P. Clawson. 2000. *The U.S. Draft Treaty for Syria–Israel Peace: A Textual Analysis*. Washingtion, DC: Washington Institute for Near East Policy. Available at ⟨http://www.ciaonet.org/pbei/sites/winep_peace2001.html⟩.

Save the Environment Afghanistan. 2000. *The Status of Environment in Afghanistan*. Kabul, Afghanistan: Government of Afghanistan.

Sayre, R., E. Roca, G. Sedaghatkish, B. Young, S. Keel, R. Roca, and S. Sheppard. 2000. *Nature in Focus: Rapid Ecological Assessment*. Washington, DC: Nature Conservancy and Island Press.

Schaller, G. 2004. The status of Marco Polo sheep in the Pamir mountains of Afghanistan. Unpublished manuscript. WCS. 30 pp.

Schellenberger, M., and T. Nordhaus. 2005. Death warmed over: Beyond environmentalism imagining possibilities as large as the crisis that confronts us. *American Prospect*, p. 29.

Schelling, T. 1960. *The Strategy of Conflict*. Cambridge: Harvard University Press.

Scholes, R. J., and R. Biggs. 2004. *Ecosystem Services in Southern Africa: A Regional Assessment*. Pretoria, South Africa: CSIR.

Schuerholz, G. 2003. Parque Nacional do Limpopo, Business Plan, Period 2004–2006, First Draft.

Schwartzman, S., A. G. Moreira, and D. G. Nepstad. 2000. Rethinking tropical conservation: Peril in parks. *Conservation Biology* 14(5): 1351–57.

Scientific Committee on Antarctic Research (SCAR). 2004a. *Strategic Plan 2004–2010*. London: SCAR Secretariat, Scott Polar Research Institute.

Scientific Committee on Antarctic Research (SCAR). 2004b. Science and implementation plan for a *SCAR scientific research program Subglacial Antarctic Lake Environments (SALE)*. Prepared by SCAR Subglacial Antarctic Lake Exploration Group of Specialists.

Scott, D. A. 1995. *A directory of wetlands in the Middle East*. Gland, Switzerland: IUCN–The World Conservation Union, International Waterfowl, and Wetland Research Bureau.

SDNPP. 2000. *Sonoran Desert National Park: A Citizens' Proposal*. Tucson: Sonoran Desert National Park Project.

Sehga, V. 1998. Demilitarization of the Himalayas as an environmental imperative. Paper presented at the India international center, New Delhi, November.

Selby, J. 2003. Introduction. In *Global Governance, Conflict and Resistance*, ed. by F. Cochrane, R. Duffy, and J. Selby. London: Palgrave/Macmillan, pp. 1–18.

Sellers, R. W. 1997. *Preserving Nature in the National Parks*. New Haven: Yale University Press.

Seo, H.-j. 2002. 6.4 million North Koreans face starvation: WFP Official. *The Korean Herald*, November 18, p. 2.

Shambaugh, J., J. Oglethorpe, and R. Ham (with contributions from S. Tognetti). 2001. *The Trampled Grass: Mitigating the Impact of Armed Conflict on the Environment*. Washington, DC: Biodiversity Support Program.

Shaplen, J. T., and G. Laney. 2005. What should US do about North Korea? [Napsnet] *Policy Forum Online*, April 27, pp. 1–4.

Sharman, J. 2001. Invasions threaten peace park. *Mail&Guardian*, October 26 to November 1, p. 17.

Shaw, M. N. 2003. *International Law*, 5th ed. Cambridge: Cambridge University Press.

Shimotomai, N. 2000. Japanese–Soviet relations under perestroika. In *Japan and Russia: The Tortuous Path to Normalization, 1949–1999*, ed. by G. Rozman. New York: St. Martin's Press.

Shoman, A. 1995. *Backtalking Belize: Selected Writings*. Belize City: Angelus Press.

Sidaway, J. 2002. *Imagined regional communities*. London: Routledge.

Simon, J. 1997. *Endangered Mexico*. San Francisco: Sierra Club Books.

Simonian, L. 1995. *Defending the Land of the Jaguar*. Austin: University of Texas Press.

Singh, J. 1999. *Study on the Development of Transboundary Natural Resource Management Areas in South Africa: Lessons Learned*. Washington, DC: Biodiversity Support Program.

Singh, J., and H. Van Houtum. 2002. Post-colonial nature conservation in southern Africa: Same emperors, new clothes? *GeoJournal* 58(4): 253–63.

Singh, K. Undated. Talk titled "The Himalayas—A Civilizational Identity." In *Himalayan Ecology*. New Delhi, India: People's Commission on Environment and Development.

Sites, W. 2004. Progressive regionalism: A "deliberative" movement? *Antipode* 36(4): 766–78.

Sivaramakrishnan, K. 1999. *Modern Forests: Statemaking and Environmental Change in Colonial Eastern India*. Oxford: Oxford University Press.

Skåre, M. 2000. Liability annex or annexes to the environmental protocol: A review of the process within the Antarctic treaty system. In *Implementing the Environmental Protocol Regime for the Antarctic*, ed. by D. Vidas. London: Kluwer Academic, pp. 163–80.

Smillie, I. 2002. *Diamonds, Timber and West African Wars*. Winnipeg, Canada: International Institute for Sustainable Development. Available at: ⟨http://www.iisd.org/pdf/2002/envsec_diamonds_timber.pdf⟩.

Smith, G., and R. D. Baldus, eds. 2005. *Environmental Baseline and Pre-feasibility Study for the Niassa Game Reserve–Ruvuma River Interface*. Dar as Salaam: Mtwara Corridor Project and GTZ. GTZ Publication.

Smith, R. D., and E. Maltby. 2003. *Using the Ecosystem Approach to Implement the Convention on Biological Diversity: Key Issues and Case Studies*. Gland, Switzerland: IUCN, 118 pp.

Smyth, L. F. 1994. Intractable conflicts and the role of identity. *Negotiation Journal* (October): 311–21.

Snidal, D. 1985. The limits of hegemonic stability theory. *International Organization* 39(4): 579–615.

Sochaczewski, P. S. 1999. Can "peace parks" help feuding countries "make wildlife, not war"? *International Wildlife* (July): 1–9.

Solomon, B. 2003. Kant's perpetual peace: A new look at this centuries-old quest. *On-line Journal of Peace and Conflict Resolution (OJPCR)* 5(Summer): 106–26.

Sonna, P. 2003. *The Status and Distribution of Eld's Deer* cervus eldi siamensis *in Preah Vihear Province, Cambodia*. Phnom Phen: Wildlife Protection Office, Forest Administration.

Sonoran Institute and International Sonoran Desert Alliance. Undated. Cooperative resource management (proposal to the Ford Foundation). Files of the International Sonoran Desert Alliance, Ajo, Arizona.

South African Tourism. 2004. *2003 Annual Tourism Report*. South African Tourism Strategic Research Unit, April 2004.

Spence, M. D. 1999. *Dispossessing the Wilderness: Indian Removal and the Making of the National Parks*. New York: Oxford University Press.

Spirkovski, Z., O. Avramovski, and A. Kodzoman. 2001. Watershed management in the lake ohrid region of albania and macedonia. *Lakes and Reservoirs: Research and Management* 6(3): 237–42.

Staff Reporter. 2004. Government relocating illegal Gonarezhou settlers. *Financial Gazette Zimbabwe*, January 22. Available at ⟨www.peaceparks.org⟩.

Stares, P. B. 1996. *Global Habit: The Drug Problem in a Borderless World*. Washington, DC: Brookings Institution.

Starr, S. F. 2005. A partnership for central Asia. *Foreign Affairs* 84(4): 164–78.

Statistics SA. 2004. Cited in van Jaarsveld 2004, op. cit.

Stattersfield, A., M. J. Crosby, A. J. Long, and D. C. Wege. 1998. *Endemic Bird Areas of the World, Priorities for Biodiversity Conservation*. Cambridge, England: Birdlife International.

Steenkamp, C. 1998. The Makuleke Land Claim signing ceremony: Harnessing social justice and conservation. *African Wildlife*, July–August, 52(4). Available at ⟨http://wildnetafrica.co.za/wildlifearticles/africanwildlife/1998/julaugust_makuleke.html⟩.

Steenkamp, C., and D. Grossman. 2001. People and parks: Cracks in the paradigm. IUCN Policy Think Tank Series 10. May.

Stein, A. 1993. *Why Nations Cooperate: Circumstance and Choice in International Relations*. Ithaca: Cornell University Press.

Stephan, J. J. 1974. *The Kuril Islands: Russo–Japanese Frontier in the Pacific*. Oxford: Clarendon Press.

Stevens, M. L. 2004. Restoration of the Mesopotamian marshes of southern Iraq: Synthesis of current restoration activities. *Canadian Reclamation*: 7–11.

Stevens, M. L., and S. Alwash. 2003a. Draft report biological characteristics, Mesopotamian marshlands of southern Iraq. In *Background Material for Technical Advisory Panel, Eden Again Project*. The Iraq Foundation, Washington, DC.

Stevens, M. L., and S. Alwash. 2003b. Draft report cultural values and traditional resource management, Mesopotamian marshlands of southern Iraq. In *Background Material for Technical Advisory Panel, Eden Again Project*. The Iraq Foundation, Washington, DC.

Stevens, S., ed. 1997. *Conservation through Cultural Survival: Indigenous Peoples and Protected Areas*. Washington, DC: Island Press.

Strauss, A. 1993. *Continual permutations of action*. Hawthorne, NY: Aldine de Gruyter.

Sundberg, J. 2003. Conservation and democratization: Constituting citizenship in the Maya biosphere reserve, Guatemala. *Political Geography* 22(7): 715–40.

Sungchun-Munhwa-Jaedan, ed. 1996. Bi Mujang Jidae: Yasaeng eu Bogo (Demilitarized zone: Wildlife report). Seoul, Korea: Hyunamsa. (In Korean)

Sungchun-Munhwa-Jaedan and B.-O. Won, et al., eds. 1996. Bi-MuJang-Jidae (The Demilitarized Zone). Seoul: Hyunam-sa, 612 pp. (In Korean)

Susilowati, I., and L. Budiati. 2003. An introduction of co-management approach into Babon River management in Semarang, Central Java, Indonesia. *Water Science and Technology* 48(7): 173–80.

Susskind, L., J. Thomas-Larmer, and S. McKearnen. 1999. *The Consensus Building Handbook: A Comprehensive Guide to Reaching Agreement*. Thousand Oaks: Sage.

Swanepoel, E. 2004. TFCA and PDF Spending 2002–2004. Report to SANParks.

Swanson, T. 1997. *Global Action for Biodiversity: An International Framework for Implementing the Convention on Biological Diversity*. London: Earthscan.

Swyngedouw, E. 1992. The mammon quest: Glocalisation, interspatial competition and monetary order: The construction of new spatial scales. In *Cities and Regions in the New Europe: The Global-Local Interplay and Spatial Development Strategies*, ed. by M. Dunford, and G. Kafkalas. London: Belhaven Press.

Swyngedouw, E. 1996. Reconstructing citizenship, the re-scaling of the state and the new authoritarianism: Closing the Belgium mines. *Urban Studies* 33(8): 1499–1521.

Tajik National Park. 2003. *Bulletin of the Tajik National Park*. August 1, Dushnabe. 4 pp.

Tanakarn, N. 2003. Management of the Pha Taem protected forest complex to promote cooperation for trans-boundary biodiversity conservation between Thailand, Cambodia and Laos (Phase I): Socio-economic technical final report. Faculty of Forestry, Kasertsart University, Bangkok.

Tanner, R. 2004. Transfrontier conservation areas of southern Africa and international law in the context of community involvement. Masters thesis. University of Montana.

Tanner, S. 2003. *Afghanistan: A Military History*. New York: Perseus.

Tanzania Wildlife Research Institute. 2001. Aerial census in the Selous–Niassa corridor, wet and dry season, 2000. Arusha.

Tarlow, L. 2000. Russia decision-making on Japan in the Gorbachev era. In *Japan and Russia: The Tortuous Path to Normalization, 1949–1999*, ed. by G. Rozman. New York: St. Martin's Press.

Tata Energy and Resources Institute. 2001. Background Paper on Himalayan ecology, main issues and concerns. Available at ⟨www.terina.org/himalayan.pdf⟩.

Taylor, M. 1987. *The Possibility of Cooperation*. Cambridge: Cambridge University.

Theron, P. 2004. Progress report. Project Management Meeting held on April 1 in Letaba, Kruger National Park, National Treasury Tourism Infrastructure Investment Programme—Proposed Projects for Funding within the Kruger National Park relating to the development of the Great Limpopo Transfrontier Park. South African National Parks.

Thesiger, W. 1964. *The Marsh Arabs*. Harmondsworth, England: Penguin.

THETA. Undated. D. The Gaza Kruger Gonarezhou TFCA: Makuleke Region of the Kruger National Park. *Skills Development Requirements*. Available at ⟨www.theta.org.za/downloads/makuleke_pilot.doc⟩, accessed November 2004.

Thorsell, J., ed. 1990. *Parks on the Borderline: Experience in Transfrontier Conservation*. Gland, Switzerland: IUCN–The World Conservation Union.

Treaty of the Ais-Ais/Richtersveld. 2003. Treaty between Namibia and South Africa on the establishment of the Ais-Ais/Richtersveld transfrontier park. Unpublished document.

Treaty of the Great Limpopo. 2002. Treaty between Mozambique, South Africa and Zimbabwe on the establishment of the Great Limpopo transfrontier park. Unpublished document.

Trisurat, Y. 2003a. Lesson learned from trans-boundary biodiversity conservation: A case of Pha team protected forests complex (Thai). *Journal of Wildlife in Thailand* 24: 194–208.

Trisurat, Y. 2003c. Defusing the trans-boundary minefield. *ITTO Tropical Forest Update*. 13: 10–12.

Trisurat, Y. 2004. *Ecosystem-based management zones of western forest complex in Thailand*. Avialable at ⟨http://www.sampaa.org/PDF/ch1/1.6.pdf⟩.

Turner, R. L. 2004. Communities, conservation, and tourism-based development: Can community-based nature tourism live up to its promise? Center for African Studies, University of California. Available at ⟨http://repositories.cdlib.org/cas/breslauer/turner2004a⟩.

US Environmental Protection Agency, National Park Service, Texas Commission on Environmental Quality. 2004. Big Bend regional aerosol and visibility observational study (BRAVO). Available at ⟨http://vista.cira.colostate.edu/improve⟩.

US FWS and DGCEUNR. Undated. *Sixty Years of Cooperation between the United States and Mexico in Biodiversity Conservation (1936–1996)*. Prado Norte, Federal District, Mexico: US Fish and Wildlife Service, Direction General for Conservation and Ecological Use of Natural Resources of Mexico, and Agrupación Sierra Madre, SC.

US FWS. Undated. *Cabeza Prieta National Wildlife Refuge*. US Fish and Wildlife Service. Available at ⟨http://www.fws.gov/southwest/refuges/arizona/cabeza.html⟩, accessed March 28, 2005.

US National Committee for the Man and the Biosphere Program. 1997. Governors Symington and Beltrones endorse biosphere reserve agreement. *US MAB Bulletin*, March 21(1). Available at ⟨http://www.state.gov/www/global/oes/bul_3_97.html⟩.

US National Park Service. 1965. *Sonoran Desert National Park, Arizona: A Proposal*. Southwest Region. Washington, DC: US Department of the Interior.

US National Park Service. Undated. *Organ Pipe Cactus National Monument*. US National Park Service. Available at ⟨http://www.nps.gov/orpi/⟩, accessed March 28, 2005.

US–Mexico Border Field Coordinating Committee. 2001. *US–Mexico Sister Areas Issue Team*. US Department of the Interior. Last revised April 1. Available at ⟨http://www.cerc.usgs.gov/fcc/letter-of_intent.htm⟩.

US–Mexico Border XXI Program. 1996. Framework document. US Environmental Protection Agency. October. EPA 160-R-96-003.

Udall, M. K. 1966. A national park for the Sonoran Desert. *Audubon Magazine*: 105–109. Available at ⟨http://dizzy.library.arizona.edu/branches/spc/udall/sonoran_htm.html⟩.

Udall, S. 1997. Stewart Udall: Sonoran Desert National Park. Interview by Jack Loeffler. *Journal of the Southwest* 39(3–4): 315–20.

UK Department for International Development (DFID). 2002. *Conducting Conflict Assessments: Guidance Notes*. London: DFID.

Underdal, A. 1992. The concept of regime effectiveness. *Cooperation and Conflict* 27: 227–40.

UNEP. 2006. Progress Report on the Capacity Building and Institutional Strengthening Programme for Environmental Management. Kabul: UNEP/PCOB, 29 pp.

UNEP. 2003a. *Post Conflict Environmental Assessment: Afghanistan.* Geneva: PCAU/UNEP, 176 pp.

UNEP. 2003b. *Post Conflict Assessment Unit: Afghanistan Update* (January). Geneva: PCAU/UNEP, 10 pp.

UNEP. 2002. *Global Environment Outlook 3 (GEO-3).* London: UNEP/Earthscan.

UNEP. 2004. Iran–Iraq Technical Meeting, May 17–18, in Geneva.

UNESCO/MAB, and IUCN. 2001. UNESCO/MAB-IUCN joint statement, January 21. Available at ⟨http://www.kurilnature.org/html/resolution.html⟩.

UNESCO-MAB Biosphere Reserve Directory. 2002. *Big Bend.* Man and the biosphere program. Last revised 28 February. Available at ⟨http://www2.unesco.org/mab/br/brdir/directory/biores.asp?code=USA+02&mode=all⟩.

United Nations Development Programme (UNDP). 2004. *Human Development Report 2004: Cultural Liberty in Today's Diverse World.* Available online at ⟨http://hdr.undp.org/reports/global/2004/⟩.

United Nations Disengagement Observer Force (UNDOF). *Golan Heights–UNDOF–Background.* Available at ⟨http://www.un.org/Depts/dpko/missions/undof/background.html⟩.

United Nations Educational, Scientific, and Cultural Organization (UNESCO). 2005. World Heritage. Available at ⟨http://whc.unesco.org/⟩.

United Nations Educational, Scientific, and Cultural Organization (UNESCO). 2005. *Man and Biosphere.* Available at ⟨http://www.unesco.org/mab/⟩.

United Nations Environment Program/World Conservation Monitoring Centre (UNEP/WCMC). 2004. *Parks for Peace.*

United Nations Environment Program (UNEP). 2004. *Understanding Environment, Conflict and Cooperation.* Nairboi, Kenya: UNEP.

United Nations Environment Programme, UNEP RRC.AP/UNDP. 2004. *DPR Korea: State of the Environment 2003.* Pathumthani, Thailand: UNEP RRC.AP.

United Nations Environmental Program (UNEP). 2001. Hassan Partow. The Mesopotamian marshlands: Demise of an ecosystem. Early Warning and Technical Assessment Report. UNEP/DEWA/TR.01-3 Rev.1, Division of Early Warning and Assessment. Nairobi, Kenya: UNEP. Available at ⟨http://www.grid.unep.ch/activities/sustainable/tigris/marshlands/⟩.

United Nations. 1959. Antarctic Treaty. Washington, DC. December 1.

United Nations. 1988. Convention on the Regulation of Antarctic Mineral Resources.

United Nations. 1991. Protocol on environmental protection to the Antarctic treaty, (Madrid Protocol) Madrid, October 4.

United States Central Intelligence Agency. 2004. *CIA World Factbook.* Available at ⟨http://www.cia.gov/cia/publications/factbook/⟩.

United States House of Representative. 2005. H.R. 418, §102.

United States National Science Foundation. 1994. *Facts about the U.S. Antarctic program* (NSF 92–134), 7 November.

United States National Science Foundation. 2002. Office of Legislative and Public Affairs, *NSF Fact Sheet*, May. Available at ⟨http://www.nsf.gov/od/lpa/news/02/fslakevostok.htm⟩.

United States National Science Foundation. 2005a. Arctic and Antarctic: An overview of NSF research. Available at ⟨http://www.nsf.gov/news/overviews/arcticantarctic/comp_q03.jsp⟩.

United States National Science Foundation. 2005b. Icebreakers Clear Channel into McMurdo Station. Press Release 05-13. Available at ⟨http://www.nsf.gov/news/news_summ.jsp?cntn_id =100849&org=NSF&from=news⟩.

UNSC (United Nations Security Council). 2001. Resolution 1343. Available at ⟨http:// daccessdds.un.org/doc/UNDOC/GEN/N01/276/08/PDF/N0127608.pdf?OpenElement⟩.

UNSC. 2003. Resolution 1478. Available at ⟨http://daccessdds.un.org/doc/UNDOC/GEN/ N03/348/12/PDF/N0334812.pdf?OpenElement⟩.

UNSC. 2004s. Resolution 1579. Available at ⟨http://daccessdds.un.org/doc/UNDOC/GEN/ N04/658/25/PDF/N0465825.pdf?OpenElement⟩.

Urrea, L. A. 2004. *The Devil's Highway*. New York: Little Brown.

USAID. 2004. Iraq marshland restoration plan. April 2004. FORWARD task order. Water Indefinite Quantity Contract US Agency for International Development 192. Available at ⟨http://www.iraqmarshes.org/Documents/Publications/ProjectPublications/Iraq%20Marshes %20Action%20Plan.pdf⟩.

USAID. 2005. Biological Diversity and Natural Resources Assessment: Afghanistan 2005–2009 Country Strategic Plan (Draft). Kabul. 39 pp.

USFWS. 2002. *El Camino Del Diablo: Highway of the devil*. US Fish and Wildlife Service, April 15. Available at ⟨http://www.fws.gov/southwest/refuges/arizona/diablo.html⟩.

Valucia, M. J. 1998. Joint utilization of strategic areas in preparation for the reunification of the two Korea. East–West Center Monograph, Honolulu.

Van Amerom, M. 2002. National sovereignty and transboundary protected areas in Southern Africa. *GeoJournal* 58(4): 265–73.

Van Amerom, M., and B. Büscher. 2005. Peace parks in southern Africa: Bringing an African renaissance? *Journal of Modern African Studies* 43(2): 159–82.

van der Linde, H., J. Oglethrope, T. Sandwith, D. Snelson, and Tessema, Y. (with contributions from A. Tiega and T. Price). 2001. *Beyond Boundaries: Transboundary Natural Resource Management in Sub-Saharan Africa*. Washington, DC: Biodiversity Support Program.

van Jaarsveld, A. 2004. Application in terms of Regulation 16.8 of the Public Finance Management Act ("PFMA"), 1999, Dealing with Public Private Partnerships, for approval of amendment and variation of agreements for the concession contracts. South African National Parks.

Vasquez, J., and M. T. Henehan. 2001. Territorial disputes and the probability of war, 1816–1992. *Journal of Peace Research* 38(2): 123–38.

Verbitsky, S. 2000. Factors shaping the formation of views on Japan in the USSR. In *Japan and Russia: The Tortuous Path to Normalization, 1949–1999*, ed. by G. Rozman. New York: St. Martin's Press.

Vetaas, O. R. 2003. The future of the Himalayan glaciers. Paper presented at workshop on Fragile mountains, fragile people? Understanding "fragility" in the Himalayas. Center for Development Studies, University of Bergen, Bergen, Norway.

Virtanen, P. 2005. Community-based natural resource management in *Mozambique*: A critical review of the concept's applicability at local level. *Sustainable Development* 13(1): 1–12.

Wagenaar, H. 2005. "Knowing" the rules: Administrative work as practice. *Public Administration Review* 64(6): 646–55.

Waldman, C. 1985. *Atlas of the North American Indian*. New York: Facts on File Publications.

Walker, S. Undated. *El Pinacate Biosphere Reserve*. San Antonio: Nature Conservancy. Available at ⟨http://parksinperil.org/files/page_4_el_pinacate_biosphere_reserve.pdf⟩.

Waltz, K. 1999. Globalization and governance. *PS: Political Science and Politics* 32(4): 693–700.

Waters, M. 1995. *Globalisation*. London: Routledge.

WCPA (World Commission on Protected Areas). 2001. Programme on protected areas: Progress and assessment report. Available at ⟨http://www.iucn.org/pareport/docs/protected_areas_report.pdf⟩.

Weed, T. J. 1994. Central America's "peace parks" and regional conflict resolution. *International Environmental Affairs* 6: 175–90.

Weiner, D. R. 1999. *A Little Corner of Freedom: Russian Nature Protection from Stalin to Gorbachev*. Berkeley: University of California Press.

Weingarter, E. 2005. Reframing the US-DPRK conflict. [Napsnet] *Policy Forum Online* July 12, pp. 1–4.

Weiss, E. B., and H. K. Jacobson. 1999. Getting countries to comply with international agreements. *Environment* 41(6): 16–20, 37–45.

Weiss, T. G. 2005. Governance, Good Governance and Global Governance: Conceptual and Actual Challenges, in *The Global Governance Reader* ed. by R. Wilkinson. London: Routledge, pp. 66–88.

Weissbert, W. 2005. Lluvias ayudan a Mexico a pagar deuda de agua a EEUU. In *El Nuevo Herald*, March 17. Available at ⟨http://www.miami.com/mld/elnuevo/news/world/americas⟩.

Wells, M. P. 2003. Protected area management in the tropics: Can we learn from experience? *Journal of Sustainable Forestry* 17(1–2): 67–90.

Welsh, M. 2002. The administrative history of big bend national park—Landscape of ghosts, river of dreams: A history of big bend national park. National Park Service, US Department of the Interior.

Westing, A. H. 1998. A transfrontier reserve for peace and nature on the Korean Peninsula. *International Environmental Affairs* Winter 10(1): 8–17.

Westing, A. H. 1998. Establishment and management of transfrontier reserves for conflict prevention and confidence building. *Environmental Conservation* 25(2): 91–94.

Westing, A. H. 1997. A transfrontier reserve for peace and nature on the Korean peninsula. In *Conference Proceedings, Parks for Peace: International Conference on Transboundary Protected Areas as a Vehicle for International Co-operation*, September 16–18. Somerset West, South Africa.

Westing, A. H., ed. 1993. *Transfrontier Reserves for Peace and Nature: A Contribution to Global Security*. Nairobi: United Nations Environment Programme.

Wikramanayake, E., E. Dinerstein, C. J. Loucks, D. M. Olson, J. Morrison, J. Lamoreaux, et al. 2000. *Terrestrial Ecoregions of the Indo-Pacific: A Conservation Assessment*. Washington, DC: Island Press.

Wildlife Conservation Society. 2005. *Biodiversity Conservation in Afghanistan: A Proposal from WCS to the USAID Global Conservation Program—Leaders with Associates Program*. New York: WCS, 32 pp.

Wildlife Division. 1998. *The Wildlife Policy of Tanzania*. Dar as Salaam: Government of Tanzania.

Wilkinson, R., and S. Hughes, eds. 2002. *Global Governance: Critical Perspectives*. London: Routledge.

Wilkinson, R., ed. 2005. *The Global Governance Reader*. London: Routledge.

Wilshusen, P. R., S. R. Brechin, C. L. Fortwangler, and P. C. West. 2002. Reinventing a square wheel: Critique of a resurgent "protection paradigm" in international biodiversity conservation. *Society and Natural Resources* 15: 24.

Wishnick, E. 2002. The regional dynamic in Russia's Asia policy in the 1990s. *Russia's Far East: A Region at Risk*, ed. by J. Thornton and C. Ziegler. Seattle: National Bureau of Asian Research/University of Washington Press.

Wolf, A. T. 2002. The present and future of transboundary water management. Paper presented at the expert workshop on Bridging the Gap between South and North, San José, Costa Rica.

Wolf, C., Jr. 2005. One Korea? *Wall Street Journal* (2005 Dow Jones & Company). [Napsnet] *Policy Forum Online*, July 8, pp. 1–3.

Wolmer, W. 2003a. Transboundary conservation: The politics of ecological integrity in the Great Limpopo Transfrontier Park. *Journal of Southern African Studies* 29(1): 261–78.

Wolmer, W. 2003b. Transboundary Protected Area Governance: Tensions and Paradoxes. Paper prepared for the workshop on *Transboundary Protected Areas in the Governance Stream of the 5th World Parks Congress, Durban, South Africa*, September 12–13.

Wolmer, W., J. Chaumba, and I. Scoones. 2003. Wildlife management and land reform in southeastern Zimbabwe: A compatible pairing or a contradiction in terms? Sustainable Livelihoods in Southern Africa Research Paper 1. Institute of Development Studies, Brighton.

Wolmer, W. 2004. Tensions and paradoxes in the management of Transboundary Protected Areas. *Policy Matters* 13(November): 137–47.

Woo, S. 2006. North Korea—South Korea relations in the Kim Jong il era. In *North Korea: The Politics of Regime Survival*, ed. by Y. W. Kihl and H. N. Kim. Armonk, NY: M.E. Sharpe, pp. 225–44.

World Bank. 1996. Transfrontier conservation areas pilot and institutional strengthening project. World Bank/GEF Project Document, Report 15534-MOZ. Washington, DC.

World Bank. 1998. Statement of work: Support for the development and management of transboundary conservation areas in Southern Africa. Washington, DC.

World Bank. 2004. Transfrontier conservation areas pilot and institutional strengthening project. Project P001759.

World Bank. 2006. Draft project concept note: Institutional development for environmental management. Kabul. 16 pp.

World Tourism Organization. 2003. *Mt. Kumgang Preparatory Assistance for Sustainable Tourism Development. Volume III: Environmental Report*. Hyundai Asan Corporation, World Tourism Organization, Korea National Tourism Organization.

World Travel and Tourism Council (WTTC). 2003. Country League Tables. *The 2003 Travel and Tourism Economic research*.

World Travel and Tourism Council. 1999. *Southern African development community's travel and tourism: Economic driver for the 21st Century*. London: WTTC.

World Wide Fund for Nature (WWF). 2005. *An Overview of Glaciers, Glacier Retreat and Subsequent Impacts in Nepal, India and China*. Khatmandu, Nepal: WWF-Nepal Program.

Wubneh, M. 2003. Building capacity in Africa: The impact of institutional, policy and resource factors. *African Development Review* 15(2–3): 165–98.

Wynne, B. 1992. Uncertainty in environmental learning: Reconceiving Science and policy in the preventive paradigm. *Global Environmental Change* 2(2): 111–27.

Yamagiwa, J. 2003. Bushmeat poaching and the conservation crisis in Kahuzi-Biega National Park, Democratic Republic of Congo. In *War and Tropical Forests: Conservation in Areas of Armed Conflict*, ed. by S. V. Price. Binghamton, NY: Haworth Press, pp. 115–35.

Yin, R. K. 1984. *Case Study Research: Design and Methods*, rev. ed. Sage, CA: Newbury Press.

Yordan, C. L. 1998. Instituting problem-solving processes as a means of constructive social change. *Online Journal of Peace and Conflict Resolution 1.4*. November, 6.

Young, G. 1977. *Return to the Marshes*. In *Life with the Marsh Arabs of Iraq*. London: Collins.

Young, O. R., and M. A. Levy. 1999. The effectiveness of international environmental regimes. In *The effectiveness of international environmental regimes*, ed. by O. R. Young. Cambridge: MIT Press, pp. 1–32.

Young, O., ed. 1997. *Global Governance: Drawing Insights from the Environmental Experience*. Cambridge: MIT Press.

Young, O. 1994. *International Governance: Protecting the Environment in a Stateless Society*. Ithaca: Cornell University Press.

Zbicz, D. C., and M. J. B. Green. 1997. Status of the world's transfrontier protected areas. *Parks* 7(3): 5–10.

Zbicz, D. C. 1999. The nature of transboundary cooperation. *Environment* 41: 15–16.

Zbicz, D. C. 2003. Imposing transboundary conservation: Cooperation between internationally adjoining protected areas. *Journal of Sustainable Forestry* 17(1–2): 21–38.

Zeckhauser, R. et al., eds. 1996. *Wise Choices; Decisions, Games and Negotiations*. Boston: Harvard Business School Press.

Zimbabwe Tourism Authority (ZTA). 2001. Tourism trends and statistics 2001. Research and development division, ZTA.

Zimmer, K. S., and K. R. Young. 1998. *Nature's Geography: New Lessons for Conservation in Developing Countries*. Madison: University of Wisconsin Press.

Index

Ab-i-Estada reserve, Afghanistan, 295
Academicians, and practitioners, 335, 339
Action, as identity need, 111
Afghanistan, 291–94, 309–11
 conservation in, 294, 303, 305, 307, 308, 310
 Environment Act (2006) in, 307–308
 and Pamirs Conservancy, 302
 peace parks for 294, 295–302, 303–304, 308
 protected areas in, 293, 295–98, 300, 303, 304–309, 310–11
 in Sistan wetlands dispute, 298, 308
 Transitional Government of (TGA), 293, 294, 297
Africa, 127. *See also* Southern Africa; West Africa; *individual countries*
 border tensions in, 75
 Great Lakes region of, 8, 37, 78
 and regionalism, 73, 77–78
African Union, 75
African Wildlife Foundation, 23
Aga Khan Development Network, 303
Aga Khan Rural Support Program (AKRSP), 302, 303
Agreed Measures for the Conservation of Antarctic Fauna and Flora (1964), 170, 173
Agreement Protocol for Transhumance (1989), 130
Agriculture
 as encroaching on PPFC buffer zone, 151
 need to improve (Afghanistan), 310

Ainu, and sovereignty over Kuriles, 265
Ais-Ais/Richtersveld Transfrontier Park, 76
Ajar Valley Wildlife Reserve, Afghanistan, 295
Akademik Sergei Vavilov (Russian ship), 174
Akagera National Park, Rwanda, 27
Al Ahwar marshes, 313–16. *See also* Hawizeh-Azim marshes
 biodiversity heritage of, 316–17
 cultural heritage of, 317–18, 320–21
 desiccation of (long-term), 319–20
 diversion of water supply from, 316, 328
 draining of, 314–15, 319–20
 heritage of (al-Khayoun), 329
 international efforts on behalf of, 326–27
 Iraqi efforts on behalf of, 327
 restoration and conservation potential of, 320
 Saddam's draining of, 319
Alaska Solution (Kuriles), 266
Al Azim marsh, Iran, 315
Albania, peace park proposed for, 75
Algiers Agreement (1975), 318
Al Hammar marsh, Iraq, 315
Al Hawizeh marsh, Iraq, 315, 320
Ali, Aamir, 284, 285
Al-Jaafari, Ibrahaim, 327
Alternative futures, for Siachen region, 284–88
Alternative livelihood initiatives, in Afghanistan, 297, 305–306
Alwash, Azzam, 322–23
Amar Appeal, 322

AMAR international charitable foundation, and Marsh Arabs, 319
American Association for the Advancement of Science (AAAS), and Karakoram Science Project, 287
Angola
  in alliance, 78
  UNITA in, 26
Antarctica, 12, 167–69
  backseat sovereignty in, 178–80
  dealing with incidents in, 172–76
  and environmental cooperation, 165–66
  and "governance without government," 163–64
  harmony-and-denial system for, 163, 171–72, 180
  management of activities in, 164–65
  permit process for, 176
  specially managed areas in, 176–77
  specially protected areas in, 176, 177–78, 180
Antarctic Conservation Act (1978), 170
Antarctic Convergence, 168–69
Antarctic Mineral Resources Commission, 170
Antarctic Science, Tourism, and Conservation Act (1996), 170
Antarctic Treaty (1959), 170, 287
  and Siachen issue, 286–87
  territorial sovereignty rejected by (Article IV), 163
Antarctic Treaty Consultative Meetings, 164
Antarctic Treaty Consultative Party (ATCP), 164, 171
Antarctic Treaty System (ATS), 166–67, 168, 169–72
  forgiving application of, 173, 174
  sovereignty lesson learned from, 180
Anthropologists, and indigenous people, 15
Aquatic management, 338
Arendt, Hannah, 261–62
Argentina, Antarctic claims of, 171, 172, 176
Armistice, vs. permanent peace, 239
Asian Development Bank (ADB), 152–53, 153, 154, 292
  Afghanistan funding by, 304, 307
  Afghanistan report of, 293–94, 297
Asian Pacific Economic Cooperation (APEC), 72
Assessment
  in Peace and Conflict Impact Assessment, 31 (*see also* Peace and Conflict Impact Assessment)
  transboundary-conservation need for, 38
Australia, and Antarctica, 164, 170, 171, 172
Autonomous region, as peace-park parallel, 178–79
Aversion, common (cooperation pathway), 277, 288, 335

Babbit, Bruce, 212, 215–16
Bahia Paraiso (Argentine supply ship), 173
Bande Amir National Park, Afghanistan, 293, 295
Bekme Dam, Iraq, 328
Belize
  Guatemala in border dispute with, 64
  Kekchi Council of, 61, 62, 63
  and peace-park financing, 62
  and transborder orchid trade, 64–65
  in tri-national peace park, 56, 59
Belize barrier reef, 60–61
Beltrones, Manlio Fabio, 215
Benin
  independence won by, 229
  sociopolitical and economic changes in, 131–32
  and W Peace Park, 7, 12, 128, 129, 130, 131
Beringian Heritage Program, 268
Bhutan, dangerous glacial lakes in, 281
Big Bend National Park (BBNP), 206–208, 221n.4
  as biosphere reserve, 210
  border crossing from, 218
  study on, 212
  as US attempt to control border (Ezcurra), 214
Big Bend Ranch State Park, 207
"Big Pamir" reserve, Afghanistan, 295

Binational Red Sea Marine Peace park, 41, 48
Biodiversity
 in Himalayan regions, 282
 of Korean DMZ, 246
 Korea's conservation of, 248, 252
 Korea's loss of, 239, 240, 241, 244, 256
 of Kuril Islands, 264, 271, 273
 in Liberia, 227, 232, 233, 234
 in Pha Taem Complex/Emerald Triangle (Thailand), 141, 144–45, 148, 150
 regional, 232
 Selous-Niassa research on, 118–19
 as threatened in Selous-Niassa region, 119–21
 in W Park, 130–31, 133, 138
Biodiversity conservation, 247–48
 in Afghanistan, 297, 304, 310–11
 of Al Ahwar marshes, 316–17
 and Ecosystem Approach, 322
 in Korea, 248
 and peace parks, 58
 and Pha Taem/Emerald Triangle, 146, 153, 155–56, 156–57, 159
 in Selous–Niassa Wildlife Corridor, 122
Biodiversity neutral site, in Liberia (Mittal Steel), 234
Biological diversity, in TBPA definition, 182n.47
Bioregionalism and bioregions, 58–59, 73, 74, 78–79, 80, 338
Biosphere reserve(s), 161n.11
 in Afghanistan, 304
 BBNP as, 210
 Carpathian, 48
 Hawizeh–Azim marshes as, 323
 and Korean DMZ, 253
 KPBRS as, 257
 in northern Mexico, 215, 220
 Pamirs-based, 302
 Sonoran Desert, 215
 UNESCO, 148
Black Gap Wildlife Management Area, 207
Bob, Clifford, 16
Border(s). *See also specific borders*
 ambiguity of (Kashmir), 279
 conservation (peace parks), 193
 re-definition of, 75–76
 Shatt al Arab as (Iraq-Iran), 313, 318
Border buffer, environmental conservation providing, 179. *See also* Buffer zone
Border conflicts, 74–76, 79
Border politics, in Emerald Triangle region, 157–58, 159
Border regions (borderlands), 70, 80
 between Tanzania and Mozambique, 115
Borders. *See also specific borders*
 conservation (peace parks), 193
 as political constructs, 28
 re-definition of, 75–76
 US–Mexico, 13, 205–206 (*see also* US–Mexico border)
Border security
 as national imperative, 13
 and US–Mexican border, 206, 217–18
 and Waterton–Glacier Park, 189, 192–96, 197
Borneo, 8
 Lanjak Entimau in, 157
Botswana, mining operations in, 5
Bowden, Charles, 216
BRAVO project, 212
Brechin, Steven, 18n.28
Britain (United Kingdom)
 Antarctic claims of, 171, 172
 locality studies in, 73
British Antarctic Survey, 164
Brown, John George "Kootenai," 187
Broyles, Bill, 216, 217
Budowski, Gerardo, 8
Buffer zone
 for Afghanistan, 293
 in Cyprus, 41, 48
 in global conservation schemes, 59
 and Hawizeh–Azim marshes, 321–23, 329
 of Korean DMZ, 245
 peace park as, 42, 45–48
Buffer zone management, in Afghanistan, 305
Buni Zom National Park, Karakoram Constellation, 301

Bun Thrik–Yot Mon proposed Wildlife
    Sanctuary, Thailand, 142, 148, 150
Burkina Faso
  on Council of Understanding, 130
  and Komoé-Leraba Forest–Warigué Forest,
    166
  sociopolitical and economic changes in,
    131–32
  and W Park, 7, 12, 128, 129, 130, 131,
    132–33, 134
Burundi
  guerrilla groups in, 26
  Kibira National Park in, 8
Bush, Ken, 30
Bwindi Impenetrable National Park, 33, 34

Cabeza Prieta National Wildlife Refuge
    (CPNWR), 213, 216
Cambodia
  in Biodiversity Initiatives, 153
  and border politics, 157
  and Pha Taem Complex/Emerald Triangle,
    8, 152, 154, 155, 156–57, 159, 160
  poverty of, 158
  and Southeastern Indochina Dry Evergreen
    Forests ecoregion, 142
Canada, local peace parks in, 70
Canada Iraq Marshlands Initiative (CIMI),
    326, 331nn.39,43
Canada–US peace park, 2, 24, 166, 183. *See
    also* Waterton–Glacier International
    Peace Park
Canadian International Development Agency
    (CIDA), 326
Canadian-Iraq Consultation Workshop on
    Iraqi marshlands, 327
Canadian Trust, 62
Canhane community (Limpopo National
    Park), 100
Canon de Santa Elena, 207, 211, 212
Capacity building or development
  in Afghanistan, 304
  in Southern Africa, 125
Capacity-Building International, Germany
    (InWEnt), 124
Cape Verde, in ECOWAS, 130

Carabias, Julia, 212, 215–16
Cardenas, Lazaro, 207–208, 213
Carnegie Endowment, 291
Carpathian Biosphere Reserve, 48
Causality, and environmental endogeneity,
    5–6, 340
Cavaly protected area, Ivory Coast, 233
Ceasefire, vs. permanent peace, 239
Ceasefire line (CFL), for India–Pakistan war
    (1948), 279
CEMEX (cement producer), 211–12, 219–
    20, 222n.32
Center for Restoration of Iraq Marshlands,
    326
Central America, peace parks in, 58–65, 66
Central American Commission on
    Environment and Development, 62
Central Asia
  and Afghanistan, 310
  West Tien Shan Mountains, 7
Central Karakoram National Park, 300, 301
Central marsh, Iraq, 315
Centro Ecológico de Sonora, 214
Chapin, Mac, 15
Charles, Graham, 167
Chihuahuan desert, 205, 206–12, 220
Chihuahuan ecoregion project, 219
Chile
  Antarctic claims of, 171
  mining operations in, 5
China, People's Republic of (PRC)
  Afghanistan suspicious of, 310
  and Antarctica, 173
  in Biodiversity Initiatives, 153
  GLOF damage in, 281
  and India, 278
  and Japan island dispute, 266
  and Korean armistice, 245
  and Pamirs Conservancy, 302
  and rapport between Koreas, 250
  and Russia-Japan settlement, 273
  in six-party talks, 240
  and Taxkorgan Wildlife Reserve, 300
Chisos Mountains, 206–207
Chitral Gol National Park, Pakistan, 301
Chitsa people, 96–97

Ch'oc, Gregory, 62, 63
Cholwon Basin, 254
Cholwon Crane Festival (2003), 254
Cisneros, Jose, 212, 220
CITES (Convention on International Trade in Endangered Species of Wild Fauna and Flora), 151, 161n.16
Civilian Conservation Corps (CCC), Liberia, 13, 235
Civilian Control Zone (CCZ), Korea, 245
Clinton, William
  and Peru–Ecuador border, 9
  and Sonoran Desert National Monument, 216
Cold war, 74–75
  and environmental security literature, 3
  and Korea, 240, 242, 249
  and Kuriles, 266, 271
Collaboration. *See also* Conservation
  lessons for, 196–197
  in Waterton–Glacier Park, 193–96, 197
Collaborative activities, soft, 154, 160
Collaborative environmental projects, and dominant economic power, 141
Collaborative management, in Afghanistan, 305
Colombia, guerrilla groups in, 26
Co-management, of Hawizeh–Azim Peace Park, 324
Common aversion, as cooperation pathway, 277, 288, 335
Common property resources (CPRs), Ostrom's model of, 48
Commons
  model of, 42
  tragedy of, 4, 174
Community-based conservation (CBC), 123
Community-based natural resource management (CBNRM), 58
  for Selous Game Reserve, 116
CONANP, 220, 225n.81
Condominium, 179
Condor–Kutuku conservation corridor (Ecuador/Peru), 8–10
Confidence-building, Siachen Peace Park for, 284

Conflict
  border, 74–76, 79
  and conservation, 23, 38
  entrapment in, 340
  vs. environmental cooperation, 166
  and environmental security, 3–5
  identity-based, 109–10, 112
  Iran-Iraq, 323
  micro-level conflicts, 12, 333, 334, 339
  and multiethnic states, 291
  and Peace and Conflict Impact Assessment, 29–37, 38
  and peace parks (transfrontier reserves), 25, 64, 127, 137
  after previous conflict, 27
  and protected areas, 25–29
  and resource abundance, 5, 64
  and resource scarcity, 4–5, 63–64
  over scarce natural resources, xiii
  in Siachen Glacier region, 280, 288 (*see also* Siachen Glacier region dispute)
  and transboundary protected areas, 29–30
  and world system, 55
  and W Park, 131
Conflict avoidance, from conservation management activities, 165
Conflict de-escalation process, as nonlinear, 6
Conflict mapping, in Peace and Conflict Impact Assessment, 31
Conflict mitigation, and environmental cooperation, 5–6
Conflict resolution
  common aversion as tactic in (Siachen dispute), 277, 288
  and conservation, xvii, 8–10, 15–16
  through focus on what is mutually valuable, 239
  in Korea, 252
  through peace parks (transboundary protected areas), xiv, 63, 75, 83, 198, 249, 333, 339
Conflict-sensitivity, 29, 35, 36
"Conflict trap," 277
Congo
  gorillas protected from violence in, 179
  mining operations in, 5

384   Index

Connectedness, as identity need, 110–11
Conservation. *See also at* Transboundary
  in Afghanistan, 294, 303, 305, 307, 308, 310
  of Al Ahwar marshes, 320–21
  biodiversity, 156–57, 247–48
  and border security, 194
  community-based, 113, 123
  and conflict, 23, 38
  and conflict resolution, xvii, 8–10, 15–16
  grassroots, 109, 112–13, 124–25
  in Korea and Korean DMZ, 239, 240, 242, 246–47, 248, 251–52, 255, 257, 285
  and Kuriles (International Peace Park), 263, 268, 269, 271, 272, 274
  and local group identity, 109, 110
  as low politics, 337, 339
  and Mesopotamia marshes, 313, 329
  in Mexico (under Cardenas), 208
  Mexico–US cooperation in, 208–209
  and Pamirs Conservancy, 302
  vs. peace-building, 339
  and peace parks, 69, 83, 198 (*see also* Peace parks)
  of Selous–Niassa Wildlife Corridor, 121–22
Conservation area design, in northern Mexico deserts, 220
Conservation of Biodiversity initiative (NAFTA), 219
Conservation Biology, 15
Conservation groups. *See* Environmental organizations
Conservation International, 6, 9, 10, 15, 16, 17n.21
  and ecoregions, 23
  in Liberia, 231, 233, 235, 236
  transboundary conservation supported by, 18n.26
Conservation interventions, and Peace and Conflict Impact Assessments, 32
Conservation management activities, conflict avoidance from, 165
Conservation programming, in situ/ex situ (Afghanistan), 305
Conservation projects, transboundary, 32–33
Conservation and Utilization of Wildlife Project, 300
Conservation work, complexity of, 37–38
Conservation zones. *See* Environmental conservation zones
Conté, Lansana, 230
Convention on Biological Diversity (CBD) (1992), 131, 232, 252, 322
  Article 8 of, 247
Convention Concerning the Protection of the World Cultural and Natural Heritage, 185
Convention for the Conservation of Antarctic Marine Living Resources (1980), 170
Convention for the Conservation of Antarctic Seals (1972), 170
Convention on Conservation of Biological Diversity (CD), 294, 301
Convention on Desertification, 131
Convention on the Regulation of Antarctic Mineral Resources (CRAMBA), 170, 180
Convention on the Safeguarding of Natural Fauna and Flora, 128
Convention of the World Heritage (1972), 130
Cooperation. *See also* Collaboration
  between Afghanistan and Pakistan, 302
  in Antarctica, 174
  common aversion as pathway in, 277, 288, 335
  and conflict mitigation, 5–6
  environmental, 155–56, 165–66
  and environmental security, 3–5
  functional (Waterton–Glacier), 186–88, 192
  for Kuril Islands, 270, 272–273
  as low politics 337, 339
  on Pha Taem Complex, 151
  peace park as zone of, 45–47, 51
  regional, 230–32
  and regional stability, 155–56
  on resource conservation, 333
  in Siachen Glacier region, 280

in synergies for peace, 269
over US–Mexican border, 212, 218–20
various forms of, 220
within Waterton–Glacier Park, 186–88
Coordination, of international efforts for Mesopotamian marshland restoration, 326–27
Cordillera del Condor, conservation and conflict resolution in, 8–10
Costa Rica, and La Amistad, 177, 179, 247
Co-stewardship, for Kuriles, 269
Côte d'Ivoire. *See* Ivory Coast
Council for the Mutual Economic Assistance (CMEA), 72
Coutada 16 (hunting zone), 97
CPNWR (Cabeza Prieta National Wildlife Refuge), 213, 216
CRAMBA (Convention on the Regulation of Antarctic Mineral Resources), 170, 180
Crown of the Continent Ecosystem, 191–92
"Crown Managers Partnership," 187, 191–92
Cultural re-connection and revitalization, through peace parks, 61
Cultural values
  conservation projects as undermining, 109
  impact of, 107
Culture, and connectedness, 111
Cyprus, buffer zone in, 41, 48

Dabelko, Geoffrey, xvii
Dag Hammarskjold International Peace Park, Zambia, 69–70
Danish Cooperation for Engineering and Development (DANCED), 153
Darfur crisis, xiii, 3
Dashte Nawar protected area, Afghanistan, 295–96
Declaration of Brasilia, 9
Deep-rooted conflict, 110
Demilitarization, 339
Demilitarized zone (DMZ)
  and Antarctica, 164
  as buffer zones, 46
  Korean, 14, 46, 48, 164, 239, 244–46 (*see also* DMZ Forum)

De-mining. *See also* Minefields
  in Emerald Triangle region, 157–58, 159
  and Limpopo National Park, 99
  in Pha Taem Complex, 152
Democratic peace theory, 166
Democratic People's Republic of Korea. *See* North Korea
Democratic Republic of Congo
  in alliance, 78
  Virunga–Bwindi region in, 33, 38
  Virunga National Park in, 8, 27
Deosai Plateau, conservation activity on, 300
Deosai Wilderness Park, Karakoram Constellation, 301
Desert ecosystems, across US–Mexican border, 205
Desertification, and Darfur conflict, xiii, 3
Development, economic. *See* Economic development
Development, sustainable. *See* Sustainable development
Diamond revenues, Kimberley process for regulation of, 5, 17n.11
Diecke Massif du Ziama forests, Guinea, 234
Disadvantaged regions, protected areas as catalyst of conflict in, 26
Dispute resolution, Korean conservation efforts as, 242
Diversity, biological, 182n.47. *See also* Biodiversity
DMZ. *See* Demilitarized zone
DMZ ecosystems
  conflict resolution through, 252
  preservation of, 246–247, 252–55, 256–57
DMZ Forum, 14, 252–53. 253, 254, 255
Doe, Samuel, 229
Dong Phayayen–Khao Yai Biodiversity Conservation Corridor, Thailand/Cambodia, 156
Donor organizations and nations, 62
  for Afghan protected areas, 292, 307, 308
  in grassroots conservation, 113, 125
  influence of, 55
  for Mesopotamian marshlands restoration, 326
  need to coordinate (GLTP), 90, 91

Donor organizations and nations (cont.)
  need for in Southern Africa, 124
  and Pha Taem Complex, 153, 154
  and Selous–Niassa Wildlife Corridor, 11
  in trust-building, 334
Drug trafficking
  in Afghanistan, 292
  and Belize, 65
  and US–Mexican border, 206, 210, 217
Dykinga, Jack, 216

Ecological geography, of Kuril Islands, 264
"Ecological pacifist," and biosphere
  (Rapaport), 21
Ecological regions, advent of, 78
Ecological resilience, 88. *See also* Resilience
Ecological restoration, in Kuriles Park, 273
Ecological systems, and state borders, 69
Ecological value, of proposed Liberian peace
  parks, 233, 234, 234–35
Ecology, and International Peace Park in
  Kuriles, 271–73
Economic/community cooperation zones, for
  Afghanistan, 305
Economic Community of West African States
  (ECOWAS), 130
Economic development(s). *See also* Logging;
  Mining; Sustainable development
  in Afghan protected areas, 309, 310
  and biodiversity loss, 240
  in Korea, 240–41, 252
  and Liberia's control of natural resources,
    237
  and linkage, 335, 337
  in Mexico and US, 209, 215
  in North Korean private sector, 250–51
  and Pamirs, 303
  South Korea's success in, 256
  and peace parks, 139
  of Thailand, 158
  and transnational forms of management,
    60–61
Economic management, and global
  governance, 57
Economic and Monetary West African
  Union (UEMOA), 140

ECOPAS (the Protected Ecosystems of
  Sahelian Africa), 131, 137–38, 140
Eco-revisionists, 15
Ecosystem(s)
  interconnections among, 333
  of Korean DMZ, 245, 245–47, 252
  regional (Mexico–US), 210
  valuing services for, 341n.4
  and Waterton–Glacier Park (Crown of the
    Continent), 191–92
Ecosystem Approach, 322
Ecosystem management, and regional
  biodiversity, 232
Ecotourism, 60. *See also* Tourism
  for Afghanistan, 297, 305
  and Antarctica, 165, 169
  and Central American peace parks, 60–61
  as economic development, 337
  in Hawizeh–Azim Peace Park, 324
  on Kuril Islands, 263, 270
  in Niger ("W" Park), 133
  in Pakistan, 300
  and Pamirs, 303
  in Pha Taem Complex/Emerald Triangle,
    150, 157, 158
  in Selous–Niassa Wildlife Corridor, 121
ECOWAS (Economic Community of West
  African States), 130
Ecuador, and Peru (Condor–Kutuku
  conservation corridor), xiv, 8–10
Eden Again/New Eden Project, 327, 331n.39
Education
  in Afghanistan, 305
  on Waterton–Glacier Park, 188–89
Effectiveness, of cooperation (Waterton–
  Glacier), 186
El Carmen Wilderness, Mexico, 220
Elephant migration routes, and Selous–
  Niassa Wildlife Corridor, 117–19, 120
El Salvador, buffer zone separating
  Honduras from, 46
Emerald Triangle, 144
  transboundary cooperation in, 156–58,
    160
Emerald Triangle Protected Forests
  Complex, 12, 153–55, 159–60. *See also*

Pha Taem Protected Forests Complex, Thailand
  tourism in, 158
  and transboundary biodiversity conservation, 155
Encroachment, of agriculture on Pha Taem Complex, 151
Endangered species, and Afghanistan, 293
Endogeneity, environmental, 5–6, 340
Engineered resilience, 88
Entrapment, avoidance of, 340–41
Environment
  and peace, xiii
  Redford on, 203
"Environment, Peace and the Dialogue among Civilizations and Cultures" (Iranian government conference), 314
Environmental Change and Security Project (ECSP), Woodrow Wilson Center, xvii–xviii, 3
Environmental conservation zones, 23
  as border buffers, 179
  and conflict resolution, 1
  peace parks as, 1–2 (*see also* Peace parks)
Environmental cooperation, 165–66
  and regional stability, 155–56
Environmental degradation
  and Afghanistan, 292
  in Himalayan ecosystem, 280–82
  in Korea, 240–42, 244, 256, 285
Environmental diplomacy (Iran-Afghanistan), 308
Environmental endogeneity, 5–6, 340
Environmental factors, in politics of war and peace, xvii
Environmental integrity, as security, 14
Environmentalism, varieties of, 16
Environmental NGP Roundtable and Regional Environmental Forum, 326
Environmental organizations, criticism of, 15–16, 18n.28, 23
Environmental Peacemaking (Conca and Dabelko), 165
Environmental planning, 334–37
Environmental problems, and global governance, 55

Environmental security
  and conflict or cooperation, 3–5
  and sustainable development (Korea), 256
Eritrea, 75
Ethiopia, 75
Euphrates River, 315, 328
European green belt, 23
European Union, 78
  in Afghanistan, 307–308
  and ECOPAS Project (W Park financed by), 137–38, 140
  and W Park in Niger, 134
Evaluation, for Liberian peace parks, 235–36
Extractive industries, and conservation organizations, 5, 18n.28. *See also* Mining
Ezcurra, Exequiel, 214

Failed states, and peace parks, 291–92
"Failed States Index," for Afghanistan, 291
Fauna and Flora International (FFI), 116, 231, 233, 236
Financial sustainability, of Afghan nature reserves, 306
Finland, and Afghanistan, 307
Fishing
  and al Ahwar marshes, 317, 320
  and aquatic systems, 338
  and Kuril Islands, 263, 272
Five State Good Neighbor Council, 209, 222n.21
Folk theorem, 49
Food and Agriculture Organization (FAO), 152, 307
Ford Foundation, 18n.28
Forest fires, in Pha Taem Complex, 152
Forestry Administration (FA) of Cambodia, 153, 159
France, and Antarctica, 170, 171, 172–73
Fridman, Susanna Nikolaevna, quoted, 261
Frontera, La. See US–Mexico border
Frontier zones, and political experimentation, 263
Functional cooperation, 186
  in Waterton–Glacier Park, 186–88, 192
Fund for Peace, 291

Gambia, in ECOWAS, 130
Game-theoretic model, of peace parks, 42–52
Game theory, 10, 335
Garden of Eden, in al Ahwar marshes, 314
Gaza–Kruger–Gonarezhou (GKG) Transfrontier Park, 84
Gaza Safaris, 98
Gbagbo, Laucent, 229–30
GEF. *See* Global Environmental Facility
Genocide, in Rwanda (1994), 27, 34, 78
Geographic(al) Information systems (GIS), 141
 and ECOPAS (W Park), 138
 for Pha Taem project (Thailand), 142, 145, 145
 on transboundary wild areas, 2
German Development Bank (KfW), 122
German development cooperation program, and Selous–Niassa Wildlife Corridor, 112
German government
 and Selous Game Reserve, 116
 and Selous–Niassa Wildlife Corridor, 121–22, 123–25
German Wildlife Program, 123
Germplasm, 245
Ghana, 229
Gill, I. S., 283
GIS. *See* Geographic(al) Information systems
Glacier Lake Outburst Floods (GLOF), 281
Glaciers, Himalayan, 281. *See also at* Siachen Glacier
Glacier–Waterton International Peace Park. *See* Waterton–Glacier International Peace Park
Gleditch, Nils Petter, 1
Global Environmental Facility (GEF), 62, 325
 and Afghanistan, 297, 304, 307, 307–308
 and Korean DMZ, 254
 and Mesopotamian marshes, 325
 and Mountain Areas Conservancy Project (Pakistan), 301
 and Pha Taem Complex, 153
 and Selous–Niassa Wildlife Corridor, 121

Global governance, 11
 and peace parks, 56–58, 66, 76–77
Globalization
 and peace parks, 64
 and scale rearrangement, 72
Global Land Ice Measurements from Space (GLIMS) Project, 287
Global politics, as multipolar, 55
Global Positioning Systems (GPS), and ECOPAS (W Park), 138
Global program leverage, for Afghan protected areas, 307
Global Transboundary Protected Area Network, 7
Global warming, and Himalayan region, 282, 287
Glocalization, and peace parks, 74–79
GLOF (Glacier Lake Outburst Floods), 281
GLTP. *See* Limpopo Transfrontier Park, Great
GMS (Greater Mekong Subregion), 154
Gojal Conservancy Area, 301
Gola Forest Reserve, Sierra Leone, 234
Gola–Lofa–Mano corridor, 234
Gonarezhou National Park, Zimbabwe, 84, 89, 90, 95–97, 101
Gondwanaland, and Antarctic Lake, 175
"Good Neighbor Fiesta," 210
Gorbachev, Mikhail, 266
Gorillas
 as environmental common bond, 179
 and International Gorilla Conservation program, 33
 and tourism (Virunga-Bwindi region), 37
"Governance without government," 163–64
Governance systems
 cooperative outcomes from, 5
 global, 11
 post-Westphalian (strengthening of), 36
GPS (Global Positioning Systems), and ECOPAS (W Park), 138
Grameen Bank, xiii
Grassroots conservation, 109, 112
 donors in, 113, 124–25
 in Selous–Niassa Corridor, 109, 112–13, 125

Great Limpopo Transfrontier Park. *See* Limpopo Transfrontier Park, Great
Grebo Park, Liberia, 233
Grinnell, George, 191
GTZ-International Services, 11, 117, 121, 123, 141
GTZ Wildlife Program, 121, 124
Guatemala
  Belize in border dispute with, 64
  and transborder orchid trade, 64–65
  in tri-national peace park, 56, 59
Guinea, 229, 230, 233, 234
Guinea Bissau, in ECOWAS, 130
Gulf of Honduras, 59
Gulf War (1991), 318–19

Haas, Peter, 338
Hakeem, Asad, 283
Hala Bala Forest (Emerald Triangle region), 142
Hamburg, Steven, xvii
Hammarskjold, Dag, 69–70
Hasnain, Syed, 281
Haute Dodo protected area, Ivory Coast, 233
Hawizeh–Azim marshes, 315. *See also* Al Ahwar marshes
  international coordination for restoration of, 326–27
  as peace park or conservation area, 323, 324, 328–29
  transboundary conservation area in, 321–23
Hawizeh–Azim Peace Park, 324
Hawr al Azim marsh, 320
Hawr al Hawizeh marsh, 320
Helvetas (Swiss NGO), 99–100
Heritage territories, 28, 39n.17
High-level (top-down) negotiation, on Korean ecosystem preservation, 255
Himalayan ecosystem, environmental degradation of, 280–82
Himalayan High Ice Symposium, 287
Hindu Kush Cultural Conferences, 302
Hindu Kush–Himalayan (HKH) region, environmental degradation of, 280

Hindu Kush Mountains, 293, 297–98
Honduras
  buffer zone separating El Salvador from, 46
  in tri-national peace park, 56, 59
Houphouët-Boigny, Félix, 229
Human needs theory, 109, 110
Human rights, and environmental protection, 338
Human rights groups, critics of, 16
Human rights institutions, 55
Hunting
  illegal (Liberia), 233
  and Sapo–Taï corridor, 233
  in W Park, 135
Hyundai Asan Corporation, 249, 250

ICF (International Crane Foundation), 254, 268
Identity
  and peace park formation, 333
  shared collective, 166
Identity, national
  fragmentation of, 291
  and Kuril Islands, 267, 268
Identity-based conflicts, 109–10, 112
Identity needs
  action, 111
  connectedness, 110–111
  meaning, 110
  as nonnegotiable, 112
  recognition, 111
  security, 111
IGCP (International Gorilla Conservation Program), 33, 34–37
IMF, 55
Immigration, illegal, and US–Mexican peace park prospects, 205–206, 210, 217
India
  in Antarctic Treaty, 287
  buffer zone separating Pakistan from, 46
  desertification in foreseen, 282
  GLOF damage in, 281
  guerrilla groups in, 26
  and Kashmir/Siachen Glacier, 41, 75, 277, 278–80, 283, 300 (*see also* Siachen Glacier region dispute)

India (cont.)
  and Pakistan, 278, 289, 337
  and Pamirs Conservancy, 302
India International Center, 284
Indigenous peoples. *See also* Local communities
  and Condor–Kutuku conservation corridor, 10
  as divided by borders, 28
  environmental organizations' disregard of, 15
  and Gonarezhou National Park, 96–97
  and Great Limpopo Transfrontier Park, 91–92, 92–93
  Marsh Arabs, 314, 317, 319, 320–21, 327, 330nn.14,15
  O'odham of Sonoran desert, 212–13, 214
Indonesia, Kayan Mentarang National Park in, 8
Institute for Zoo Biology and Wildlife Research of Berlin, 117–18
Institutional adjustment, for protected areas (Afghanistan), 306
Inter-American Development Bank, 62
Interdependence, of US and Mexico, 219
Inter-Korean (horizontal) approach, to ecosystem preservation, 255
Internally displaced people (IDP), and protected areas, 27
International Association of Antarctic Tour Operators (IAATO), 164
International boundaries, in transboundary conservation areas, 28
International Conference of the Conservation of African Fauna (London, 1933), 128
International Conservancy, Pamirs, 302–303
International cooperation, for Afghan Protected areas, 306–307
International Council for Science, and Antarctic research, 164
International Crane Foundation (ICF), 254, 268
"International free zones" (alternative name for peace parks), 220
  in Big Bend–Sierra del Carmen area, 222n.20

International Gorilla Conservation Program (IGCP), 33, 34–37
International Institute for Peace through Tourism (IIPT), 69
International Institute for Sustainable Development (IISD), 34
International Karakoram Science Project (IKSP), 287–88
International Mountaineering and Climbing Federation, 289
International Peace Park (IPP), for Kuril Islands, 262–63, 267–74
International Peace Park Foundation, 268
International regimes, 186
International Snow Leopard Conference, 300–301
International Sonoran Desert Alliance (ISDA), 215
International Sonoran Desert Biosphere Reserve, 215
International space, Kuriles as, 269, 270
International system. *See* Global politics
International Tropical Timbers Organization (ITTO), 10, 12, 141, 142, 153, 154, 156, 159–60, 160n.2
International Union for Nature Conservation (IUCN), 132, 154
International Year of the Mountains (2002), 288, 303
International Year for Water (2004), 288
Interpretation, on Waterton–Glacier Park, 188–89
Interstate Committee against Drought in the Sahel, 130, 140
Iran
  Afghanistan suspicious of, 310
  and al Ahwar marshes, 313–14 (*see also* Al Ahwar marshes)
  Al Azim marsh in, 315
  in environmental dialogue, 328
  and Hawizeh–Azim marshes, 326, 328–29
  Iraq as adversary of, 14, 313, 323
  Iraq in environmental conversation with, 337
  and Ramsar Convention, 325
  in Sistan wetlands dispute, 298, 308

wildlife areas in, 297–98
and World Heritage Convention, 325
Iran–Iraq War (1980s), 315, 318, 319
Iraq
  and al Ahwar marshes, 313–314 (*see also* Al Ahwar marshes)
  and Hawizeh–Azim marshes, 328–329 (*see also* Hawizeh–Azim marshes)
  Iran as adversary of, 14, 313, 323
  Iran in environmental conversations with, 337
  and Ramsar Convention, 325
  uprising in (1991), 319
  water management in, 322
  and World Heritage Convention, 325
Ireland, peace park in, 69
Israel
  and Binational Red Sea Marine Peace Park, 41, 48
  and Jordan, 337
Issue-linkage, peril of, 277
Italian government, and Mesopotamian marshlands restoration, 326, 328
ITTO (International Tropical Timbers Organization), 10, 12, 141, 142, 153, 154, 156, 159–60, 160n.2
IUCN. *See* International Union for Nature Conservation; World Conservation Union
IUCN Pakistan, 300, 301
IUCN World Conservation Congress, 301
Ivory Coast (Côte d'Ivoire), 229–30
  on Council of Understanding, 130
  and Komoé–Leraba Forest–Warigué Forest, 166
  Nimba corridor in, 233
  private security forces' attacks on, 227
  Rapid Biological Assessment in, 233
  Sapo–Taï corridor in, 233
  UN presence in, 229

Japan
  and Kuril Islands, 13–14, 261, 263, 268, 270 (*see also* Kuril Islands)
  and Mesopotamian marshlands restoration, 326

  and minefield defusing (Emerald Triangle), 157
  peace park in, 69
  and rapport between Koreas, 250
  in six-party talks, 240
  and South-East Asia region, 72
Johnson, Lyndon B., 213, 223n.47
Joint Management Board (JMB), for Great Limpopo Transfrontier Park, 85, 90, 91
Joint sovereignty, 179. *See also* Co-management; Co-stewardship
Jordan
  and Binational Red Sea Marine Peace Park, 41, 48
  and Israel, 337
Jordan, Z. Paulo, xiii–xiv
Juridicial solutions, for buffer zones, 48
Justice, and Kuriles dispute, 267, 269, 273–74

Kaeng Kachan Forest, 142
Kaeng Tana National Park, Thailand, 142, 145, 148
Kali Cuqmasti Nature Reserve, Afghanistan, 301
Kanwal, Gurmeet, 283
Karakoram Constellation, 300–302
Karakoram Mountains, 277, 279, 286, 287
Karkheh Dam, 328
Karkheh River, 315, 320
Kashmir dispute, 75, 277–278, 279–80
Kayan Mentarang National Park, Indonesia, 8
Kazakhstan, 7
Ke Chung Kim, 285
Kemkar, Neal, 285
Kemp, Geoffrey, 323
Keumgang, Mount, 249
KFEM (Korean Federation of Environmental Movement), 254
Khattaq, Khushhal Khan, 291
Khayoun, Rasheed Bander al-, 320–21, 329
Khorog Declaration on the Pamirs Initiative, 302
Khunjerab National Park, Pakistan, 300, 301

Kibira National Park, Burundi, 8
Kimberley Process for regulating diamond revenues, 5, 17n.11
Kim Dae-jung, 241, 243, 249, 250
Kim Il-Sung, 243
Kim Jong-il, 241, 243, 250, 251
Kim Young Sam, 253
Koithara, V., 283
Koizumi, Junichiro, 266
Kole Hashmat Khan protected area, Afghanistan, 296
Komoé–Leraba Forest–Warigué Forest, West Africa, 166
Korean demilitarized zone (DMZ), 14–15, 46, 48, 239, 244–46
 and peace parks, 246, 249
 as transboundary nature reserves (TBNRs), 248, 251–54
 and trust-building, 337
 as UNESCO World Heritage Site, 254
Korean peninsula, 239–40. *See also* North Korea; South Korea
 environmental degradation in, 240–42, 244, 256, 285
 Inter-Korean Summit (2000) on, 241, 255, 256, 257
 nuclear crisis in, 240, 242, 243, 249, 257
 population of, 244
Korea Peace Nature Reserves System (KPNRS), 246–47, 248, 257
Kosovo, peace park proposed for, 75
KPBRS, as Biosphere Reserve or World Heritage Site, 257
Kpelle National Forest Reserve, Liberia, 234
Krasin (Russian icebreaker), 174
Krill, 168, 173
Kristeva, Julia, 292
Kruger National Park, South Africa, 77, 84, 90, 92–95
Kubo community (Limpopo National Park), 100
Kunduz Province Wildlife Reserve (Tajikistan–Afghanistan), 298
Kuril Island Network, 268
Kuril Islands, 13, 261
 ecological geography of, 264
 geopolitical history of, 263, 264–67
 International Peace Park (IPP) for, 262–63, 267–74
Kurilsky Reserve, Russia, 272
Kuwait
 and Gulf War, 318–319
 and World Heritage Convention, 325
Kuwait Institute of Scientific Research (KISR), 326
Kyrgyz Republic, 7

La Amistad International Park and World Heritage Site, 177, 179, 247
Land conservation, contention over, 6–7
Landmines. *See* Minefields
Lanjak Entimau, Borneo, 157
Laos (Lao PDR)
 in Biodiversity Initiatives, 153
 and border politics, 157
 and Pha Taem Complex/Emerald Triangle, 8, 151, 152, 155, 156–57, 159, 160
 poverty of, 158
 and Southeastern Indochina Dry Evergreen Forests ecoregion, 142
Le Blanc, Steven, quoted, 203
Leopold, Aldo, xvii
Letter of Intent for Adjacent Protected Areas (LOI), Mexico–US, 212, 215–16, 220
Liberia, 227, 229
 biodiversity in, 227, 232, 233, 234
 and common aversion, 335
 in ECOWAS, 130
 and Guinea, 230
 logging in, 228, 233
 natural resources illegally controlled and exploited in, 237
 peace parks for, 227–28, 230, 232, 233–36, 237
 and regional cooperation, 230–32
 Sapo National Park in, 13, 231, 233, 236
 Taylor coup in, 26–27
Liberia Protected Areas Trust, 235
Limpopo National Park, Mozambique, 90, 97–100

Limpopo Transfrontier Conservation Area, 41
Limpopo Transfrontier Park, Great (GLTP), Mozambique/South Africa/Zimbabwe, 77, 84–87, 90–91, 101
  in Mozambique, 97–100, 101
  as social-ecological system, 87–92
  in South Africa, 92–95
  in Zimbabwe, 77, 84, 95–97, 101
Line of control (LOC), in Kashmir, 278
Linkage of issues, peril of, 277
Local communities. *See also* Indigenous peoples
  in Afghanistan, 293, 303, 305–306
  in Benin, 132
  and biodiversity in Emerald Triangle (Thailand), 153
  in Burkina Faso, 133
  conservation policies as changing lives of, 109
  and economic development, 337
  and Emerald Triangle tourism, 158
  and Great Limpopo Transfrontier Park, 101
  involvement of needed, 124
  and Kruger National Park, 94
  and Kuril Islands, 263, 270
  and Liberian peace parks, 231, 236
  in Limpopo National Park, 100
  and Pamirs Conservancy, 302
  and parks vs. militarization, 231
  in peace-park implementation, 339
  and peace park initiatives, 61–63
  and Selous Game Reserve, 116
  and Selous–Niassa Wildlife Corridor, 121
  and W Park, 134, 135, 137, 138
Locality studies, in Britain, 73
Lofa National Forest Reserve, Liberia, 234
Logging
  and Liberian peace parks, 228, 232, 233
  and Liberia under Taylor, 227
  and Pha Taem Complex, 148, 151
  and Sapo–Taï corridor, 233
Low politics, 12, 16, 337, 339
Lusk, Gil, 210

Maathai, Wangari, xiii, 1
MAB (Man and Biosphere) Program (UNESCO), 130, 131, 185, 215, 222n.24, 271
MACP (Mountain Areas Conservancy Project), 301, 302
Madagascar, 235
Maderas del Carmen, Mexico, 207, 209, 211, 212, 220
Madrid Protocol, 165, 170, 170–71
  Committee for Environmental Protection created by (Article 11), 177
  on emergency response issues (Annex IV, Article 12), 173–74
  on liability rules and procedures (Article 16), 173
  on specially protected areas (Annex V), 176, 177
Makuleke people, 92–94
Malawi, and Mtwara Development Corridor, 115, 122
Mali, on Council of Understanding, 130
Mallaby, Sebastian, 16
Management information system design, for Afghan conservation activities, 304
Management Protected Areas, categories of, 160n.3
Manatees, protection of (Gulf of Honduras), 59
Mandela, Nelson, 25
  and Korean peace park, 249
  quoted, 21
Marine biodiversity, and Kuril Islands dispute, 271–72
Maritime Zones, Special, 247
Marsh Arabs, 314, 317, 319, 320–21, 327, 330nn.14,15
Marshlands, of Mesopotamia, 14, 313–16. *See also* Al-Ahwar marshes; Hawizeh–Azim marshes; Marsh Arabs
Matswani Safaris, 93
Maxwell, Gavin, 324
Meaning, as identity need, 110
Mediterranean Action Plan, 338
Mekong River Commission (MRC), 152, 153, 154

Memoranda of Understanding (MoUs), 77
 on Mtwara Development Corridor, 122
 on Sonoran Biosphere Reserves, 215
 on US–Mexican technical cooperation, 218
 on Waterton–Glacier, 187
Meso-American biological corridor, 23, 62
Meso-American Reef System Project, 62
Mesopotamia, 313, 318
Mesopotamian marshes, 14, 313–16, 317. *See also* Al Ahwar marshes; Hawizeh–Azim marshes; Marsh Arabs
Mexican Revolution, 208, 221n.11
Mexico, and Belize barrier reef, 60–61
Mexico–US border. *See* US–Mexico border
Military, in peace-park implementation, 339
Military conflicts
 and militia cover in conservation zones, 127
 and protected areas, 26
 in southern Africa, 26–27
 in West Africa, 227, 229–30, 231
Millennium Development Goals of United Nations, 113
Minefield(s)
 in Afghanistan, 295
 de-fusing of
  in Emerald Triangle region, 157–58, 159
  and Limpopo National Park, 99
  in Pha Taem Complex, 152
 in Pha Taem area, 148, 152
Mining
 and Antarctic Treaty System, 170–71
 and Kuril Islands, 263
 in Liberia, 227, 231, 233, 234
 looting of (vs. Kimberley process), 5, 17n.11
 and Sapo–Taï corridor, 233
 in Selous–Niassa region, 121
 small- vs. large-scale, 5
Mittal Steel, and biodiversity neutral site in Liberia, 234
Monitoring, for Liberian peace parks, 235–36

Montenegro, peace park proposed for, 75
Mountain Areas Conservancy Project (MACP), 301, 302
Mountain Societies Development Support Program (MSDSP), 303
Mozambique
 and Great Limpopo Transfrontier Park, 77, 83, 84, 89, 97–101 (*see also* Limpopo National Park; Limpopo Transfrontier Park, Great)
 history of, 114–115
 insurgent financing in, 26
 and Mtwara Development Corridor, 122
 Niassa Reserve in, 116
 in SADC, 115
 and Selous–Niassa Wildlife Corridor, 112
 and Tanzania, 125
 Tanzania border of, 115
MRC (Mekong River Commission), 152, 153, 154
Mtwara Development Corridor, 115, 120, 121, 122–23
Mugabe, Robert, 84
Multiethnic states, with ethnic majority in neighboring state, 291
Mumbai bombings (2006), 289
Musharraf, Pervez, 284,
Muslims, Shi'a-Sunni division of, 318
Mweka College of African Wildlife Management, 124
Myanmar, in Biodiversity Initiatives, 153

Nabhan, Gary Paul, 205
NAFTA (North American Free Trade Agreement), 210, 218, 219, 222n.29
NAFTA Commission on Environmental Cooperation, 13, 219
Nagasaki, Japan, "peace park" in, 7, 69
Namibia
 in alliance, 78
 peace park in, 76
Nanga Parbat Conservancy Area, 301
Narcotics trade. *See* Drug trafficking
National Adaptation Plan of Action, Afghanistan, 308

National Aeronautics and Space Administration (NASA), and Karakoram project, 287
National Biodiversity Strategy and Action Plan (NBSAP), 308
National Capacity Self-Assessment (NCSA), Afghanistan, 308
National identity. *See* Identity, national
Nationalism, territorial (and Kurils), 267
National Science Foundation (NSF), Office of Polar Programs (OPP) of, 164, 170, 175
Native Americans, and US national park, 6
Nature Conservancy, 6, 23, 62, 214, 219
Nature conservation. *See* Conservation
Nature Iraq, 322
Nature Iraq/New Eden Project, 322, 326
Nature protection. *See also* Conservation
  Fridman on, 261
  Russian approach to, 272
Nature reserves, 247. *See also* Transboundary nature reserves
  in Afghanistan, 305
  Korean DMZ as, 246, 257
  in Russia, 271–72
Nature villages, 247
Neoliberalism, and global governance, 57, 60
Nepal
  dangerous glacial lakes in, 281
  desertification in foreseen, 282
  guerrilla groups in, 26
New Eden/Eden Again, 327, 331n.39
New Eden project, 322, 326
New Keum Su Gang San (NKSGS), Korea, 248
New Zealand, Antarctic claims of, 171
NGOs. *See* Nongovernmental organizations
Niassa Reserve, Mozambique, 112, 116
  as elephant range, 119
  in transboundary cooperation, 125
Niger
  on Council of Understanding, 130
  sociopolitical and economic changes in, 131–32
  and W Park, 7, 12, 128, 129, 130, 131, 133–34, 134
Nigeria, 130, 229, 230
Nimba corridor, Liberia/Guinea/Ivory Coast, 233–34
Nimba Nature Reserve, Liberia, 233–34
1956 Declaration (Russia), 266
Nobel Peace Prize (2004), xiii, 1
Nobel Peace Prize (2006), xiii
Nongovernmental organizations (NGOs). *See also* Donor organizations and nations; *specific organizations*
  criticisms of, 16, 63
  and GLTP, 91
  influence of, 55
  and program sponsorship, 307
North American Agreement on Environmental Cooperation (NAAEC), 210
North American Free Trade Agreement (NAFTA), 210, 218, 219, 222n.29
Northern Areas Strategy for Sustainable Development, 300
North Korea (Democratic People's Republic of Korea, DPRK), 239, 242
  and Basic Agreement (1992), 253
  as Convention on Biological Diversity signatory, 252
  in Inter-Korean Summit (2000) on, 241, 255, 256, 257
  nuclear standoff forced by, 240, 249
  in six-party talks, 240
  South Korean business enterprises in, 241
  suffering of, 249–50
  "Sunshine Policy" of, 243, 245, 250
North Korean Nature Conservative Union (or Federation), 253
Northwest Afghanistan (Iran-Turkmenistan-Afghanistan) wildlife reserve, 298
Norton, Gale, 220
Norway
  Antarctic claims of, 171
  and minefield defusing (Emerald Triangle), 157
Nuclear crisis, Korean, 240, 249

Nuristan National Park, Afghanistan, 301
Nyerere, Julius, 114

O'odham people, 212–13, 214
Orchid collection, on Belize–Guatemala border, 64–65
Organization of African Unity (OAU), 75
Organ Pipe Cactus National Monument (ORPI), 213, 214, 216, 217

Pakistan
 Afghanistan suspicious of, 310
 buffer zone separating India from, 46
 conservation and development program for, 300
 desertification in foreseen, 282
 GLOF damage in, 281
 and India, 278, 289, 337
 and Karakoram Constellation, 300
 and Kashmir/Siachen Glacier, 41, 75, 277, 278–80, 283 (*see also* Siachen Glacier region dispute)
 and Pamirs Conservancy, 302
 protected areas in, 301
 and state failure, 291
Pamir-i-Buzurg Wildlife Reserve, Afghanistan, 295, 301
Pamirs International Conservancy, 302–303
Pamirs National Park, Tajikistan, 301
Panama, and La Amistad, 177, 179, 247
Park "W." *See* "W" International Peace Park
PCOB (Postconflict Branch) of UNEP, 294, 339
Peace
 and environment, xiii
 hard and soft, 12
 synergies for, 262, 269
Peace agreements, and environmental issues, 15
Peace-building
 bioregional approach to, 78–79
 vs. conservation, 339
 environmental issues in, 337–38
 and International Gorilla Conservation Program (IGCP), 34, 37
 Selous–Niassa Wildlife Corridor in, 115
 in Siachen Glacier dispute, 278
 transboundary conservation areas for, 32–33
Peace and Conflict Impact Assessment (PCIA), 29–33, 38, 333
 on transboundary conservation projects, 32–33
 of Virunga–Bwindi region, 33–37
"Peace corridor," suggested for Mexican deserts, 220
Peace park(s) (Parks for peace), xiv, 1–2, 197–98, 247, 248–49, 324, 341. *See also at* Conservation; Nature; Transboundary; *individual parks*
 advocacy needed for, 220
 for Afghanistan, 294, 295–302, 303–304, 308
 in al Ahwar marsh area, 314, 329
 alternative names for, 220
 and backseat or joint sovereignty, 180
 and borderlands, 70
 in Central America, 58–65, 66
 challenges to, 339–40
 claims on behalf of, 38
 collaboration in, 196–97
 and conflict, 25, 64, 127, 137
 conflict resolution through, xiv, 63, 75, 83, 198, 249, 333, 339
 and conservation, 83, 198 (*see also* Conservation)
 conservation balanced against conflict reduction in, 83
 conservation-induced conflicts ameliorated by, 16
 contributions of, 24, 25
 criminal networks in (Central America), 64–65
 cultural re-connection and revitalization through, 61
 definitions of, 8, 24, 180n.6
 as economically self-sustaining, 60–61
 and economic developments, 139
 as environmental conservation zones, 1–2 (*see also* Environmental conservation zones)
 evaluation of progress on, 38

as exemplars, 6–8
as exit strategy, 14
and failed states, 291–292
first example of (Waterton–Glacier), 2, 24, 56, 69, 183
formation of, 11, 334
frontcountry and backcountry areas in, 193, 198n.15
game-theoretic model of, 42–52
and global governance, 56–58, 66, 76–77
and glocalization, 74–79
implementation of, 339
and Korean DMZ, 246, 249
as laboratories for peace, 41
lack of consideration of, 4
and Liberia, 227–28, 230, 232, 233–36, 237
measuring performance of, 23
and Nagasaki, 7, 69
and national pride or prestige, 14
nontransborder, 69–70
and Peru–Ecuador conflict, 9
planning for, 334, 340–41
and policy makers, 339
and political interventions, 59, 61, 64, 66
political process in, 333
proposed for Kosovo, Montenegro, and Albania, 75
and Red Sea marine peace park, 338
and regionalism, 70–71, 74, 80
as re-ordering, 69
requirements for effectiveness of, xiv
resistance to, 333–34
and scale, 70, 71, 79
for Siachen Glacier region, 284–85, 288–89
and specially protected areas (SPAs) on Antarctica, 176, 177
between states with positive and conflictual relations, 11
and transnational governance, 55
for US–Mexico, 214 (see also US–Mexico border)
vs. wild reserve, 137, 139
Peace Parks Foundation (PPF), South Africa, 7, 63, 77, 84, 97

Peace park threat assessments, call for, 127
Peru, and Ecuador (Condor–Kutuku conservation corridor), xiv, 8–10
Pha Taem National Park, 142, 148, 150
Pha Taem Protected Forests Complex (PPFC), Thailand, 142–45, 150, 159
  and Emerald Triangle, 153–55
  management plan and objectives for, 146, 148
  management programs for, 150–51
  opportunities for, 152–53
  threats to, 151–52
  wildlife distribution in, 145–46
  zoning for, 148–50
Pha Taem Transborder Initiative, 8
Phouxeingthong National Biodiversity Conservation Area (NCBA), Laos, 144, 146, 148, 151
Phu Jong–Na Yoi National Parks, 142, 145, 146, 148
Pinacate region, 214, 219
Poaching
  in Gonarezhou National Park, 97
  and Great Limpopo Transfrontier Park, 92
  on Kuril Islands, 263, 268, 272–73
  and Limpopo National Park, 99
  in Niassa Reserve, 116
  and Pha Taem Complex/Emerald Triangle, 148, 151, 155, 157
  in Selous Game Reserve, 115–16, 117
  and Selous–Niassa Wildlife Corridor, 112, 118, 119–20
  and transboundary conservation areas, 29
  and W Park, 129, 131, 135, 138, 139
Polar Sea (US icebreaker), 174
Political interventions, in peace-park formation, 59, 61, 64, 66
Postconflict Branch (PCOB) of UNEP, 294, 339
Postconflict reconstruction
  and peace park formation, 13, 13
  and resource exploitation, 27
Post-Westphalian governance, strengthening of, 36

Poverty alleviation
  through Afghanistan parks and protected areas, 297, 301, 303, 305–306, 310
  through Emerald Triangle projects, 158, 160
  through Mtwara Development Corridor, 115
  through wildlife management (Selous Game Reserve), 116
  in W Park, 135
Powell, Colin, 237
Practitioners, and academicians, 335, 339
Praxis, 335
Project-based (bottom-up) approach, to ecosystem preservation (Korea), 254
Project/program mapping, in Peace and Conflict Impact Assessment, 31
Pronatura Noreste, 219
Property rights, and environmental peace-building, 4–5
Protected areas (PAs), xiii, 247. *See also* Transboundary protected areas
  for Afghanistan, 293, 295–298, 300, 303, 304–309, 310–11
  benefits of (World Conservation Union), 285
  and conflict, 25–29
  Selous Game Reserve as, 115
  and US–Mexican border, 207, 211, 212, 220, 220
Protected Forest for Conservation of Genetic Resources of Plants and Wildlife, Cambodia, 144
Protected landscapes or seascapes, 247
Protocol on Environmental Protection to the Antarctic Treaty. *See* Madrid Protocol
Public-private partnership, and US–Mexico border, 13
Public safety, in Waterton–Glacier Park, 189, 191
Public trusteeship, 107
Putin, Vladimir, 266, 268

Qashqar Conservancy Area, Karakoram Constellation, 301

Raghavan, V. R., 283
Ramsar Convention on Wetlands (1971), 130, 131, 323, 324–325, 327, 331n.42
Ramsar Wetland of International Importance. *See* Wetlands of International Importance
Rapaport, Anatol, quoted, 21
Rapid Biological Assessment (RAP), 233
Rapid ecological assessment (REA), 145, 161n.8
Rationality, in peace-park model, 42, 43, 51
Recognition, as identity need, 111
Redford, Robert, quoted, 203
Red Sea marine peace park, 338
Regional cooperation, and Liberia, 230–32
Regional diversity, and ecosystem management, 232
Regional ecosystems (Mexico–US), 210
Regional hegemonic powers, and environmental cooperation, 141
Regionalism
  in Africa, 73, 77–78
  and economic globalization, 72
  new awareness of (US), 73
  and peace parks, 70–71, 74, 80
Regional Organization for the Protection of the Marine Environment (ROPME), 324, 326, 327
Regional stability, and environmental cooperation, 155
Regionness, stages of, 72–73
Relationship-based model, peace park in, 52
Replicability, in protected-area system (Afghanistan), 306
Republic of Korea. *See* South Korea
Reserva de la Biosfera Alto Golfo de California y Delta del Rio Colorado, Mexico, 215
Reserva de la Biosfera El Pinacate y Gran Desierto de Altar, Mexico, 215
Resettlement, 27
  for Liberian peace parks, 236
  from Gonarezhou National Park (Zimbabwe), 96
  in Limpopo National Park, 90, 98
  World Bank policy on, 98

Resettlement Working Group (RWG), for Limpopo National Park, 98
Resilience, 88, 187, 197–98
  of cooperation (Waterton–Glacier), 186, 187, 197
Resource abundance, and conflict, 5, 64
Resource scarcity
  and conflict, 4–5, 63–64
  and environmental peace-building, 4–5
Reychler, Luc, 30
Reynolds, Henry "Death on the Trail," 187
Rio Convention on Biological Diversity (CBD). *See* Convention on Biological Diversity
Rio Grande River, 207, 221n.6
  water quality in, 210
Rio Grande Wild and Scenic River (protected area), 207
Robustness, 88, and Great Limpopo Transfrontier Park, 92
Roh Moo-hyun, 243
Roosevelt, Franklin D., 207, 213, 235
ROPME (Regional Organization for the Protection of the Marine Environment), 324, 326, 327
ROPME/UNEP High-Level Meeting on the Restoration of the Mesopotamian Marshlands (al Ahwar) in Bahrain (2005), 331n.35
Rosado, Natalie, 60
Ross Ice Shelf, 167
Rotary Club International
  and first peace park, 69, 185, 198n.1, 206
  and US–Mexico Park proposals, 209, 219
Royal Forest Department (RFD), Thailand, 142, 153, 159
Rubin, Jeffrey, 340
Rupert, Anton, 7, 84, 87
Rupert Nature Foundation, 84
Russia
  Afghanistan suspicious of, 310
  and Antarctica, 175
  and Kuril Islands, 13–14, 261, 263, 268, 270 (*see also* Kuril Islands)
  and rapport between Koreas, 250
  in six-party talks, 240

Russian Academy of Science, 268
Rwanda
  genocide in (1994), 27, 34, 78
  gorillas protected from violence in, 179
  IDP settlement in, 27
  swath through bamboo forest cut in, 26
  Virunga–Bwindi region in, 33, 38
  Volcanoes National Park in, 8

SADC (South African Development Community), 115
Saddam Hussein, uprising against, 319
Sahel, transhumance in, 137
Sandia National Labs, and Siachen dispute, 283–84
Sands, Peter, quoted, 107
SANParks (South African National Parks), 94, 101
Sapo National Park, Liberia, 13, 231, 233, 236
Sapo–Taï corridor, 233
Sarambwe Conservation Area, 33
Sarstoon-Temash transfrontier conservation area (Belize/Guatemala), 60, 61, 62, 63, 64–65
Sasawara Forest Reserve (Tanzania/Mozambique), 121
Scalar transformation, 71–74
Scale analysis, 71
Science
  and Antarctic treaty, 12
  in peace-building, 334
Science center, for Siachen region, 285–87
Scientific Committee on Antarctic Research (SCAR), 164, 177
Scientific management, 58
Security
  as identity need, 111
  redefinition of, 14
Selous Conservation Program (SCP), 116, 116–17, 117
Selous Game Reserve, Tanzania, 112, 115–16
  as elephant range, 119
  and village wildlife management areas (WMA), 116

Selous–Niassa Wildlife Corridor (SNWC), 11, 117
  beginnings of, 112–13
  community-based conservation in, 121–22, 123–25
  grassroots efforts in, 109, 112–13, 125
  and Mtwara Development Corridor, 122
  in peace-building process, 115
  research on, 117–19
  threats to 119–21
Seoul Environmental Declaration, 253
September 11 terrorist attack, 193, 194
Shackleton, Ernest, 167
Shaller, George, 299–300, 308
Shandur–Handrup National Park, 301
Shared collective identity, 166
Shared Peace Park, Hawizeh–Azim marshes as, 323, 324
Shatt al Arab waterway, 313, 315, 317, 318, 329
Shi'a Muslims, 313, 318
Siachen: Conflict without End (Raghavan), 283
Siachen Glacier region, environmental degradation of, 280
Siachen Glacier region dispute, 41, 277–78, 288
  environmental impact of, 282–84
  and Karakoram Constellation project, 300
  origins of, 278–80
Siachen Peace Park, 284–85, 288–89
Siachen Science Center, 285–87
Sierra Club, 213
Sierra del Carmen, 209
Sierra Leone, 230
  in ECOWAS, 130
  guerrilla groups in, 26
  and Guinea, 230
  independence won by, 229
  mining operations in, 5
  Revolutionary United Front (RUF) in, 227, 230
  and Taylor, 230
  UN in, 229
Silk Road, 298–99
Singh, Manmohan, 289

Sistan Baluchistan/Hamun-i-Puzak wetlands area (Iran/Afghanistan), 297, 308
"Sister parks" (alternative name for peace parks), 220
Six-Party Talks (Korean nuclear crisis), 14–15, 240, 242, 249, 251, 257
Small business assistance programs, for Afghanistan, 305
Smuggling
  in Liberia, 234
  and Sapo–Taï corridor, 233
  and transboundary conservation areas, 29
Social-ecological system, 83
  Great Limpopo Transfrontier Park as, 87–92
Social identity, 112
Soft collaborative activities, 154, 160
Sokoine University of Morogoro/Tanzania, 117–18
Sonoran desert, 205, 212–17
  Western, 212
Sonoran Desert National Monument, 216
Sonoran Desert National Park Friends (SDNPF), 216
"Sonoran Desert Peace Park," 217
Sorak National Park, South Korea, 249
South Africa
  conflict resolution attempted by, 78
  and Great Limpopo Transfrontier Park, 84, 89, 92–95 (*see also* Limpopo Transfrontier Park, Great)
  Limpopo Transfrontier Conservation Area in, 41
  majority government established in, 115
  and Mtwara Development Corridor, 115
  peace parks on border of, 76, 77
  Peace Parks Foundation in, 7, 77
  transboundary conservation promoted in, 83
  wildlife conservation zones in, 6
South African Development Community (SADC), 115
South African National Parks (SANParks), 94, 101
South America, conservation tensions in, 6
Southeast Anatolia Project (GAP), 319

Southeastern Indochina Dry Evergreen
    Forests ecoregion, 141–42
Southern Africa, 83, 84–87
  capacity building in, 125
  stronger government capacity needed in,
    124
  transboundary projects in, 109, 249
Southern African Development Community
    (SADC), 77, 78
  Wildlife Policy of, 85
South Korea (Republic of Korea, ROK), 239,
    243–44
  and Basic Agreement (1992), 253
  as Convention on Biological Diversity
    signatory, 252
  economic development and environmental
    degradation in, 240–42
  in Inter-Korean Summit (2000), 241, 255,
    256, 257
  and Japan island dispute, 266
  and rapport with North Korea, 250
  in six-party talks, 240
Sovereignty, 172
  and Antarctica, 163, 166, 169–70, 172–73
  back seat for, 178–80
  over borderlands, 76
  and change of scale, 70
  vs. common environmental sensitivities, 338
  dilution of discussed, 4
  and global governance (peace parks), 57
  joint, 179
  and Kuril International Peace Park, 263,
    267, 268
  and transboundary conservation areas, 29
  and Westphalian divisions, 166
Soyinka, Wole, quoted, v
Specially managed areas, Antarctica, 176–
    77
Specially protected areas, Antarctica, 176,
    177–78, 180
Special Maritime Zones, 247
Species extinctions, 273
Spence, Mark David, quoted, 107
Staff training, in Afghanistan, 304
Sudan, Darfur crisis in, xiii, 3
Sunni Muslims, 313, 318

Sustainable development
  and Emerald Triangle initiatives, 155
  and environmental security (Korea), 256
  International Institute for (IISD), 34
  Northern Areas Strategy for, 300
  protected areas in, xiii
  in West Africa, 237
  World Summit on (2003), 303
  and W Park, 134
Sustainable Development Institute, Liberia,
    231
Sweden, and Greater Nimba Highlands
    Peace Park, 234
Swiss Agency for Development and
    Cooperation (SDC), 300
Symington, Fife, 215
Syria
  and dams on Tigris and Euphrates, 328
  and World Heritage Convention, 325

TAAF (Terres Australes et Antarctiques
    Françaises), 172
Taï Park, Liberia, 233
Taiwan, and Japan island dispute, 266
Tajikistan
  and Pamirs Conservancy, 302
  wildlife areas in, 298
Tajik National Park (TNP), 302
Takhar Province Wildlife Reserve
    (Tajikistan/Afghanistan), 298
Tanzania
  community-conservation approaches in,
    113
  history of, 114–15
  and Mozambique, 125
  Mozambique border of, 115
  and Mtwara Development Corridor, 122
  in SADC, 115
  Selous Game Reserve in, 112, 115–16,
    119
  and Selous–Niassa Wildlife Corridor, 112,
    117 (*see also* Selous–Niassa Wildlife
    Corridor)
Tanzania Wildlife Research Institute, 118
Taxkorgan Nature Reserve, China, 300, 301
Taylor, Charles, 26–27, 227, 228, 229, 230

TBPA. *See* Transboundary protected areas
Technical assistance, importance of, 124
Terres Australes et Antarctiques Françaises (TAAF), 172
Territoriality, 41
Terror, war on, 13
  and terrorist infiltration from Mexico, 217
Terrorist attack on Mumbai (2006), 289
TFCA. *See* Transfrontier conservation area
Thailand
  in biodiversity Initiatives, 153
  and border politics, 157
  economic development of, 158
  immigrants and refugees in, 158
  and Pha Taem Complex/Emerald Triangle, 8, 12, 142, 152, 154, 155, 159, 160
  as regional power, 141
  and Southeastern Indochina Dry Evergreen Forests ecoregion, 142
Thesiger, Wilfred, 324
Tibet, GLOF damage in, 281
Tienshansky, Simon, 299
Tigris-Euphrates catchment area, 315. *See also* Al Ahwar marshes; Hawizeh–Azim marshes; Mesopotamian marshes
Tigris River, 315, 328
Tilaboy NT, Afghanistan, 301
Tirich Mir Conservancy Area, Pakistan, 301
Togo, 229
Tolbert, William R., 229
Toledo Institute for Environment and Development (TIDE), 59
Touré, Sékou, 230
Tourism. *See also* Ecotourism
  in Antarctica, 164, 171
  and Central American peace parks, 60–61
  in Emerald Triangle, 158
  in Gonarezhou National Park, 95–96
  and gorillas (Virunga-Bwindi region), 37
  in Great Limpopo Transfrontier Park, 83, 87, 88, 89, 90, 101
  in Kruger National Park, 93–95
  in Limpopo National Park, 98–100
  in North Korea, 250
  and peace-park formation, 337
  and peace-park impact, 340

  in Selous–Niassa Wildlife Corridor, 121
  in W Park, 135
Tragedy of the commons, 4, 174
Transborder protected landscapes, at US–Mexican border, 206
Transborder transhumance, in W Park, 136–37
Transboundary biodiversity, and Emerald Triangle (Thailand), 153
Transboundary biodiversity conservation area (TBCA), Emerald Triangle as, 160
Transboundary Biosphere Reserve
  Hawizeh–Azim marshes as, 323
  W Park as, 130
Transboundary community-based wildlife management initiatives, 113
Transboundary conservation, 23–24. *See also* Conservation
  evaluation of progress on, 38
  origin of, 24
Transboundary conservation initiatives, 23, 36
  and Emerald Triangle, 159
Transboundary conservation projects
  and high-level political movement, 109, 125
  Peace and Conflict Impact Assessments on, 32–33
Transboundary conservation zones or areas, 2
  contributions of, 24
  in Hawizeh–Azim marshes, 321–323
  incorporation of international boundaries in, 28
  informal establishment of, 29
Transboundary cooperation
  in Afghanistan, 304
  Antarctica as model for, 178–80
  in Emerald Triangle, 156–58
  environmental, 165–166
  between Tanzania and Mozambique, 125–26
Transboundary initiatives, proliferation of, 25
Transboundary National Park, Siachen Peace Park as, 288

Transboundary natural resource
    management (TBNRM), 58, 63
Transboundary nature reserves (TBNRs),
    247–48. *See also* Nature reserves
  Korean DMZ as, 248, 251–54
Transboundary parks, multilevel governance
    of, 90
Transboundary protected areas (TBPA), xiii–
    xiv, 7. *See also* Protected areas
  conflict mitigation through, 314
  definition of, 182n.47
  examples of, 166
  increasing interest in, 142
  in Karakoram Constellation (Afghanistan/
    Tajikistan/China), 301
  and Korean DMZ, 248
  lessons on, 124
  Peace and Conflict Impact Assessment
    (PCIA) for, 29–33
  peace parks as, xiv, 324 (*see also* Peace
    parks)
  replication of (Afghanistan), 306–307
  and specially protected areas (SPAs) on
    Antarctica, 176, 177
  types of, 7–8
  workshops on, 124
Transboundary Protected Areas Network, of
    World Conservation Union, 1–2, 8
Transboundary wild areas, 2
Transboundary wildlife reserves
    (Afghanistan), 297–300
Transfrontier conservation area (TFCA), 7,
    55, 314. *See also* Peace park(s)
  and local communities, 61
Transhumance, in W Park, 136–37
Transhumance Agreement (Africa) (1991),
    130
Transitional Government of Afghanistan
    (TGA), 293, 294, 297
Trilateral Ministerial Committee (TMC), for
    Great Limpopo Transfrontier Park, 85,
    91
Trophy hunting, and Makuleke people
    (Kruger National Park), 93
Trust-building, 334
  in Antarctica, 165

  calibration of, 337
  improper sequencing of, 277
  in synergies for peace, 269
Tugitang Mountains (Turkmenistan/
    Uzbekistan/Afghanistan) wildlife reserve,
    298
Turkey
  dams built in, 319, 328
  and World Heritage Convention, 325
Turkmenistan, wildlife areas in, 298
Turner, Ted, 14, 255

Udall, Morris, 213, 223n.47
Udall, Stewart, 213
Uganda
  gorillas protected from violence in, 179
  Virunga–Bwindi region in, 33, 38
UICN
  and ECOPAS program, 140
  and W Park in Niger, 134
UNDP (United Nations Development
    Program), 62, 121, 153, 297, 300, 307,
    325, 326
UNEP. *See* United Nations Environment
    Program
UNEP Iran–Iraq Dialogue, 325
UNESCO
  and ECOPAS program, 140
  and Himalayan region, 282
  on Kuril Islands, 268
  and Mesopotamian marshland restoration,
    326, 327
  and W Park in Niger, 134
UNESCO Biosphere Reserves, 148
UNESCO MAB (Man and Biosphere)
    Program, 130, 131, 185, 215, 222n.24,
    271
UNESCO World Heritage and Biospheres
    Reserves, 325
UNESCO World Heritage Convention, 285
United Kingdom. *See* Britain
United Nations
  and common aversions outcomes, 335
  influence of, 55
  in Ivory Coast (UNOCI), 229
  and Korean armistice, 245

United Nations (cont.)
  in Liberia (UNMIL), 229, 231, 236
  Millennium Development Goals of, 113
  in Sierra Leone (UNIOSIL), 229
United Nations Commission on Global Governance, 57
United Nations Convention on Biological Diversity. *See* Convention on Biological Diversity
United Nations Development Program (UNDP), 62, 121, 153, 297, 300, 307, 325, 326
United Nations Environment Program (UNEP), 3, 121, 290n.16, 292, 325
  in Afghanistan, 297, 304, 307, 307, 308
  on desiccation of al Ahwar marshes, 314–15
  and Mesopotamian marshlands restoration, 326, 327
  Postconflict Branch (PCOB) of, 294, 339
United Nations General Assembly, Special Session on Environment and Development of, 253
United Nations Security Council Resolution 1647, 227–28
United States. *See at* US
University of Nebraska IKSP Expedition, 287
UN Office for Project Services (UNOPS), 235
UN-World Bank Reconstruction and Development Conference, 237
US
  and International Peace Park in Kuriles, 271
  and Korean armistice, 245
  national parks vs. Native Americans in, 6
  and rapport between Koreas, 250
  in six-party talks, 240
US Agency for International Development (USAID), 235, 303, 304, 308, 326
US–Canada peace park, 2, 24, 166, 183. *See also* Waterton–Glacier International Peace Park
US Department of State, and DMZ conservation, 255
US Geological Survey (USGS), and Karakoram Science Project, 287
US Institute of Peace, on Iran vs. Iraq, 323

US–Mexico border, 13, 205–206
  and border security, 217–18
  and Chihuahua Desert, 206–12
  cooperation across, 218–20
  and Rio Grande water-quality management, 210
  and Sonoran Desert, 212–17
Utilization of Waters bilateral treaty (US–Mexico; 1944), 205, 209
Uzbekistan, 7, 298

Vajpayee, Atal Bihari, 284
Vegetation restoration, in Waterton–Glacier, 188
Vietnam
  in Biodiversity Initiatives, 153
  Cat Loc reserve in, 141
  and Southeastern Indochina Dry Evergreen Forests ecoregion, 142
Violence. *See* Conflict; Genocide; Warfare
Virunga–Bwindi region, 29, 38
  Peace and Conflict Impact Assessment of, 33–37
Virunga Massif, 34
Virunga National Park, Democratic Republic of Congo, 8, 27
Virunga volcanoes, 26
Visas, and access to peace park areas, 341
Visitor access, within Waterton–Glacier Park, 192–96
Volcanoes National Park, Rwanda, 8, 27
Vostok, Lake (Antarctica), 175

Wakhan Corridor, 298–300, 303
  and Karakoram Constellation, 300–302
  and Pamirs International Conservancy, 302–303
Warfare, 239
  Le Blanc on, 203
War on terror, 13
Water (fresh)
  cooperation over, 4, 337
  diversion of from Mesopotamian marshes, 328
  and Himalayas environmental degradation, 280–81, 287

Iran-Afghanistan contest for, 298
peace parks as sources of, 231–32
proportion of in Antarctica, 167
from Rio Grande, 210
and US–Mexico border, 205
Water management, in Iraq, 322–23
Waterton–Glacier International Peace Park (Waterton–Glacier), 2, 24, 69, 166, 183–86, 197
  border security in, 192–96, 197
  cooperation and collaboration within, 186–88, 193–96, 197
  education and interpretation in, 188–89
  public safety in, 189, 191
  size of, 207
  and US–Mexico border, 206, 212
  vegetation restoration in, 188
  visitor access in, 192–96
  wildfires in, 189
West Africa, 127–28, 139. *See also* "W" International Peace Park; *individual countries*
  forests reduced in, 232
  wildlife reserves in, 127
  "W" International Peace Park in, 7, 11–12, 128–31, 139 (*see also* "W" International Peace Park)
Western Forest Complex, Thailand, 142, 153, 158
Westphalian sovereign-state divisions, 166
West Tien Shan Mountains, 7
Wetlands areas
  Ramsar Convention on (1971), 130, 131, 323, 324–25, 327, 331n.42
  reduction of (Afghanistan), 296
  Sistan Baluchistan/Hamun-i-Puzak, 297–98, 308
Wetlands of International Importance (Significance)
  Hawizeh–Azim marshes as, 323, 324–25, 327
  W Park registered as, 130
Wetzler, Richard, xvii
Wilderness Safaris, 93–94
Wildfire, in Waterton-Glacier Park, 189
Wildlife Conservation Society, 299–300, 308

Wildlife conservation units (WCU), in Burkina Faso, 133
Wildlife corridor, for Pha Taem Complex, 150, 159
Wildlife management areas (WMA), village
  in Selous Game Reserve, 116
  in Selous–Niassa Wildlife Corridor, 121, 122
Wildlife reserves, in Afghanistan, 297–98
Wildlife sanctuary, and Limpopo National Park, 99
"W" International Peace Park, 7, 11–12, 128–31, 139
  institutional comparisons of three countries involved in, 131–34
  institutional evolution of, 128–31
  management methods for, 134
  physical constraints on, 136
  and Project ECOPAS, 137–38, 140
  protection measures for, 135, 137
  research and ecological follow-up on, 136
  and transborder transhumance, 136–37
  valorization of, 135
Wirth, Conrad, 221n.19
Wologizi protected area, Liberia, 234
Wonegizi protected area, Liberia, 234
World Bank, 62
  and Afghanistan, 307
  and al-Hawizeh/al-Azim marshlands, 326–27
  influence of, 55
  and Limpopo National Park, 97, 98
  and northern Mexico, 219
  and resettlement, 236
  and Selous–Niassa Wildlife Corridor, 121
  on transfrontier conservation, 63
  and West Tien Shan Mountain project, 7
  and Wetland of International Significance designation, 325
World Commission on Protected areas, 2, 7, 306–307
World Conservation Congress (IUCN), 301
World Conservation Monitoring Center (WCMC), 297
World Conservation Union (IUCN), xiii, xiv, 3, 7

World Conservation Union (IUCN) (cont.)
  and Afghanistan conservation programs, 306–307
  on categories of Management Protected areas, 160n.3
  Environmental Law Center of, 124
  and Kuriles, 267–68
  "peace park" defined by, 180n.6
  on protected areas, 285
  and Siachen Peace Park, 289
  "transboundary protected area" defined by, 24, 182n.47
  Transboundary Protected Areas Network of, 1–2
World Heritage Areas Program, 215
World Heritage Convention, 325
World Heritage Site
  Korean DMZ as, 254
  KPBRS as, 257
  Hawizeh-Azim marshes as, 323, 325
  La Amistad as, 177
  Waterton–Glacier International Peace Park as, 185
World Parks Congress, 7, 303
World Summit on Sustainable Development (2003), 303
Worldwatch magazine, 15
World Wide Fund for Nature (WWF), 281, 307
World Wildlife Fund (WWF), 6, 15
  and Chihuahuan ecoregion project, 219
  transboundary conservation supported by, 18n.26
World Wildlife Fund for Nature, 23
World Wildlife Fund (WWF)–South Africa, 84

Yale School of Forestry and Environmental Studies, and Leopold, xvii
Yeltsin, Boris, 266
Yot Dom Wildlife Sanctuary, Thailand, 142, 148
Young, Gavin, 324
Young, Oran, 163–64
Yunus, Muhammad, xiii

Zaïre, 78
Zambia
  Dag Hammarskjold International Peace Park in, 69–70
  and Mtwara Development Corridor, 115, 122
Zanzibar, 114
Zimbabwe
  in alliance, 78
  and Great Limpopo Transfrontier Park, 77, 84, 89, 90, 95–97, 101 (*see also* Limpopo Transfrontier Park, Great)
  majority government established in, 115
Zoning, for Pha Taem Complex, 148–50

Global Environmental Accord: Strategies for Sustainability and Institutional Innovation

Nazli Choucri, series editor

Nazli Choucri, ed., *Global Accord: Environmental Challenges and International Responses*

Peter M. Haas, Robert O. Keohane, and Marc A. Levy, eds., *Institutions for the Earth: Sources of Effective International Environmental Protection*

Ronald B. Mitchell, *Intentional Oil Pollution at Sea: Environmental Policy and Treaty Compliance*

Robert O. Keohane and Marc A. Levy, eds., *Institutions for Environmental Aid: Pitfalls and Promise*

Oran R. Young, ed., *Global Governance: Drawing Insights from the Environmental Experience*

Jonathan A. Fox and L. David Brown, eds., *The Struggle for Accountability: The World Bank, NGOS, and Grassroots Movements*

David G. Victor, Kal Raustiala, and Eugene B. Skolnikoff, eds., *The Implementation and Effectiveness of International Environmental Commitments: Theory and Practice*

Mostafa K. Tolba, with Iwona Rummel-Bulska, *Global Environmental Diplomacy: Negotiating Environmental Agreements for the World, 1973–1992*

Karen T. Litfin, ed., *The Greening of Sovereignty in World Politics*

Edith Brown Weiss and Harold K. Jacobson, eds., *Engaging Countries: Strengthening Compliance with International Environmental Accords*

Oran R. Young, ed., *The Effectiveness of International Environmental Regimes: Causal Connections and Behavioral Mechanisms*

Ronie Garcia-Johnson, *Exporting Environmentalism: U.S. Multinational Chemical Corporations in Brazil and Mexico*

Lasse Ringius, *Radioactive Waste Disposal at Sea: Public Ideas, Transnational Policy Entrepreneurs, and Environmental Regimes*

Robert G. Darst, *Smokestack Diplomacy: Cooperation and Conflict in East-West Environmental Politics*

Urs Luterbacher and Detlef F. Sprinz, eds., *International Relations and Global Climate Change*

Edward L. Miles, Arild Underdal, Steinar Andresen, Jørgen Wettestad, Jon Birger Skjærseth, and Elaine M. Carlin, *Environmental Regime Effectiveness: Confronting Theory with Evidence*

Erika Weinthal, *State Making and Environmental Cooperation: Linking Domestic and International Politics in Central Asia*

Corey L. Lofdahl, *Environmental Impacts of Globalization and Trade: A Systems Study*

Oran R. Young, *The Institutional Dimensions of Environmental Change: Fit, Interplay, and Scale*

Tamar L. Gutner, *Banking on the Environment: Multilateral Development Banks and Their Environmental Performance in Central and Eastern Europe*

Liliana B. Andonova, *Transnational Politics of the Environment: The European Union and Environmental Policy in Central and Eastern Europe*

David L. Levy and Peter J. Newell, eds., *The Business of Global Environmental Governance*

Dennis Pirages and Ken Cousins, eds., *From Resource Scarcity to Ecological Security: Exploring New Limits to Growth*

Ken Conca, *Governing Water: Contentious Transnational Politics and Global Institution Building*

Sebastian Oberthür and Thomas Gehring, eds., *Institutional Interaction in Global Environmental Governance: Synergy and Conflict among International and EU Policies*

Ronald B. Mitchell, William C. Clark, David W. Cash, and Nancy M. Dickson, eds., *Global Environmental Assessments: Information and Influence*

Helmut Breitmeier, Oran R. Young, and Michael Zürn, *Analyzing International Environmental Regimes: From Case Study to Database*

J. Timmons Roberts and Bradley C. Parks, *A Climate of Injustice: Global Inequality, North-South Politics, and Climate Policy*

Saleem H. Ali, ed., *Peace Parks: Conservation and Conflict Resolution*